America's Snake

America's SNAKE

THE RISE AND FALL OF THE TIMBER RATTLESNAKE

Ted Levin

Illustrations by Alexandra Westrich

The University of Chicago Press   *Chicago and London*

TED LEVIN is a veteran naturalist and award-winning writer whose essays have appeared in *Audobon*, where he's a frequent contributor, *Sports Illustrated, National Wildlife, Sierra, National Geographic Traveler, Yankee, OnEarth, Nature Conservancy, Attaché, Boston Globe Sunday Magazine, Chicago Sun-Times, Newsday,* and numerous other print publications. His writing has been included in numerous anthologies, including *The Best American Sports Writing 2003*. He is the author of three critically acclaimed books: *Backtracking: The Way of a Naturalist, Blood Brook: A Naturalist's Home Ground,* and *Liquid Land: A Journey through the Florida Everglades,* for the last of which he was awarded the 2004 Burroughs Medal for distinguished nature writing.

The University of Chicago Press, Chicago 60637
The University of Chicago Press, Ltd., London
© 2016 by The University of Chicago
All rights reserved. Published 2016.
Printed in the United States of America

25 24 23 22 21 20 19 18 17 16   1 2 3 4 5

ISBN-13: 978-0-226-04064-6 (cloth)
ISBN-13: 978-0-226-04078-3 (e-book)

DOI: 10.7208/chicago/9780226040783.001.0001

Library of Congress Cataloging-in-Publication Data

Names: Levin, Ted, 1948– author.
Title: America's snake : the rise and fall of the timber rattlesnake / Ted Levin.
Description: Chicago ; London : The University of Chicago Press, 2016. | Includes bibliographical references and index.
Identifiers: LCCN 2015037816 | ISBN 9780226040646 (cloth : alk. paper) | ISBN 9780226040783 (e-book)
Subjects: LCSH: Timber rattlesnake. | Snakes—United States.
Classification: LCC QL666.O69 L495 2016 | DDC 597.96/38—dc23
LC record available at http://lccn.loc.gov/2015037816

♾ This paper meets the requirements of ANSI/NISO Z39.48–1992 (Permanence of Paper).

FOR ALCOTT

For our kind, rattlesnakes are coiled, tail vibrating, fangs at ready to poison us. It is as if we formed our entire knowledge of automobiles from head-on collisions.

CHARLES BOWDEN

Contents

Etymology

venom
A poisonous fluid secreted by certain animals, as in the viper, in a state of health, and which they preserve in a particular reservoir, to use as a means of attack or defense.

ROBLEY DUNGLISON, 1857

The venom immobilizes the prey quickly and in circulating through the bloodstream prepares it for the "subsequent" action of the snakes' digestive juices.

JAMES R. KINGHORN, 1929

Venom in snakes almost certainly evolved as an adaptation for subduing prey. This remains its primary role in nearly all species; it serves secondarily as a defensive adaptation.

SHERMAN A. MINTON, *Venom Diseases*, 1974

It is important to again re-emphasize that the coincidental medical effects of snake venoms should have no role in their definition, as these were evolved long before humans.

SCOTT A. WEINSTEIN, 2015

venomous
An epithet applied to animals which have a secretion of venom, as the viper, rattlesnake ... as well as to the venom itself; and, by some, to liquids in the animal body, which have been perverted by previous disease, that their contact occasions serious mischief in sound individuals.

ROBLEY DUNGLISON, 1857

rattlesnake | ˈratlˌsnāk|
noun
a heavy-bodied American pit viper with a series of horny rings on the tail that,
when vibrated, produce a characteristic rattling sound as a warning. • Genera
Crotalus *and* Sistrurus, *family* Viperidae: *several species.*

Apple online dictionary

Rattlesnake
n: any of numerous New World pit vipers that have a series of horny interlocking
joints at the end of the tail which make a sharp rattling sound when vibrated,
*that comprise two genera of which one (*Sistrurus*) contains small snakes (as the*
massasaugas and ground rattlesnakes) having the head covered with symmetri-
*cal plates and the other (*Crotalus*) contains usu. larger snakes that have scales*
instead of headplates, are rather thick-bodied, large-headed snakes of sluggish
disposition which seldom bite unless startled or pursuing prey, and occur across
most of America from southern Canada to Argentina—SEE CANEBRAKE
RATTLER, DIAMONDBACK RATTLESNAKE, PRAIRIE RATTLESNAKE,
SIDEWINDER, TIMBER RATTLESNAKE, WESTERN DIAMOND RATTLE-
SNAKE

Webster's Third New International Dictionary, 1986

But above all, the crucial characteristic that distinguishes rattlesnakes from all
other snakes—even from other pit-vipers—is the possession of the rattle.... All
rattlesnakes have rattles, and no other kind of snake has them. No snake is a
rattlesnake because it is shaped like a rattler, or because it has blotches like those
of a rattler, or because it is venomous, or because it is found among rattlers, or
because it will coil like a rattler, or because it will vibrate its tail as does a rattler.
Many harmless or venomous snakes have some or all of these characteristics, but
lacking rattles, they are not rattlesnakes.

LAWRENCE M. KLAUBER, *Rattlesnakes:*
Their Habits, Life Histories, and Influence on Mankind, 1956

Rattlesnakes are distinctly American pit vipers that probably developed on the
Mexican plateau and dispersed chiefly northward. Of the twenty-seven species
of Crotalus *and three of* Sistrurus, *only one has an extensive range south of*
Mexico, while fifteen occur in the United States.

SHERMAN A. MINTON AND
MADGE RUTHERFORD MINTON,
Venomous Reptiles, 1969

klapperschlange	German
ratel-slange	Dutch
casabel	Spanish
cascavel	Portuguese
strepito	Italian
caudason	Latin
serpente á sonnettes	French
sin ta dah (yellow snake)	Sioux
sesek	Narragansett and Natick
sizikwa	Abenaki
sheehhkwaa	Lenape
un-cu-gi-sheki	Osage

American colloquial
sizzle-tail
buzz-worm
chatter viper
snattle-rake

James Fenimore Cooper, 1827, was the first to use the term *rattler* in print in his book *The Prairie*.

Prologue

Because I think they are beautiful.

CARL KAUFFELD

Why would Alcott Smith, at the time nearly seventy, affable and supposedly of sound mind, a blue-eyed veterinarian with a whittled-down woodman's frame and lupine stamina, abruptly change his plans (and clothes) for a quiet Memorial Day dinner with his companion, Lou-Anne, and drive from his home in New Hampshire to New York State, north along the western rim of a wild lake, to a cabin on a corrugated dirt lane called Porcupine Hollow? Inside the cabin fifteen men quaffed beer, while outside a twenty-five-inch rattlesnake with a mouth full of porcupine quills idled in a homemade rabbit hutch. It was the snake that had interrupted Smith's holiday dinner. Because of a cascade of consequences there aren't many left in the Northeast: timber rattlesnakes are classified as a threatened species in New York and an endangered species everywhere in New England except Maine and Rhode Island where they're already extinct. They could be gone from New Hampshire before the next presidential primary. Among the cognoscenti it's speculated whether timber rattlesnakes ever lived in Quebec; they definitely did in Ontario, where rattlesnakes inhabited the sedimentary shelves of the Niagara Gorge but eventually died off like so many failed honeymoons consummated in the vicinity of the falls.

That rattlesnakes still survive in the Northeast may come as a big surprise to you, but that they have such an impassioned advocate might come as an even bigger surprise. Actually, rattlesnakes have more than a few advocates, both the affiliated and the unaffiliated, and as is so often the case, this is a source of emotional and political misunderstandings, turf battles and bruised egos. As you may have guessed already, Alcott Smith is a timber rattlesnake advocate, an obsessive really, who inhabits the demilitarized zone between the warring factions. How else to explain this spur-of-the-moment, four-hour road trip?

By the time Smith arrived, the party had been percolating for a while. Larry Boswell opened the door. As he spoke, a silver timber rattlesnake embossed on an upper eyetooth caught the light. Boswell owned the cabin and access to a nearby snake den, a very healthy one, where each October the unfortunate rattlesnake outside, following its own prehistoric biorhythms, had crawled down a crevice and spent more than half the year below the frost line dreaming snake dreams. Porcupines also favor sunny slopes, which likely is how the two met, one coiled and motionless and the other blundering forward. You'd think that after thousands of years of cohabitation on the sunny, rocky slopes of the Northeast, rattlesnakes and porcupines might have worked things out, but not so. No doubt, both animals instinctually took a defensive stance, and whether the snake struck and quills came out, or the startled porcupine lashed the snake with its pincushion tail, both had been severely compromised.

Without Smith's help, the rattlesnake might have been doomed to starve as the quills festered. Ailing snakes die slowly, very slowly. One western diamondback is reported to have survived (and grown longer) in a wooden box for eighteen months without food and water, and a timber rattlesnake from Massachusetts lived twelve months (in and out of captivity) with its face consumed by a white gelatin-like fungus, a Quasimodo in the Blue Hills.

The cabin was small, dank, poorly lit. There wasn't a sober individual in the group. Lou-Anne thought of *Deliverance,* and all evening she stood by the front door. Smith examined the snake and found fifteen quills embedded inside its mouth, which curled back a corner of the upper lip and perforated the margin of the glottis, gateway to the lungs, compromising both the snake's breathing and its eating while protecting the outside world from the business end of the fabled, hollow (and grossly misunderstood) fangs. Essentially, the snake's mouth had been pinned open.

Although this was a rattlesnake-tolerant (if not friendly) group, Smith wasn't about to trust any of their less-than-steady hands to hold the animal. With imaginary blinkers on, Smith worked on a cleared-off coffee table in the middle of the cabin, with the overly supportive crowd keyed to every nuance. Smith gripped the head with one hand and pulled quills with the other, while the snake's dark, thick torso sluggishly undulated across the coffee table. Slowly, methodically, he plucked each quill with a hemostat, and the men, who had tightened into a knot around the coffee table, cheered, toasted, chugged. After the last quill was pulled, the ebullient crowd roared approvingly, and the snake was returned to the hutch. Eight-years later, Lou-Anne, still jazzed by the potpourri of emotions, intensity, and images of that night, remembers feeling "relieved to have left there alive" as the couple returned home on the morning side of midnight.

The timber rattlesnake had been discovered several days before the tabletop surgery. Three of the unaffiliated herpetological adventurers—a couple from Connecticut and a man from northern Florida—had concluded an annual spring survey of the bare-bone outcrops behind the cabin. There, in the remote foothills above the shores of a narrow valley, where a wild brook strings together a run of beaver ponds, is one of the most isolated series of rattlesnake dens in the Northeast, perhaps in the entire country. (The word *infested* might come to more discriminatory minds.) For me, seeing those small, gorgeous pods of snakes basking in the October sunshine is stunning, a natural history right of passage, sort of like a bar mitzvah without the rabbi.

Beside the rattlesnakes, the trio found a fresh porcupine carcass in the rocks, unblemished, and on their way back down the mountain, they found the quilled snake, coiled loosely in a small rock pile one hundred fifty feet behind the cabin, last snake of the afternoon. The rock pile was at the base of a corridor, a bedrock groove in the side of the mountain that rattlesnakes use as a seasonal pathway from the den to the wooded shore and back. The cabin's unkempt backyard is a veritable (and historic) snake thoroughfare. One of these three, a man who calls himself Diamondback Dave, thought he could pull the quills. Well known in the small, fervid circle of snake enthusiasts, Diamondback Dave maintains the website Fieldherping.com, where, among scores of photographs posted of himself (and a few friends) holding various large and mostly venomous snakes, you can view a full-frame picture of his bloody hand, the injury compliments of a recalcitrant banded water snake. You can also read synopses of field trips and random journalistic entries like this one:

> I had a meeting with the director of a wildlife conservation society to discuss strategies on protecting rattlesnake populations in Eastern North America. What turned out was a weird combination of trespass warnings and a lengthy and unnecessary lecture on going back to school and finishing my degree, so that I could make 80,000 a year ... welcome to the new age of Academic Wildlife Exploitation! ... Business as usual.

Although in the spring of 2003, Diamondback Dave had never "pinned" a snake, a term that means immobilizing a venomous reptile's head against the ground using any of a number of implements — snake hook, snake stick, forked branch, golf putter, and so forth — he convinced his two friends that he knew what he was doing. He did. Once the rattlesnake was pinned, Diamondback Dave directed his female companion to hold the body. Three visible quills protruded several inches from a corner of the snake's mouth, fixed like miniature harpoons with their barbed tips. Dave's efforts to pull them proved fruitless, however; not wanting to risk further injury to the snake, he released it.

On their way back to the car, they reported the incident to Boswell, who returned the following day and transferred the rattlesnake from rock pile to rabbit hutch. In his spare time, Boswell taught police officers and game wardens how to safely catch and relocate nuisance snakes, and he had been issued a permit by New York Department of Environmental Conservation (DEC) to harbor them on a temporary basis. This snake needed more than he could offer, though, so the next afternoon, Boswell phoned Alcott Smith.

After surgery, the timber rattlesnake recuperated in the hutch on Larry's side of the bed. Three weeks later, when it was able to swallow a chipmunk, the snake was returned to the rock pile, where it immediately disappeared into a jumble of sun-heated stones. Today, the quilled snake can be found on Dave's glitzy website among a host of other photographs. Just scroll down to the image labeled "Spike."

The thing between timber rattlesnakes and me started in the late fifties. Although there used to be colonies from Brooklyn to Sag Harbor, by the end of World War II the only Long Island rattlesnakes were hanging as trophies in old farmhouses or submerged in specimen jars, their glassy eyes dulled by formalin. The last record of a Long Island rattlesnake came from the pine barrens of East Moriches in 1962, the year after my bar mitzvah. According to a 1915 article in *Copeia*, the journal of the American Society of Ichthyologists and Herpetologists, the rapid decline of the island's rattlesnakes closely followed the eastward extension of the Long Island Railroad in 1895. The snakes adopted the fatal habit of sunning themselves on railroad embankments and lounging on the heated rails. As a kid, I laid pennies on those same tracks, under those same wheels, perhaps, innocent of the fate of these prior denizens.

There were other snakes in my neighborhood, all quotidian— garter snakes, ribbon snakes, northern water snakes, little brown snakes, milk snakes, black racers, even theatrical hognose snakes—

every one of them cool in their own right, but certainly without the aura of a husky rattlesnake, dark as night or sulfur yellow and longer than I was tall. That timber rattlesnakes see in the dark through pits in their face and have rattle-tipped tails that send out a loud, monotonous insect-like buzz when they are afraid or annoyed, quickening one's senses, as would a gunshot in the woods, adds to the allure. And, of course, there's the venom.

My home on the South Shore was once part (a very small part) of a sprawling kingdom of potato farms. Before that, it had been a mixed coastal hardwood forest—black oak, scrub oak, black cherry, sweet gum, tulip poplar, tupelo; in low-lying areas grew Atlantic white cedar and red maple, and in flat sandy stretches, pitch pine. I lived a few miles from the ocean, closer than that to the salt marsh. Below a thin layer of humus, the soil was mostly sandy. Near the ocean, trees pruned by the salty onshore wind often assumed cartoonish shapes, their branches swept back from the sea like spiked hair. Nowhere on the island were there jumbles of sun-heated rock. No ledges. No scree or talus slopes. In fact, the only rocks I recall were boulders strewn about beaches on the North Shore or embedded in lawns and woodlands on the estates of northern Nassau County. Long Island may be composed of ground-down rock from mountain ranges to the north, but who or what would ever mistake its mostly flat and extremely permeable epidermis for the Adirondacks or the Taconics? Certainly not the island's rattlesnakes. They likely lived like their kindred that still survive in the New Jersey Pine Barrens, having pared their social urges down to the basics: mating and sunbathing. There were no large communal dens on Long Island like the one behind Larry Boswell's cabin. Coastal timber rattlesnakes hibernate in wetlands instead of on sunny, rock-strewn slopes, often alone in rodent burrows or in tunnels left by decayed tree roots. Getting below the frost line is all that really matters.

Wherever Long Island rattlesnakes passed the winter, they were long gone before I ever knew they were there. To see one, I'd go to the Staten Island Zoo. Carl Kauffeld, curator of reptiles and later director, had transformed the small zoo into a rattlesnake emporium. Kauffeld exhibited the finest rattlesnake collection in the world, all

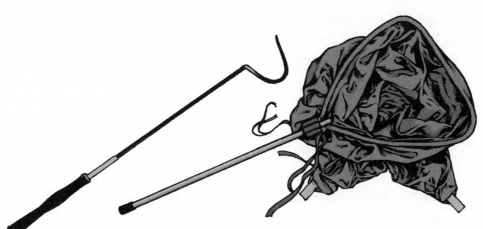

thirty-two species and subspecies occurring within the United States. I saw my first timber rattler there, a dark-phased adult collected in the mountains north of New York City. The snake was shedding (ecdysis, biologists call it), and I stood by the glass-fronted cage as it crawled out of its skin, one, long serpentine sleeve peeling back, inch by inch, inside out, like taking off a sock.

Kauffeld was well known outside Staten Island. The *New Yorker* had twice covered his exploits in Talk of the Town, and he himself had written three books, of which two, *Snakes and Snake Hunting* (1957) and *Snakes: The Keeper and the Kept* (1969), were eye-openers for a boy naturalist. Reading them, I realized that Major League baseball players were not the only men who made a living doing boy things. In these books, an exuberant Carl Kauffeld caught snakes across the ridges and ravines of North America. Unlike birders, who comb the countryside for birds, recording their achievements in "life lists," "state lists," even "backyard lists," or watch familiar birds doing unfamiliar things, many snake enthusiasts hunt snakes and then take them home.

Kauffeld's stories launched an army of youthful collectors, of whom some became prominent herpetologists, while others simply collected and collected and collected. Dealers ravaged the rattlesnake dens Kauffeld had so lovingly described. With a memorable chapter in *The Keeper and the Kept*, titled "Life and Death in Okeetee," Okeetee, South Carolina, became an ophidian Mecca. The first March after the book's publication, the unwashed began to arrive in Okeetee and bag every gorgeous corn snake they found—"it is happily quite numerous," Kauffeld had written. No longer. Reptile clubs chartered buses, flooding Okeetee in such numbers that landowners posted their properties. Thirty years earlier, in 1939, a young Kauffeld had gained notoriety when he took a handful of snake people on an annual autumn collecting trip to the southerly slopes of Listening Mountain in the Ramapos. Over the years, the "Hunt," as reported in the *New Yorker*, had produced fifty-nine specimens, including thirty copperheads, five blacksnakes, and fifteen timber rattlesnakes. All were brought back to the Staten Island Zoo to be traded like baseball cards to other institutions.

ॐ

When it comes to eliciting empathy, it's the back of line for rattlesnakes, creatures seemingly with, face it, not much personality. One could argue that our squeamishness at the sight of a snake began with the story of the Garden of Eden in the book of Genesis, but it also may be coded in our genes, suggests Harvard biologist E. O. Wilson. Humans, says Wilson, could be hardwired to fear snakes. In Africa, where our closest primate kin have multiple predators to fear, chimpanzees have been observed shadowing dangerous snakes at a safe distance, staring and hollering. Charles Darwin even weighed in on the issue of ophidiophobia: "I took a stuffed snake into the monkey-house, and the hair on several of the species instantly became erect," he wrote in 1872 in *The Descent of Man*. Though timber rattlesnakes rarely harm humans or domesticated animals, Americans nevertheless have a long history of organized efforts to collect and eliminate them.

In 1680, a Massachusetts hunter could earn two shillings a day killing timber rattlesnakes, and beginning in 1740, Massachusetts chose one day each fall for a community-wide hunt, called a rattlesnake bee, which took place in towns across the state. In 1810, hunters in Pennsylvania strapped powder horns to rattlesnakes, lit them, and released them back into their dens; in 1849, in Madison County, Iowa, teams competed for the most snakes killed. The prize for the winning team: two bushels of corn. Bounties were paid for rattlesnakes in New York and Vermont into the early 1970s.

Twenty-five years ago, I visited a Vermont town clerk to examine old bounty records. "Why," she asked, "would anyone care?" That was a hard question to answer. I had just driven an hour and a half to learn something about the snakes and the people of western Vermont, maybe something about the hard-rock ledges. I found it difficult to articulate what I was after. She pressed me again.

"It's not every day someone comes here to talk about snakes. I don't even know where that book *is*."

She apparently found it hard to say the word *rattlesnake*. "I saw one this spring, crossing the road near the Blatsky River. I can't stand to look at 'em."

A man in a three-piece suit walked into the clerk's office. He was in a hurry.

"Hey, Bob," the clerk said, "this guy wants to know about rattlesnakes." Finally, she had said the word, hanging on to the *a*'s and *t*'s as though she were shaking a castanet. (Until that moment, I hadn't thought of the word *rattle* or *rattlesnake* onomatopoetically.) Bob apparently didn't like rattlesnakes, either. He said he had killed one in East Steeple, not far from Crystal Lake, a couple years previously. Whacked off its head with a hoe.

No one wanted to touch the bounty book, so I collected it myself. What I found was that between 1899 and 1904, two hundred forty-one timber rattlesnakes were bountied, a dollar a piece. The earliest bounty was paid on May 9, and the latest on October 19. Of the two hundred forty-one snakes listed, sixty-two were killed between May 9 and May 31, and one hundred fifty-four after August 21, when the snakes, including the young-of-the-year, had returned to their dens. This seasonal pattern confirmed that timber rattlesnakes go to bed early and wake up late.

One snake hunter, Andy Howard, collected the one-dollar bounty on one hundred ninety-six rattlesnakes during that five-year period. According to the town clerk, Andy liked liquor, and the bounty payments warmed the long, cold winters, so he made it his business to find snake dens. On September 13, 1902, he killed thirty-seven rattlesnakes.

Only twenty-five snakes were bountied from early June to mid-August. This is not too surprising. Timber rattlesnakes need the ice to melt and the soil to warm before they are ready to expend energy on growth, to leave the vicinity of their dens for the wooded ridge, where they lie in wait for mice and chipmunks. To find one in summer is a matter of chance. Great chance.

There were no records from 1905 through 1947. After 1947, sixty-four snakes were killed in a twenty-year period, ending in 1967. With so few snakes to record, the bounty book began noting the length of each snake and the number of rattles segments: the longest was four-and-a-half feet.

In some regions of the country snake killing is still sanctioned. As recently as 1989, Clairemont, Texas (now a ghost town), held its forty-first and final Peace Officers Rattlesnake Shoot, in which law-enforcement officials and other contestants competed for points by shooting live rattlesnakes. A shooter was awarded ten points for a head shot, five for a body shot; prizes were given for five categories: masters, first place, second place, third place, and guest.

Several years ago, Dartmouth College's Hood Museum of Art in Hanover, New Hampshire, exhibited the watercolors of George Catlin, a native Pennsylvanian who traveled throughout the nineteenth-century American West painting the lives of Plains Indians. Catlin's subjects engaged the landscape—hunting bison, praying and dancing, preparing food, pitching tepees in the shadows of great mountains and along the shores of winding rivers.

Not all the watercolors in the exhibit celebrated the West, however. In one painting, Catlin depicted his own home ground, the green woods and rocky ledges above the Susquehanna River in central Pennsylvania. The time is early May. Dozens of timber rattlesnakes lounge on the rocks, basking in the sunshine, while two men attack them with clubs and guns. A boy, perhaps the artist himself, stands in the background, screened by foliage, watching, waiting his turn. Catlin's chance eventually came. On another occasion, the artist, known for his sensitivity to vanishing cultures, allegedly destroyed a Pennsylvania den by strapping a powder horn to the tail of a rattlesnake. He lit the fuse and released the snake into the talus, towing the bomb behind it.

More than one hundred fifty years have passed since Catlin painted the snake hunt, yet these timid serpents still evoke the same fear and loathing that motivated the destruction of America's other predators. We've since made our peace with most of these—bald and golden eagles, wolves, and catamounts, the alligators and crocodiles, the silver-tipped grizzlies. Why not with the rattlesnake?

Decades ago, we stopped slaughtering hawks and owls. We welcomed gray wolves back to Yellowstone, red wolves to South Carolina, and black-footed ferrets to the Northern Plains. Today, we celebrate jaguars in Arizona, ocelots in South Texas, and great white sharks off Cape Cod, and we commiserate with the plight of polar bears swimming to exhaustion in the Beaufort Sea. But when the subject turns to timber rattlesnakes, we are collectively and decidedly pigheaded about their future; trying to sell an ophidiophobe the merits of rattlesnakes is as difficult as trying to convince a member of Red Sox Nation on the merits of the Yankees. Timber rattlesnakes are perceived as bad to the bone. Even those who care can't agree on the best way to ensure survival of the snakes; worse, it is difficult for the different factions even to hear each other's concerns.

Forty years after Kauffeld's death, timber rattlesnakes, which are not inherently aggressive—just unforgiving of being mishandled—are still pursued by both collectors and persecutors, and face a litany of other problems ranging from isolated colonies, depleted gene pools, and inbreeding—a prescription for local extinction—to fatal fungal infections, climate change, automobile traffic, and political paralysis. Timber rattlesnakes, which are as American as apple pie, still live a short drive from Boston, Hartford, New York, Philadelphia, Pittsburgh, Baltimore, Richmond, Saint Louis, and Minneapolis, which says something about their passive nature, their secretive ways, and the breadth of their evolutionary adaptations, which allows them to count among their immediate neighbors animals as geographically disparate as peccaries, alligators, and moose.

The story of the timber rattlesnake (America's snake) is as much a story of human attitudes—good and bad, but rarely indifferent—and of places—pockets of wildness between the Atlantic and the west bank of the Mississippi—as it is the story of a snake.

In the autumn, I go to a certain talus slope above a wild brook to watch rattlesnakes, and I stay until cool weather ushers them underground for the winter. I wish I could lead a field trip to a timber rattlesnake den. While trekking around the remote pockets of wildness where they live, I want to educate people about the *true* nature of the snake, gregarious and docile, and to share the sense of wonder I feel as I watch the last of the Northeast's apex predators. Sadly, this isn't possible.

Lethargic and predictable, timber rattlesnakes remain vulnerable to vandals and collectors, and to a New Age group called "field herpers," who handle snakes, digitally photograph them, and then post their exploits over the Internet. Even scientific research and repeated visits by naturalists (like myself) may cause snakes to bask less or to use less-than-ideal birthing sites. With the aid of a GPS and a subsequent website announcement, an Appalachian Trail hiker who stumbles onto a pod of rattlesnakes and then electronically broadcasts exuberance could be the unwitting agent of the snakes' demise.

And that would be heartbreaking, because timber rattlesnakes are breathtakingly beautiful. They vary in base color from the blackest black to golden yellow. Some are mustard-colored; others are olive or brown, tawny or twilight gray. Neonates are the pinkish-grayish shade of exfoliated granite. Timber rattlesnakes have crossbands or chevrons or blotches (sometimes all three) that may be faintly rimmed in yellow or white, and range from black to charcoal, chocolate to tan or olive-yellow. Some snakes have a broken, rust-colored, dorsal stripe. Others are patternless black. Coiled in a bed of October

leaves, a timber rattlesnake is hidden in plain sight unless it rattles, which is electrifying.

Here in the Northeast, den-site fidelity is the hallmark of their survival. Each fall, rattlesnakes return to their maternal den as precisely and directly as a Bicknell's thrush might return to a particular hillside forest in Hispaniola. When a well-muscled rattlesnake migrates home it doesn't undulate in loops and curves as it does when it's swimming; it flows in a straight line rather like melting candle wax, belly scales caressing the ground, a thousand little pseudo-feet. Slow ... slower ... slowest. On a windless afternoon the vague sound of scales brushing against leaves gives them away.

I keep vigil at one particular den, counting, always counting snakes—a yellow morph, a black morph, a young-of-the-year, a three-year-old, an adult female with a broken ten-segment, untapered rattle, that sort of thing. I note air temperature, rock temperature, snake temperature, cloud cover, and wind speed and direction. This year, a few snakes returned to the den in late August; more arrived in September. And the number peaked in early October, when I tallied more than eighty.

One day, I followed two very *big* snakes through rock-studded woods to the base of a towering ledge and then watched them disappear down a crevice. Later that afternoon, I stood quietly in front of the main portal as more than a dozen snakes slowly passed by me and poured themselves over the stone rim, braided together in the foyer, and then one by one vanished into the abyss. Two weeks later, I found only three, including a newborn en route to the slumber party. And two days hence, after fallen leaves had slicked the rocks, two, of which one, a black male, coiled in the foyer. At fifty-four degrees Fahrenheit, he was the temperature of the rock, two degrees cooler than the air. And he moved his tongue in slow motion.

I don't pass winter underground and I stopped basking decades ago, though sun-warmed rocks feel good to me, particularly when the air is cool and the day short. I go to talus slopes to watch rattlesnakes, and I stay until the rocks cool off and autumn's last whit of heat draws them down below the surface. Like a rain of maple leaves or a flock of migrating geese, the doings of rattlesnakes in October mark a season in transition, the subtlest of autumnal tides.

The snakes at my study site ignore me, and I never touch them. I bear witness; my movements ratcheted down to mere heartbeats and breaths. For the most part, they treat me with indifference. One even glided over my boot. A few rattlesnakes living here were born the summer the Beatles released "Hey Jude"; at least one forty-five-year-old still bears young. For their continued survival I did for this book what never occurred to Kauffeld; I fictionalized the names of several people and places, especially recognizable roads, bodies of water, and mountains. I also changed the names of a few towns and states. To paraphrase the narrator of the 1950s television show *Dragnet*: Ladies and gentlemen, the story you are about to read is true. Only a few names and locations have been changed to protect the innocent: in this case, the timber rattlesnake.

1

An Introduction to *Crotalus horridus*

She strongly resembles America in this, that she is beautiful in youth and her beauty increaseth with her age, her tongue also is blue and forked as the lightning, and her abode is among impenetrable rocks.

BENJAMIN FRANKLIN

I have seen timber rattlesnakes before, mostly on the sun-baked talus of western Vermont — dark, with vague markings, or mustard-colored and distinctly banded to merge into the forest floor, but never an incandescent yellow one in the Northeast. The snake stretches out in the morning light, basking on the stone foundation of a building preempted by trees in a forested hillside in western New York. Chocolate-colored bands stain the yellow areas and loop its body from head to vent; the tail is as black as obsidian. And like most timber rattlesnakes I've seen, this one is content. No rattling. No threatening coil. No retreat. Not so much as a glance in my direction. Lidless eyes focus on the all or nothing of a New York morning, golden spheres each slashed by a vertical black pupil — eyes of the night.

At fifty-two inches and just over three and a half pounds, Hank is big. Of all the rattlesnakes I've encountered in the Northeast, only Travis, Hank's hillside neighbor, a black morph recumbent beneath the overhang of a bramble a quarter of a mile away, is bigger: fifty-two inches, four and a half pounds, and thick as the sweet spot on a baseball bat.

I'm in the field with Rulon Clark, a Utah native who completed his doctorate at Cornell University in nearby Ithaca with a dissertation on the communal habits of timber rattlesnakes and is now assistant professor in the biology department at San Diego State University. Although it was once widely believed that snakes led rudimentary lives of solitude, Clark discovered that timber rattlesnakes lead surprisingly rich social lives. Because they dwell in a different temporal realm than people, very slow and methodical, rattlesnakes live within a formerly unmapped wilderness of patience.

To understand timber rattlesnakes, you must learn to think like them, which means spending an inordinate amount of time on the ledges and in the woods. And you must also be a master of technology. According to Clark,

> Rattlesnakes have adapted to live in such a way that they do things more slowly than we do. We're too quick for them in some ways; we don't recognize important things that could be going on. You really have to either be very patient or set up experiments in such a way that you see results regardless of your timescale.

During the six years that he studied rattlesnakes in Pennsylvania and New York, Clark unveiled a sisterhood of snakes, in which female timber rattlesnakes from the same litter entwined with each other more than with unrelated females, the first demonstration of kin recognition for any species of snake. In fact, Clark determined that female littermates separated for two years after birth immediately (in snake time) recognized each other.

As a child, Rulon Clark was attracted to non-fleeing predators (like rattlesnakes) and sponsored public feedings with his menagerie of reptiles and arthropods. "I had tolerant parents," he recalls. I had tolerant parents as well, but their level of tolerance was governed by the social codes of suburban Long Island. They accepted my boyhood passion for bringing home snakes, and would find me whenever anything unusual happened in a terrarium. Once my father prevented a garter snake from consuming its own young by tapping the snake on the head with the buckle on the dog's leash. My parents would likely have drawn the line against keeping a venomous pet, but since Long Island timber rattlesnakes had already been extirpated, this never became an issue.

Through the years, Clark's boyhood obsession grew. He examined eleven hundred timber rattlesnake museum specimens from collections all over the United States, and summarizing their dietary information made him an expert in the identification of half-digested mammal parts. "Don't let anyone tell you that you can't get practical skills out of a PhD."

Clark's audience also grew. The March 2005 issue of *Natural History* featured his essay "The Social Lives of Rattlesnakes," which was how I tracked him down. The week before I arrived in Ithaca, David Attenborough and a BBC film crew had visited Clark to film the hunting strategy of a wild timber rattlesnake (essentially waiting) for the final episode of the landmark two-part series *Life in Cold Blood*.

When I google "Rulon Clark" I find newspaper stories about pregnant rattlesnakes sun-bathing at a picnic area along a major highway in south-central New York. Clark advised the state to close the rest area until the snakes either give birth (sometime in late summer) or are relocated to an alternative basking site. When asked if the snakes pose a threat to travelers, Clark responded, "People are more likely to be killed by a drunken driver veering into the rest area than they are by a basking rattlesnake."

Radio-tracking enables Clark to eavesdrop on timber rattlesnakes during the four or five months when they're away from the den and fanning out for miles across the countryside. Earthly pulses and species-specific biorhythms, formerly understood only by the snakes themselves, govern the movement of these reptiles. Using a remote-sensor video, Clark films rattlesnakes as they wait hours, or even days, to ambush the small mammals that are their principal food. He learns where they eat, when they eat, what they eat, and how they eat, as well as what they do between meals (lounge and digest and eventually wander to their next ambush station). Clark's work dispels myths, confirms scientific speculation, and illuminates unknown aspects of rattlesnake behavior. It also adds a sense of wonder to the forested hillsides of the Northeast, a glimpse into the secret life of a reptile whose continued presence is a marvel of adaptation and concealment, for timber rattlesnakes are the very last of a short list of potentially dangerous pre-Columbian predators that still survive in the twenty-first century on the virtual (and sometimes real) doorstep of the urban Northeast.

§

Rattlesnakes belong to the subfamily Crotalinae, the pitvipers, of the family Viperidae, perhaps the most advanced family of reptiles in the world. Of the approximately three hundred species of viperids, one hundred ninety-nine are pitvipers, the most advanced of the vipers. And rattlesnakes, the most recently evolved of the pitvipers, sit alone on the pinnacle of serpentine evolution, cold-blooded "state of the art." The timber rattlesnake is one of thirty-eight species (and eighty recognized subspecies) of rattlesnakes, all restricted to the Western Hemisphere, as emblematic of the New World as maize and beans. Rattlesnakes evolved two to five million years ago on the grassy plains of north-central Mexico, still their epicenter of diversity, and have left their fossils—mostly vertebrae, but with occasional ribs, fangs, and skulls bones—in the gypsum and limestone caves of the Southwest and the tar pits of California. Thirty-six species belong to the genus *Crotalus*, the mailed rattlesnakes, the most recently evolved genus. Each has numerous small scales (as well as a few large ones) on the top of the head suggesting the overlapping rings or loops of chain mail worn by Elizabethan knights. The remaining two species, in the genus *Sistrurus*, sport nine large skull plates instead of tiny scales: the pygmy rattlesnake of the Southeast, and the massasauga, whose odd diagonal distribution runs from northern Mexico to the southern edge of the eastern Great Lakes, where they barely survive in wetlands outside Rochester and Syracuse, New York.

Arguably the most novel appendage in the animal kingdom, the rattle evolved after gradual changes to the distal tail spine sparked the retention of the cone-shaped, terminal scale whenever the snake shed its skin. Made of keratin (like our fingernails and hair), the rattle is laterally flattened, thick and hard, hollow and musical when vibrated, and is so similar among rattlesnake species that biologists believe it arose just once from the common ancestor of the entire tribe. If you examine the shed skin of a rattlesnake, you'll see an opening at the tail tip where the terminal scale stayed behind with the snake to become the first rattle segment. In all other species, the shed terminal scale remains with the old skin. The base of the rattle—called the matrix—is the living end of the snake's tail. When a rattlesnake sheds, the skin of the old matrix pushes back to become the newest segment of the rattle, though the length of the rattle is not a precise indication of its of age. Each pagoda-shaped rattle segment, pinched into two- or three-tiered lobes, interlocks with its neighbor; the narrow end of one segment fit loosely into the wide base of the distal segment. The deep, transverse constrictions that create the lobes and the shallower, longitudinal grooves that run laterally along both sides of the rattle internally fasten the segments together. The entire rattle is referred to as the *rattle string*, which can be either complete or broken, tapered or untapered.

A newborn rattlesnake, or *neonate*, has a *prebutton* "rattle," a single, unconstricted lobe that covers the tail tip and is lost during the snakelet's first shed. Once that initial shed has been completed, the neonate is now a *young-of-the-year*, and its tiny rattle, or *button*, is a single constricted segment that looks like an itsy-bitsy replica of Abraham Lincoln's hat, and will remain the last segment of a complete string until broken off. As a snake grows, so too the matrix, and up to a point (about the tenth segment) each new rattle segment is slighter bigger than the one that preceded it. Because older snakes grow less dramatically than younger snakes and because rattles break, matriarchs and patriarchs sport untapered rattles. An ideal rattle length is approximately ten segments; many more would make the resonance of the rattle less efficient. The largest rattle string I've ever seen was eighteen segments, a broken, untapered rattle carried around by a big yellow male, but I've heard of wild snakes with more than twenty segments, which would be like lugging a tuba around when a bugle would be sufficient.

Like an asymmetric eight, a rattle is smaller and tilted forward above the longitudinal grooves, larger below, which prevents it from drooping, lessoning the opportunity for abrasion on rocky terrain.

Instead, the rattle is held either parallel to the ground or tilted upward when the tail is lifted slightly, the standard crawling posture. In front of my keypad sits a ten-segment, complete rattle string, an inadvertent souvenir from a New York snake. The tiny button is top-hat obvious; the gradual increase in the size of successive segments, also obvious, echoes the growth of the snake. I see the asymmetry on either side of the longitudinal grooves; and when I hold the rattle right side up (the way the snake would), it extends straight out, parallel to the ground; when I turn it upside down, it droops like a flaccid hose when the water is shut off. Sound is produced in the larger, lower loop of the figure eight, where so much empty space amplifies rattling. Using my entire arm as a lever, no matter how vigorously I shake the rattle, it sounds no louder than an anemic cricket or a couple of pebbles bouncing around in a tin can.

The timber rattlesnake, of course, doesn't have this problem and is equipped to play the instrument. The muscle that shakes the rattle (the shaker muscle) is richly endowed with blood and oxygen and capable of sustained contraction as a human heart in fibrillation. The warmer the snake, the more rapidly the shaker muscle contracts. The faster the contractions, the faster the rattle vibrates. A rattlesnake warmed to ninety-five degrees Fahrenheit can vibrate its rattle eighty-six cycles per second, a visual blur capable of running uninterrupted for up to three hours; chill the snake to fifty-five degrees, however, and the vibrations slow to twenty cycles per second. According to a University of Washington physiologist, if human leg muscles were as efficient at using oxygen as a rattlesnake's shaker muscle, we could complete a twenty-six-mile marathon in less than nine minutes.

Hearing an unexpected rattlesnake is a full-body experience, and the chills that metastasize down our spines give significance and density to "the fear of God." One evening on the outskirts of Tucson, in the late eighties, when my son Casey was not yet three years old, I carried him upside down as I followed behind a squat, lumbering Gila monster that had just emerged from a pile of rust-colored rocks. The lizard was in no particular hurry. Consequently, neither were we. Where it went, we went. Over rocks, around cacti, across an arroyo. Suddenly, after the sky had darkened lumen by lumen from rose to violet, a chunk of reddish sandstone came alive beneath the lizard's feet. The delirious rattlesnake went crazy, buzzing. Never flinching, the Gila monster plodded on. I straightened as though I had touched an electric fence, and by the time I regained my composure, the lizard had vanished, and Casey, whose face had been a mere two feet or so above the ground, was totally jazzed as though I had planned the diversion. Driven by demonical fury, the snake held its ground, threatening to strike—its upper coil rising, its tail a wild blur. I stepped back instinctively, levering Casey into the upright position, and then followed the arroyo back down the canyon, accompanied by the sound of an acutely disturbed rattlesnake.

In the Northeast, rattlesnakes are just as easy to overlook, though more equanimous, often not as quick to rattle. The first time Alcott Smith scrambled up a rockslide in search of a den, he had followed the directions of an armed acquaintance, who, for fear of snakes, proceeded no further than the base of the slide. Scrambling around the rocks, Alcott stumbled on to the threshold of Vermont's largest den, setting off a chorus of half a dozen snakes before he had laid eyes on one. He froze, stone stiff for some time until he figured where the snakes were, half hidden in rocky alcoves. Hearing one is a bit unnerving the first time. And the second ... the third ... and so on and so forth, which, of course, is why the rattle evolved in the first place. It's a sound you never fully get used to, a sound that demands attention, like an unexpected blast of thunder or the roar of a hidden lion.

Timber rattlesnakes that live on hard rock granite of the Adirondack foothills tend to have shorter rattles than those that live on soft sandy loam of the New Jersey Pine Barrens. No matter where the snake lives, however, rattles are fragile. They wear and tear and pop off like snap-on beads. Alcott Smith once knocked the rattle off a snake he was examining and snapped it back on. When Smith realized the drooping rattle was upside down, he quickly caught the snake and made the necessary adjustment. And, one hot July afternoon, in lower New York State, Randy Stechert, who monitors rattlesnakes for the state's Department of Environmental Conservation (DEC), drove past a thirty-nine-inch snake with an eight-segment, complete rattle string. Since Stechert hadn't planned on marking snakes and didn't have a snake bag, he caught the rattlesnake and held it out the open window of his car, pinching its head between his thumb and forefinger, wedging its flagellating body under his left arm, and drove home past houses, farms, and fields. En route, the rattle broke off, temporarily vanishing in the darkness beneath the front seat. Stechert eventually recovered the rattle and, a year later, the snake, which he brought back home, again; then, like the fitting of Cinderella's glass slipper, he snapped the lost rattle back in place.

The more a snake eats, the more it sheds, and a timber rattlesnake, honed in the boom or bust cycles of the deciduous forest, can depress its metabolism and go without food for more than a year or can gorge on rodents, feeding every four or five days. I once watched a well-fed western diamondback on exhibit in the Arizona-Sonora Desert Museum molt its skin (ecdysis), a delicate process that took hours (snake time). Bearing witness to the event was more interesting than watching my laundry dry, but still required downtime for both the observer and the observed. The scales of a snake and the scales of a fish are not homologous. Snake scales originate in the epidermis (the outer layer of skin) and form a seamless covering (a sleeve) of textured folds; whereas fish scales are individual out-growths of the dermis (the inner layer of skin) and can be removed one at a time. (Snake scales do not come off on your fingers as fish scales do.) Chafing against a rock, the diamondback split the skin around his snout and lips, and then moved about his enclosure. Whenever he snagged the loose flap on something rough he crept forward, literally oozing out of his skin one micrometer at a time. Hours later, when the snake had finished, he left behind a perfect grayish casting of himself, an inverted replica.

I've found timber rattlesnake skins on talus slopes above western Vermont, stuck to bluish-gray rocks like strips of cellophane. Except for the rattle, each scale and topographic feature of the snake's body was there: the lining of the pit cavities and the nostrils, the semicircular, transparent scale that covers the eye—the spectacle, an embryonic fusion of the upper and lower eyelids. Each skin had three openings: one each for the mouth, the cloaca, and the end of the tail, where the terminal scale stayed behind with the snake to become a new rattle segment. Old snakeskins are dry and brittle and crumble easily, but freshly shed skins like the one in the museum are moist and pliable (at least for the moment), and glisten in the incandescent light. And the snake itself is bright and gorgeous, its new rattle segment shiny.

Big rattlesnakes have big rattles (unless broken) that issue loud, chilling warnings. Smaller species (and young of larger species), like the pigmy and twin-spotted rattlesnakes, have small rattles that emit an insect-like buzz barely audible from more than a few feet away. I've mistaken the sound of both for that of a trilling cricket. There's no mistaking the rattle of a large, edgy timber. It's unnerving, sometimes commencing with little provocation as though the snake itself were sizzling like a downed power line.

During the initial wave of rattlesnake evolution, the interior grasslands of northern Mexico were crowded with prehistoric beasts—horses, camels, elephants, sloths, and numerous species of bison, pronghorn, and deer—that moved together like a wave across the southern plains. Most viperids and many species of nonvenomous snakes (like the milk snake in my stone wall) vibrate the tips of their tails when startled; if against dry leaves or grasses, the sound is a barely audible buzz. Rattlesnakes, of course, have taken this one giant step further. Originally, herpetologists were moored to the idea that nature, which Darwin referred to as "prodigal in variety, but niggard in innovation," fitted the rattlesnake, evolving amid all that foot traffic, with a resonator to amplify the vibratory noise to keep from being stepped on. Now, the putative theory suggests that the rattle evolved to startle midsized predators like coatimundi and ringtail cats, which found (and still find) rock-dwelling, fat-bodied rattlesnakes tasty. The undisputed fact is the sound of the rattle is primordial and demands a response. In an evolutionary reversal, one species, the Catalina Island rattlesnake, dispensed with rattles, which atrophied like the hind limbs of a whale or the wings of a penguin during thousands of years of isolation on Santa Catalina in the Sea of Cortez, where the snake crawls noiselessly along branches feeding on birds (among other things).

Seventeen species of rattlesnakes live in the United States, including twelve in Arizona, four in the East, and two in the Northeast. Until 1973, herpetologists subdivided the timber rattlesnake, *Crotalus horridus*, the most widely (though discontinuously) distributed rattlesnake of the eastern deciduous forest, into two subspecies: the timber rattlesnake of the Appalachian highlands and northern and central deciduous forest, *Crotalus h. horridus*, and the canebrake rattlesnake of the southern coastal plain and the Mississippi valley, *C. h. atricaudatus*. DNA testing has confirmed that although north-to-south, as well as east-to-west, geographic variation occurs, both are a single race of the same species. (More recent results of DNA testing, reported in a 2006 issue of *Herpetological Monograms*, also confirm that the western populations of timber rattlesnakes, which live on the bluffs above the Mississippi in southeastern Minnesota and southwestern Wisconsin and on the eastern threshold of the prairie, differ more from both the coastal lowland and the highland timber rattlesnakes than either does from the other.) The highland timber rattlesnake, the one I see most frequently, is brown or gray to yellow or black, its front half stippled in blotches, an iterating pattern that fuses to crossbands midway down the back. The head is unmarked. The canebrake (a reference to the low-lying wetlands where it often lives) is more colorful, so colorful that its painted image graces the cover of the four-hundred-fifty-page *Biology of Pitvipers*: yellowish, pinkish, or brownish with a reddish-brown back stripe that bisects the crossbands and a dark stripe behind each eye. Most timber rattlesnakes (no matter where they live or what they're called) have velvet-black tails.

The timber rattlesnake was the first snake encountered by Puritans and the first New World reptile classified by Linnaeus, who, in

1758, named a Manhattan specimen *Crotalus horridus*, a reference to the animal's "musical rattle" or "castanet" (*crotalus*) and "bristling" or "scaly texture" (*horridus*), the thin median ridge, called a keel, that projects from all but the wide belly scales. (Contrary to popular belief, the name has nothing to do with its reputed temperament.) The specimen Linnaeus examined was initially housed in the King Adolph Frideric collection and later sent to the Royal Zoological State Museum in Stockholm, where it vanished into the ether like the rest of Manhattan's once robust rattlesnake population.

In 1825, the French-American naturalist Charles Lesueur painted a drop curtain at the New Harmony cooperative in Indiana that highlighted the two most characteristic features of wild America: Niagara Falls and a rattlesnake, the timber rattlesnake to be precise. For the colonist, the rattlesnake was the most novel creature in the New World and the amount of writing about it far exceeded its usefulness.

In 1612, describing clothing worn by Virginia Indians, Captain John Smith made the first reference in English to a rattlesnake: "Some on their heads weare the wing of a bird or some large feather, with a Rattel. Those Rattels are like the chape of a Rapier but lesse, which they take from the taile of snake." In 1624, Edward Winslow, in *Good News from New England*, coined the name "rattle snake," also in connection with Indian custom. Many early accounts of timber rattlesnakes focused on their bite, their size, and their appetite, as well as their musical appendage: "teeth are as sharp as needles, and although the neck is no thicker than a man's thumb, the snake can swallow a squirrel"; or, describing a snakebitten man as turning the color of the snake, "blew, white, and green spotted." Colonists reported with exaggeration if not outright untruth, that rattlesnakes charmed their prey (not true), sported seventy or eighty rattles on their tails (ten is a lot), and had a smattering of gray hairs on their bodies (also not true).

Even John James Audubon got rattlesnakes wrong, which is rather surprising for such an astute observer of birds and mammals. His painting of the northern mockingbird (captioned *Mocking Bird*) in *Birds of America* features a sinister looking yellow-phase timber rattlesnake looped around the crotch of a small tree, a bird nest in the background. Three mockingbirds hurl passerine expletives at the gape-mouthed snake, which appears intent on eating one of the mockingbirds or the unseen chicks in the nest. That timber rattlesnakes rarely climb trees and infrequently eat birds is a minor point; Audubon's snake has round pupils and a set of teeth in the upper jaw, both anatomically incorrect. A creature of the night, the rattlesnake has pupils that are vertically slit, and except for a pair of fangs, the upper jaw is as toothless as that of an infant. Audubon's rattlesnake faux pas extended beyond art. In an Edinburgh lecture on rattlesnakes, in 1827, he dazzled his audience by claiming that while lying on his back in the Louisiana woods he watched a timber rattlesnake chase a gray squirrel through the treetops, tree to tree like a limbless acrobat. After a long, spirited chase over ground and through the woods, the snake finally caught the squirrel and then dispatched it by constriction, rather than injection (pure fiction). In an account published in a distinguished Scottish philosophical journal, Audubon referred to den-site basking as a "disgusting" ophidian orgy. He also reported that roving timber rattlesnakes hold their heads high off the ground, swinging back and forth, to scan tree limbs for vulnerable bird nests and the sky for dangerous birds of prey. Rulon Clark's research confirms that timber rattlesnakes are deliberate, not spirited, and that they read their world with bifurcated tongues and a pair of heat-sensitive facial pits, and at close range with their jeweled, laterally facing, myopic eyes; a human with vision equivalent to a rattlesnake's would be declared legally blind.

Astonishingly accurate, facial pits are a second set of eyes that read the long wavelength of the electromagnetic spectrum: infrared, which radiates off both warm-blooded animals and the landscape. As you may have already surmised, the presence of facial pits differentiates a pitviper from other classifications of snakes in the same way that the presence of a rattle differentiates a rattlesnake from other pitvipers. Located midway between each eye and nostril, the pits guide a timber rattlesnake's strike on an overcast, moonless night and sense the presence of a mammal predator like a coyote or a bobcat or a skunk. Think infrared night-vision goggles. Of equal importance, pits enable a timber rattlesnake to read a thermally patchy environment,

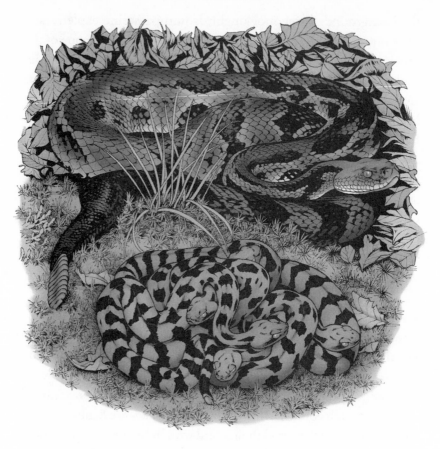

which may have been the reason for their development; knowledge of the thermal landscape is paramount for an adult snake as thick as your wrist, which is why some shelter and basking rocks are popular and others are not (because they're smaller, juvenile rattlesnakes have more thermal options). And then there is the choice of a thermally stable den, an ancestral inheritance along the matriarchal line of descent, perhaps the only reason timber rattlesnakes have been able to successfully colonize and persist in the Northeast.

In 1683, Edward Tyson, a landmark British anatomist (more on him later) who wrote the first scientific report on timber rattlesnakes, was the first person to officially misidentify the pits. He thought they were ears. In the centuries that followed, observers credited the pits with various other tasks from lubrication and washing of the cornea to breathing and smelling, as though they were an extra pair of nostrils—pitvipers are still known as *cuatro narices* in Mesoamerica. Then, in 1934, a biologist named Noble realized that they functioned as infrared retinas, the serpentine version of the pinhole camera. Each forward-facing pit (the aperture) rests against a large cavity in an upper jawbone, on either side of the head, and is divided into two chambers by a thin diaphragm that is lined on the outer surface with heat-sensitive epithelial cells. A pore in front of the eye connects the smaller, inner chamber of the pit to the outside, balancing air pressure on both sides of the diaphragm. When we stare into a rattlesnake's blank face, we see the larger, outer chamber. I have read in several scientific papers that the shape of the pit is species specific, but after dissecting and preserving several dozen timber rattlesnakes I can report that I have *no* idea what shape their pits are. You can't miss it . . . and I can't describe it.

Pits lack lenses and their heat receptors number only in the thousands, in contrast to the millions of light-sensitive rods and cones in a rattlesnake's eyes. The trigeminal nerve, largest of the cranial nerves, delivers the fuzzy, heat-driven image directly to the eyes, where it overlays the sharper visual image. The result: at *very* close range a timber rattlesnake is lethally precise, even in the dark, capable of detecting a minute temperature difference of 0.002 degrees Fahrenheit, a realm unavailable to our own naked senses. In one laboratory experiment, a blindfolded rattlesnake successfully struck a mouse that was eighteen degrees Fahrenheit warmer than its surroundings and more than two feet away forty-eight out of forty-nine times. A snake with occluded pits was successful four of fifteen times. Sensing heat works both ways: a timber rattlesnake detects warm mice against cool backgrounds and cool mice against warm backgrounds. Thermosensitivity tests performed in a forked maze, each fork at a different temperature, demonstrated that rattlesnakes consistently took the comfortably cooler path through the maze. Snakes with pits plugged were successful fifty percent of the time. It's all about the minuscule temperature differences.

Benjamin Franklin promoted the snake as a cultural icon in 1751, when he wrote "Rattlesnakes for Felons," an essay that suggested the colonies export the snakes to London in retaliation for the king's unloading British felons on the colonies. Three years later, in the *Pennsylvania Gazette,* Franklin published the first political cartoon in an American newspaper. Triggered by the French and Indian War, the cartoon featured a timber rattlesnake chopped into eight parts— with New England at the head and South Carolina at the tail—each labeled with a colony's initials, accompanied by the slogan "Join, or Die." As the Revolution approached, Franklin considered the timber rattlesnake a fitting symbol of America.

> I recollected that her eye excelled in brightness, that of any other animal, and that she has no eye-lids. She may therefore be esteemed an emblem of vigilance. She never begins an attack, nor, when once engaged, ever surrenders: She is therefore an emblem of magnanimity and true courage. As if anxious to prevent all pretensions of quarreling with her, weapons with which nature has furnished her, she conceals in the roof of her mouth, so that, to those who are unacquainted with her, she appears to be a most defenseless animal; and even when those weapons are shewn and extended for defense, they appear weak and contemptible; but their wounds however small, are decisive and fatal: Conscious of this, she never wounds till she has generously given notice, even to her enemy, and cautioned him against the danger of stepping on her. Was I wrong, Sir, in thinking this a strong picture of the temper and conduct of America?

John Paul Jones, "Father of the American Navy," wore brass rattlesnakes on the lapels and buttons of his uniform. And from 1776 to 1778, Esek Hopkins, the first commander-in-chief of the U.S. Navy, flew a yellow flag with a coiled timber rattlesnake that bore the now-famous motto "Don't Tread on Me." The flag, known as the Gadsden

Flag, was the gift of Continental Colonel Christopher Gadsden of South Carolina, and is still flown in Charleston, the city where Gadsden first presented the flag. (A symbol of uncompromising patriotism, the Gadsden Flag has enjoyed a recent renaissance as a motif of the Tea Party—you can buy traditional yellow "Don't Tread On Me" T-shirts and bumper stickers on Amazon.com and elsewhere on the Web.)

The timber rattlesnake as a positive symbol for youthful America did not last long. Originally found in thirty-one states, from east Texas to Minnesota and from northern Florida to southern Ontario (and possibly southwestern Quebec—they may have occurred at a couple of sites over the New York and Vermont line, east of Mystic and at Covey Hills), timber rattlesnakes have been extirpated in Canada (1941), Maine (mid-nineteenth century), Rhode Island (1970), and Delaware. No one knows for certain if they ever occurred in Michigan. They're on the verge of vanishing from New Hampshire, where the use of snake oil as a panacea, rendered from autumnal fat deposits in the body cavity, led to their extermination before 1920 on Rattlesnake Island in Lake Winnipesaukee, the state's largest lake. According to Clifford Pope, curator of the Division of Reptiles and Amphibians at the Chicago Natural History Museum and the author of several popular "snake books" published in the thirties and forties, "Rattlesnake oil was once sold throughout the United States as a home remedy for numerous diseases, deafness, lumbago, toothache, sore throat, and other ailments. Later, as scientific remedies slowly replaced it, peddlers recommended it chiefly for rheumatism, and not without justification. It served as a lubricant for rubbing an affected part, the high price insuring much rubbing and, no doubt, better results."

Besides an island, New Hampshire has seven hills and one point with the name Rattlesnake, but only one active den, and that lies just outside a state park wedged between Concord and Manchester, the state's rapidly metastasizing urban center. Although they are now protected in New Hampshire, it may already be too late to save them. One early May, nearly thirty years ago, together with Jim Taylor, a University of New Hampshire herpetologist, I spent a fruitless afternoon searching for an old den on an outcrop slope in Bear Brook State Park. We found nothing.

In 1842, Zadoc Thompson wrote that timber rattlesnakes were nearly gone from my home state of Vermont, and just a century ago, fewer than two thousand crept among the ledges in all of northern New England. Far fewer creep there today. Of New York's approximately two hundred extant dens, fewer than sixty-five interconnect genetically.

Vermont, an acknowledged center for forward thinking and environmental justice, paid a bounty for timber rattlesnakes from 1894 to 1971—a dollar a tail—and then listed them as an endangered species in 1987. So deep runs the fear of rattlesnakes that a Vermont town clerk refused my telephone request to discuss the town's rattlesnake bounty records, reacting as though the book itself was toxic. Our own attitude toward rattlesnakes is more venomous than the snakes themselves. In the recorded history of Vermont, only five snakebites have been reported, and four were nonfatal: Bristol in the early 1800s, Ludlow in 1959, Blatsky in 2003, and East Steeple in 2010, when a middle-aged white male was bitten twice on the hand while attempting to move a snake off the road as his wife watched from the front seat of their car. What tool did he use to move the snake with? A tongue depressor. The fifth, Vermont's only fatality, marked by a weathered tombstone in Putney that allegedly reads, "Killed by a Serpente." In New York there has been one documented death in the twentieth century; in 1929, Charles Snyder, the former head keeper in the Reptile House at the Bronx Zoo, was fatally bitten while collecting rattlesnakes in the Hudson Highlands.

Timber rattlesnakes live at the tribe's northern edge and continue to roll evolution's dice, gambling that in a landscape with less competition they can survive and reproduce during a truncated growing season. They feed only six to twenty times a summer, mostly on white-footed mice and chipmunks, but a Methuselah timber rattlesnake may live to be more than forty years old, grow nearly five feet long, and support a spade-shaped head nearly the size of a baseball. For a rattlesnake to prosper in the Northeast, the number of frost-free days is not nearly so important as the number of days the snake's body temperature is above seventy degrees Fahrenheit, the lower end of their optimal range. If reptiles were classified according to growing zones, like garden vegetables, northern timber rattlesnakes would be Zone 4 snakes, found in Zone 3. To carry this analogy one step further, if you grew them in your garden, you would have to start them indoors under grow lights, transfer them to a cold frame on the first of May, and then plant them sometime after Memorial Day. They would be harder to start than melons, slower growing than eggplants, and, like September basil, need to be covered by newspaper at the first hint of frost.

In the Northeast, timber rattlesnakes den in precise locations—microclimates where the heat of the sun warms southeast- or southwest-facing rock outcroppings, usually steep talus or scree slopes below exposed ledges, sometimes in vertical or horizontal crevices. Deep fissures among the fragmented rocks allow the snakes access to frost-free subterranean chambers, where they ball together (sometimes by the hundreds in pristine dens like the one beyond Larry Boswell's cabin) to prevent both freezing and desiccation. Den sites, or hibernacula, are the limiting factor in their highland distribution, and these long-lived, slow-reproducing reptiles are homebodies, not likely to repopulate depleted dens.

Even Carl Kauffeld, who had visited dozens of dens repeatedly, wondered about the lack of magnetic snake appeal at a seemingly good-looking den site.

> There is a point on the Appalachian Trail Herb and I visit very near the New York-Connecticut boundary—where we would find one, two, or three snakes, never more. The trail crosses through very characteristic "den rocks," but the number of snakes hardly warrants considering this a den. Could this be an incipient den—one that is just beginning to be used by the local rattler population—or is it an alternative for a few stragglers who fail to reach the main dens before cold weather? Whatever the explanation, it is a lovely spot—one that I would approve very highly if I were a snake.

At the northeast extreme of their range, in the foothills of the Adirondacks, based on still unpublished research by W. S. Brown, most female timber rattlesnakes produce young for the first time at nine or ten, occasionally as early as seven, and sometimes not until eleven or twelve, usually reproducing at only three-, four-, or five-year intervals. Less than half of adult females live long enough to breed more than once in their lifetimes. (For timber rattlesnakes calibrated for life on the High Allegheny Plateau of West Virginia the margin between survival and extirpation is razor thin; these hardy snakes have the latest age at first birth, sometimes as early as nine, but typically ten or eleven, and the longest average interval between births, typically four or five years, of any known species of snake ... anywhere.) Gravid snakes stay near the den all summer, basking and fasting, living like monks off the proceeds of long-forgotten meals. It takes years to recover from the pregnancy. Six to ten very vulnerable young, born in late summer, remain near their mother for a week or more until they shed, and then the crisp air of October ushers them all below ground.

🐍

Hank, named for country singer Hank Williams, soaks up sunshine, yolk yellow on gray rocks, while an irate catbird torments the snake from the crown of a nearby sumac. The Finger Lakes Land Trust purchased the eight-hundred-acre hillside preserve in 2001, in part to protect a rattlesnake den; in 2006, photogenic Hank put the preserve on worldwide television when David Attenborough's BBC crew filmed him catching a chipmunk and a white-footed mouse, the first capture-and-feeding sequence of a wild rattlesnake without manipulation of either prey or predator. Carrying a tiny, surgically implanted radio transmitter in his peritoneal cavity makes Hank available for daily observation. When the rattlesnake finds a potential ambush site, a log or runway rife with the odor of prey and, as Clark has recently discovered, the smell of success—Hank determines from chemical clues where and when a neighboring rattlesnake has scored a meal—Clark sets up a motion-sensitive surveillance camera to film any encounters that may occur. During this two-year project, Clark spied on seventeen different rattlesnakes (eleven females and six males)—a total of eight hundred forty hours of ophidian eavesdropping—and recorded eighty-seven encounters with prey, eleven of which, a mere thirteen percent, ended in a meal for the snake.

Today, Hank sunbathes, his coils loosely stretched across the stone foundation, seemingly oblivious to our presence. When he resumes serial foraging, the weft of odor trails that streak the hillside—those of potential prey as well as those of other rattlesnakes—will be his signposts. The wooded hillside is etched with species-specific chemical conveyances, an arabesque of molecular messages both inadvertent and intentional, whose meaning Clark has begun to interpret, revealing a world far beyond the threshold of both our senses and our sense of time.

§

On a warm day in late April or early May, I might see twenty or more Vermont timber rattlesnakes charmed by the sun, as they loiter around the mouth of one of the state's six dens, waiting with the patience of Job for nights to warm. They then disperse up the ledges or down the valleys and wander in serpentine loops across miles of woods and wetlands, feeding and slithering, blazing species-specific odiferous paths, until the chill of late August lures the snakes back toward the den. Big males leave first, then nonbreeding females and young. Pregnant females stay behind and bask and birth at specific sites, often on the ledges above the dens.

This ability to forecast the snake's behavior presents a problem. Northern timber rattlesnakes sell for a hundred dollars or more on the black-market pet trade. Collectors, who easily sack ten or more snakes in a day, pillaged a Vermont den I've visited numerous times, owned and protected by The Nature Conservancy. The same trait that made the snakes susceptible to bounty hunters and wanton killers, their predictability, makes them an easy target also for collectors. If you know where a den is, snakes will be there in season, entwined and basking, benign and vulnerable.

§

Vermont's surviving snake dens are in the southwestern part of the state, where Vermont's border, having run north from the Massachusetts line for fifty perfectly straight miles, doglegs toward New York. These relic dens were once part of a larger metapopulation that includes many more (like Boswell's) in the Adirondack foothills. The ledges of the Vermont dens face southwest, rise from the valley for several hundred feet, and support, in addition to six snake dens, a disjunct population of five-lined skinks. Vermont's only lizard, the skinks share a similar demography with the rattlesnakes: widely scattered colonies, vestiges from a prehistoric warm spell five thousand years ago, called the Hypsithermal Interval, when both crept (the snake) and scurried (the lizard) over much of the Northeast. A wet meadow stretching from the base of the bluffs marks the former floodplain of an Ice Age river, which once drained an enormous Ice Age lake. Champlain's long tail, which dominates the view from one of the dens, follows the river's old drainage.

This is an isolated corner of Vermont, a rural farming community cut off from the rest of the state by New York, which surrounds it on three sides—in fact, you can watch the sun rise over New York. To reach two of the dens, I drive on an unimproved, unmarked beaver-flooded dirt road that sees more muskrats than cars. Of the more than twenty known dens in Vermont that once supported rattlesnakes, only six still do. The others suffered the fate of the den in the George Catlin painting of the Susquehanna snake hunt.

Two of the dens are remote; four are along a ledge above a busy state highway not far from a Walmart. On Sunday afternoons, the drone of shoppers fills the valley. In late April and early May, wildflowers pepper the woods—white trillium, Dutchman's-breeches, corydalis, hepatica, spring beauty. Trees are leafless and full of birdsong, tattered branches often raking a cloudless sky. One site supports ledge-nesting peregrines. From another, I have watched ospreys build a huge nest in a marsh along the Blatsky River. The valleys below the ledges hold leopard frogs and wild turkey, while above, ravens, kvetching and proactive, dive and roll and glide on gusts of wind that deflect upward off the rock faces. A warm late-April sun brings out the snakes.

I heard my first Vermont rattlesnake more than twenty-five years ago, long before I saw it, a visceral message that shot up my spine, a loud, whirring buzz, like a deeply disturbed hornet. At the time, I wondered if I should have stuck to the trail, my back to the valley, and minded my own business. It was late morning; the rocks were warm, and I was told that the snakes would soon be emerging from the den to raise their body temperature with a round of catalytic sun worship, making them active and alert, ready.

I knew from numerous treks in the Southwest that a rattlesnake, bearing the color and markings of home, remains magically camouflaged, even when coiled at your feet. Alcott Smith, the veterinarian, who has recently spent more time in the field with rattlesnakes than in the clinic, once accidentally stood on a rattlesnake until the snake bumped his calf with the side of its head, advising him that there was a problem.

I continued roaming around the warm slabs in a heightened state of alert, studying each rock and shadow, every fallen branch as if it were a decryptable atlas. Scrambling around rattlesnake country in the Northeast, particularly when snakes aren't carrying a Rulon Clark radio transmitter, is different than wandering through the desert, where paths are open and snakes are not apt to be out in the heat of the day. In a hickory-oak glade, grasses, sedges, and blueberry bushes mask the ground. Only the rocks are bare. And since rocks shelter snakes, I walked (and still do) with considered steps.

I heard a second snake, a light buzz, more like a cicada than a hornet. In both cases, I momentarily froze. To my right, less than three feet away, a small timber rattlesnake, its two-segment rattle deliriously shaking, coiled in self-defense. The yellow and brown-banded snake, about eighteen inches long, was less than two years old. Feeling the vibrations of my footfalls, it had gathered into coils and raised its head, forked tongue suspended in midflick gathering odors for the Jacobson's organ, a chemoreceptor on the roof of the mouth. Although its blood ran cold, the rattlesnake's tongue revealed its mood. A provoked snake flicks more; an excited snake yawns to flood its mouth with stimuli. A placid snake, like central New York's Hank, remains uncoiled and untroubled, its tongue inside its mouth.

For the moment, that small rattlesnake had a split personality. While its tongue lazily arced up and out, slightly quivering, its tail vibrated so quickly that the rattles blurred. The May woods absorbed the buzz, barely audible beyond twenty feet. When I reached the summit, I saw the first snake I had heard, a big, black male recumbent on a flat-rock bed, the sweep of the wet meadows behind him.

More recently, on a pair of warm mornings in late April, I visited Vermont's snake dens with Smith, who had been monitoring rattlesnakes for the state for twenty-four years and has an encyclopedic knowledge of his subject built from firsthand, sometimes hands-on experience. Smith knows the rattlesnake hangout spots: basking and birthing sites, den portals, and the avenues of egress, those vertical paths the snakes follow from the den up the ledges and into the spring woods. Smith, who strides up broken terrain like a bighorn, is even tougher to keep up with in the woods, where he glides over the ground like the ultra-marathoner he used to be. He trained by running a weekly fifty-mile loop, beginning his route shortly after midnight. As a boy, whenever ice glazed southern Vermont, Smith and his twin brother, Avery (also a veterinarian), skated cross-country, over mountains and through woods. That he's well over seventy is of little consequence.

At all four dens, we visited snakes either basking or moving unseen through the talus, scales scraping rock and brittle leaves. At one site, the slope above the marsh, snakes were tinged with orange, stained by a veneer of iron oxide—rust—that coated the iron-rich talus. The snakes were adults, smaller than Hank and Travis, but large by Vermont standards, thirty-five to forty inches. In a week or two, the yearlings and pregnant females would be out.

Once, in 2003, on a warm, cloudless afternoon in midsummer,

when the ledges were too hot for sunbathing snakes, in the company of Smith, I counted eleven snakes, mostly gravid females in the shade of overhanging rocks. One was as dark as coal; another, dull yellow and banded brown. Some had recently shed, their eyes crystal bright. Others appeared ready to burst their worn, dull outer skin. Two were braided together on a mat of oak leaves beneath a rock awning, a living rope that sighed and heaved in almost imperceptibly slow, rhythmic pulses that, by the following August, might yield a half dozen snakelets, soft, pinkish, and vulnerable ... the tenuous future of this species. Altogether the snakes—the couple and the singletons—lay coiled at the apex of reptile evolution, forged in the furnace of a bygone age, and, like Greta Garbo wanting only to be left alone.

I returned to the ledges the following August, after a cool spring and summer—it rained the first thirteen days of the month. Because of the unseasonable weather, rattlesnake birthing was late throughout the eastern mountains from Virginia north. According to Smith, whenever the summer temperature drops an average of one degree Fahrenheit, parturition is delayed eight days. At the second birthing site Smith and I visited, five neonates, beige with pink tinges—the color of native rock—and each about the width of my thumb, basked in the late September sunshine. Four of them draped their mother, while the fifth, somewhat bolder, coiled a few feet away in the open on the ledge above and behind its siblings. And then yawned, a wide, rubbery smile.

Around the den, we counted nine adults, six males and three barren females. Using aluminum snake tongs, a utilitarian cross between a golf putter and a pair of pliers, Smith removed a dark-phased female, wedged tightly in the rocks, for closer inspection. Holding her head securely in his fingers, he gently squeezed her mouth open to demonstrate how the fangs remained sheathed until just before biting. I ran my fingers down the snake's rough back and then, squelching her buzz, I held her vibrating rattle in my hand.

On the cliff above the den, a three-foot snake, black etched with broken yellow bands, shiny as a new car, coiled almost invisible in a fractured beam of sunlight between a hickory and an oak, screened by tufts of sedge. The snake was coming home. Two days later, slowed by hard, cool rain, the rattlesnake had progressed a mere twenty feet.

After our two-day April survey, I returned with Smith on May 28 and found only small rattlesnakes, a year or two old, and one pregnant female, lounging in the sunshine beneath a balcony of rock. Carefully, gently, Alcott caught the snake and then secured her head between thumb and forefinger. I rubbed my fingers along her sides. Eight oval embryos, like beans in a pod, firm and growing, crowded her abdominal cavity. There was little room for food.

In the woods beyond the ledge, we found a three-foot-long rattlesnake coiled against the trunk of a red oak. The snake was outside a denning zone, en route to life obscured by the oak-hornbeam woods of western Vermont, feeding and moving and basking. If someone (like us) found this snake, it would be quite by accident, which is why Rulon Clark's research in central New York so intrigues me. Clark, carrying a collapsible antenna, radio-tracks the snakes as day length and scent trails guide them. After watching Hank sunbathe, which is analogous to watching a length of thick, resplendent cable, Clark carefully captures the snake with snake tongs. He coaxes Hank into a Plexiglas tube. I hold the rear half of the snake, rough and muscular, while Clark examines the front half, safely stowed in the see-through tube. A scar on the lower jaw, likely the mark of an aggressive and desperate mouse or chipmunk, and the black stitching like aberrant hairs that hides the radio transponder just inside the coelomic cavity are all that seem out of order. We return Hank to the foundation, where he pours down the stones, disappearing into a cowlick of brambles.

The catbird becomes apoplectic again.

Back in a delicatessen in Ithaca for lunch, Clark opens a laptop, downloads the snake-surveillance videos, and then plays snippets of the hidden lives of timber rattlesnakes for me. The sequences are short, seconds really. A mouse runs over the snake, peeks into the camera lens, and then hops away, unscathed. A gray squirrel runs past the snake, leaps away, and then returns to torment its incubus, jumping, chattering, flicking its bushy tail, just beyond strike range—half the length of the snake. As long as the squirrel hurls its insults, the serpent, cryptic or no, will surprise nothing. The footage reminds me of crows or jays mobbing an owl, only the squirrel is a lot closer to the business end of the rattlesnake.

Of the eighty-seven encounters with potential prey that Clark clandestinely videoed for two years, he recorded antipredator behavior twelve times, which had been unknown previously in the life of woodland rattlesnakes. Seven of these incidents were promulgated by chipmunks, four by gray squirrels, and one by a wood thrush, which took time off from making beautiful music to pester a snake (just as the catbird did to Hank). Harassed snakes gave up hunting and moved some distance away. Because foraging rattlesnakes (except when striking) respond with glacial slowness, in all twelve cases, the time between discovery and departure took over an hour.

Next clip, head against a log perpendicular to the runway, a rattlesnake waits in ambush, choosing a potential feeding site by using its forked tongue to read the odor of prey. (All snake species have sensitive tongues. A single tongue flick, for instance, alerts a garter snake to the size, sex, and species of another snake, whether it has recently mated or eaten or shed, whether it's healthy or not.) If log or runway or burrow smells (tastes) promising, the rattlesnake stays. Instead of a mouse, a long-tailed weasel appears on the screen, perhaps attracted by the same cocktail of odors, and is struck on the hip—a blur even when the video is slowed down. The stricken weasel then leaps out of view, also a blur. A strike covers a distance of a few inches to more than a foot. The sequence from start to finish takes less than half a second, less than a heartbeat. After several minutes the rattlesnake slowly unravels, tongue extended, reading the air, then moves off-screen to swallow the weasel, which is already being digested from the inside out by enzymes in the venom. Like most things a rattlesnake does, consumption of prey, usually head first, is slow and methodical. After a meal, the snake relaxes and then, driven by hunger, moves to another promising site. And waits. And

waits. For mice, usually—white-footed and deer mice, meadow voles, red-backed voles, pine voles—but also shrews, chipmunks, squirrels, and cottontails. And sometimes nosy birds, like catbirds. When a timber rattlesnake strikes a bird, it hangs on until the bird stops struggling, since unlike a fleeing mammal, a bird flies away before dying and would thus be difficult, if not impossible, to track.

On one of Clark's videos, in a case of turnabout, a horned owl checks out a rattlesnake as *its* possible meal. A Pennsylvania biologist once tracked the beeps of a radioed rattlesnake to a squirrel's nest in the crown of a chestnut oak, a very unlikely spot for a rattlesnake. A day or two later, a beeping transponder was found in a coughed-up owl pellet at the base of the tree. Accept for red-tailed hawks and great horned owls (both serious rattlesnake predators), a full-grown snake has relatively little to fear. For the neonate, however, there are many pitfalls, particularly on the road to the den, the den it has never seen. Recently, I found a horned owl pellet packed with tiny rattlesnake scales on a slope above Vermont's most active den. A New York naturalist watched a strategy-savvy adult black racer, *Coluber constrictor*, a reptilian MacArthur, attack a pod of neonate timber rattlers. The racer disabled four or five with a bite and a chew to prevent their escape, and then ate each one in succession, headfirst. In the Northeast, the gauntlet also includes wild turkey, opossum, raccoon, red and gray fox, long-tailed weasel, and coyote.

On another of Clark's videos, a deer vents, prancing and pawing the ground, utterly disrupting the snake's solitude. On my favorite video, an unsuspecting hiker finds the remote camera, performs for the lens, his wide-angle, full-faced smile like the Cheshire Cat's, his left foot inches from a coiled, yet placid, rattlesnake.

Part One

EGRESS

THE DANGERS
OF LEAVING HOME

2
A Quirky Subculture

The time you enjoy wasting is not wasted time.

BERTRAND RUSSELL

On the shore of the Hudson River, on a late September afternoon, at a field station in northern New York, four timber rattlesnakes lounged in the foyer: two adults, a black and a yellow morph, and two post-shed babies, also a black and a yellow morph, each in its own bulb-heated, pad-locked aquarium. The snakes belong to William S. Brown, a retired Skidmore College biologist, who lives across the river and over the corrugated hills, not too far from a several thriving snake dens. Brown is a rattlesnake legend; he is to the timber rattlesnake as Ted Williams is to baseball: exquisitely skilled, self-absorbed, and disposed by experience to be surly.

I opened the heavy oak door and stepped inside. Stacked like a garden hose, the big black morph buzzed, its tail a whir of motion, a reptilian doorbell somewhat out of sync with my entry. Except for the kitchen and bathroom, the institute's downstairs walls were decorated in academic-rattlesnake motif with charts, maps, graphs, photographs, posters, and newspaper articles, some going back more than ten years. The most inspiring of these was an obituary from the *New York Times*, which covered a corner wall in the recreation room above a pool table supporting photographs of fungus-riddled rattlesnakes. The obit chronicled the life of the late Roy Pinney, who died in 2010 a month shy of his ninety-ninth birthday. A Brooklyn

Jew of Polish decent, Pinney dashed off to the jungles of South America alone at the age of seventeen, where he married a native named Mina, wore a loincloth, ate monkey meat, illustrated himself with traditional tattoos, and "danced to delirium with the Gods." After his young bride died of malaria, Pinney returned to New York City where he did a good many more unusual things, among them making a hands-on study of timber rattlesnakes, which remained a lifelong passion. Pinney was the naturalist who had watched a black racer chew through a pile of baby rattlesnakes in the Hudson Highlands. After the last neonate was paralyzed the racer swallowed five in succession, nullifying a mother rattlesnake's three- or four-year reproductive effort. Although Pinney adored the baby rattlesnakes, he *never* interfered. Of a length reserved for diplomats, his obituary omitted the story of the cannibalistic racer, but did mention that Pinney kept the company of rattlesnakes in his Manhattan flat until the end of his life.

Brown was responsible for the décor, carefully selected to highlight the life and times of both his favorite reptile and those who share his passion.

As I mentioned previously, I saw my first rattlesnake in the late fifties behind glass in the Staten Island Zoo, one of Carl Kauffeld's Ramapo

Mountain trophies. In front of my eyes a thick, black morph had crawled out of its old skin and had left behind an inverted sleeve of itself. For a young boy, watching a snake was an awakening of sorts—when the snake was finally through casting off its worn skin what had been a taut and muscled, but somber-looking rattlesnake was as bright as a new penny. I was ineluctably hooked. For decades my interest in them flickered and flared, but it has recently been reignited. For the past six years, I've been on their trail full time in the field, in labs, zoos, libraries, and museums, over the phone, on the Internet, through letters, e-mails, and conversations—even in my dreams.

Because my mania leaked to William S. Brown (whose friends call him Bill), in 2011, I was invited to join the Bashers, a quirky subculture of timber rattlesnake biologists and enthusiasts, for their annual autumn gathering at the riverside field station. I was *not* invited as a Holo-Basher, a title reserved for tenured members like W. H. Martin (Marty to friends), Randy Stechert, Tom Tyning, and Alcott Smith among others. I was an incipient member called a Hemi-Basher (Latin for *half* a Basher), a carefully crafted nickname that makes reference to a prong of the male snake's cloven penis or hemipenis. Each member at the meeting would present an informal lecture on one of a range of topics related to timber rattlesnakes: conservation, evolutionary genetics, distribution, sex life, snakebite mistreatment, and regional extinction, as well as the snake's relationship to acorns, gypsy moths, gas pipelines, and hydrofracking the Marcellus Shale, the gas-bearing rock deep in the basement of Pennsylvania and western New York. Although six of the sixteen Bashers who attended the meeting with me had been bitten by timber rattlesnakes sometime during their careers—some more than once—I found it profoundly reassuring that the fate of this rare and reclusive reptile (and of future Basher reunions) rested largely in their hands.

Several Bashers gave artful PowerPoint presentations; two described in lurid details a near fatal case of anaphylaxis suffered during a recent envenomation; and I read excerpts from my work in progress. "To keep spirits from lagging" and to add zest to the discussions, a quarter keg of solar-brewed beer was on tap in the kitchen sink. The beer was from the Barrington Brewery, compliments of Anne Stengle, a graduate student at the University of Massachusetts, Amherst, her thank-you to the Bashers for supplying her with small pieces of snakeskins gathered from all over the Northeast for DNA analysis, the essence of her doctoral research. Libations were a *big* hit and kept the proceedings bubbling along. One Basher, who had never matriculated but was nevertheless up on adolescent pedagogical customs, washed down breakfast with beer. Our dress code was equally informal; an e-mail from Brown suggested that anything was appropriate—modest fig leaves, shorts, air-conditioned T-shirts, and ragged jeans. More than half the group wore rattlesnake T-shirts. Some wore custom-made Basher T-shirts.

The group had first assembled in 2007 for "The Big Book Bash," in celebration of the publication of Jon Furman's *Timber Rattlesnakes in Vermont and New York*. Many of the original Bashers had guided Furman through the quagmire of rattlesnake natural history that he needed to complete his book. (Brown had served as midwife.) Other members were luminous add-ons and in-the-trenches commissioned ophidian biologists, mostly from New York State. Everyone had so much fun they decided to meet annually, in the process truncating their name and expanding their membership.

In addition to lectures, readings, multimedia presentations, and beer, the weekend agenda also included field trips, a tricky seventy-four-question ID quiz on reptiles and amphibians of New York (I got sixteen wrong), a roast-and-toast pictorial history of the previous four gatherings—nobody was immune from Brown's jesting—catered dinners, and undiminished rattlesnake-oriented schmoozing. Brown is an iconic scientist who has achieved near rock-star status in the world of herpetology. There has been no end to the visitors (myself included) who want, and in some cases demand, Brown's time. For decades he has attracted groupies, with some of whom he has become attached more significantly as research associates, a girlfriend or two, a wife, and enemies. A few malfeasants have become saboteurs who have rearranged snake rocks and posted taunting notes in nearby trees. Some have removed research animals, an unforgivable transgression. In the days before GPS, one of Brown's guests, who had attended the biologist's third wedding but later fell out of favor, flagged a mile-long trail to a remote den so that he could return there on his own. Another rattlesnake-hotshot told Brown that he'd tallied more than two dozen rattlesnakes at a den and would continue the study when the biologist becomes too old to scramble across the rocks. Even though the gregarious biologist celebrated his seventieth birthday during the Bashers with a cake that had a timber rattlesnake embossed in the icing, losing his footing on loose rock appears to be a long way off for Bill Brown—he's as nimble as an Iberian ibex and as tenacious as a badger. In an effort to stop den-side vandalism, Brown distributed "wanted" posters throughout the rattlesnake zones of the Northeast—a hand-drawn sketch on one had the harried, over-the-shoulder look of Bigfoot.

I first encountered the name William S. Brown in the biological literature in the eighties while at work on my first book. Several years later, I interviewed him over the telephone for a magazine piece I was writing; although we spoke then for nearly an hour, he doesn't recall our conversation. Brown's work in the Adirondack foothills is the longest, largest continuous capture-recapture study of any snake in the Western Hemisphere, possibly of any reptile on the planet. He began his fieldwork when Jimmy Carter was president and has marked and released a *substantial* number of timber rattle-snakes since then (he says that the number is neither relevant to the theme of his study nor to my book). Brown has written a score of erudite articles published in peer-reviewed scientific journals, on topics such as ecology, behavior, denning, movement, temperature relationships, hibernation, scent trailing, reproduction, and genetic diversity. When he finally quits the hills, his protégé Matt Simon will carry on, and his field notebook, an academic work of art, should be placed on exhibit in the Smithsonian. The notebook's computer-coded data are replete with numerous neat, small-print columns including details of each snake's specifically assigned scale-clip num-ber, color morph, sex, length, weight, date, sequence number for the day, catch site, recapture history, release date, and hibernaculum (snakes from five adjacent dens received a dab of paint on their rattle that corresponded to each of their home dens: red, orange, green, blue, or yellow). There are columns for rattle structure, injury status, reproductive state (if female), shedding state, body condition, and so forth. In the back of the notebook is a reference series of labeled, glued-down pieces of snakeskin that demonstrate how the scale-clip marks can be read as a permanent scar on ventral scales of a shed skin. "So far," he told me, with a sense of paternal pride, "I've gotten about four dozen records based on marked skins."

In addition to his own writing, Bill Brown is an editor possessed, the prototypical schoolmarm. When he has been tasked with reviewing drafts of articles submitted for publication in scientific journals, reader-friendly magazines, chapters, and even full-fledged books, each manuscript is so densely critiqued that plumes of red corrections erupt from page margins and between the lines. One colleague who had a lot of Brown's red ink to wade through referred to the experience as Brownamous. He makes unsolicited edits to e-mails, mine included.

"Bill taught me two things," said Randy Stechert, who worked closely with Brown for four years in the early eighties, helped jump-start his Adirondack project, and has remained a good friend: "the importance of keeping meticulous notes and the importance of self-promotion." Brown has taught courses on snake ecology, behavior, and life history, including two courses specifically embracing timber rattlesnakes. He has also authored the only scientific monograph on the biology and management of the northeastern timber rattlesnake and in 1987 wrote a feature for *National Geographic* based largely on his own Adirondack research, which was how I became aware of him. Much of what is known about the snake begins with the work of William S. Brown. In fact, I haven't read a journal article on timber rattlesnakes that didn't cite him at least three or four times.

Bill Brown is old school: full white beard and wavy white hair combed straight back like a whaler aboard the *Pequod*, what my boys refer to as *flow*. He's intellectually agile, theatrical, opinionated, obdurate, funny, and sometimes pompous—earlier in the year, when he invited me for the first time to join him in the field, I was interrogated about my intentions; until he trusted me, he was chary about divulging too many secrets. Brown's capacity for grudges, particularly when it concerns rattlesnake vandals, is the stuff of herpetological legend. In 2008, when notorious rattlesnake poacher Rudy Komarek died, a Basher colleague claimed that the biologist called Rudy's next of kin and demanded proof. "Are you sure?" His humor borders on self-deprecating, and he is a target of both wit and adulation among the Bashers, which he relishes.

Bill Brown grew up in rural Pennsylvania rooting for the Phillies. As a boy, the Wild West, a theme revisited throughout his life, enchanted him; he earned bachelor's and master's degrees from Arizona State and a doctorate from the University of Utah. His mother drove a tractor. His father killed a water snake, which kick-started an inchoate zoological inquiry. The adolescent Brown unwound at the Philadelphia Zoo, in the reptile house mostly, where the curator, Roger Conant, was solicitous of his curiosity. He began his professional herpetological career as a teenager guiding tourists at the New York Serpentarium, a roadside attraction in the Adirondacks, where he played the "great white hunter," replete with pith helmet, vest, and boots, and entertained crowds by handling the "world's most dangerous reptiles."

Brown married for the first time during graduate school. His second wife he met at Skidmore. They divorced in 1987, several months after their only child, Bonnie, died of brain cancer before her eleventh birthday. Brown dedicated his landmark publication *The Biology and Management of the Timber Rattlesnake (Crotalus horridus): A Guide for Conservation* to Bonnie and prefaced one of his informal Basher presentations with her full-screen image. Then, after losing a herpetological-groupie girlfriend to a terminal illness, he married Sheila, a turtle biologist, in the mid-nineties, with whom he still resides.

Bill Brown is the Basher fulcrum. He determines the roster. He's not a misanthrope, but because the research center is short on space and he's selective about confidants, invitations are limited. Diamondback Dave and Kevin McCurley, a well-known herpetoculturist with whom the survival of New Hampshire rattlesnakes rest, have made cameo appearances at previous meetings, but were *never* invited back. The 2011 Bashers came from Vermont, Massachusetts, Connecticut, New York (mostly), Kansas, Minnesota, West Virginia, and Ontario—more or less outlining the historic northern range of the timber rattlesnake. A number of Bashers had collaborated with Brown for three decades. To them, he's a beer-drinking bud, the focus of good-natured ribbing. At the gathering I attended, almost anything he said evoked a wisecrack. After an innocent statement triggered a swift and jocose rebuttal, Brown followed.

"I should know if you try to say anything with this group, you come out a loser."

"Only if you start out a loser," someone rejoined without a moment's hesitation.

"That's what I mean. That's what I … mean."

Brown ran an entertaining meeting. After pizza Friday night, the quiz's top three scorers each received a fine-art reptile print, and a jar of homemade grape jelly was given to the person who had seen the first herp of the day, a leopard frog. Other jars were awarded for the first snake sighting—a garter snake—the first rattlesnake, and the first rattlesnake at the second den.

"Did you strain out the seeds?" a winner replied, looking incredulously at the jelly.

Saturday morning breakfast conversation touched on arcane details of milk snake taxonomy, with as much zest and commitment as if the biologists had been debating the Talmud; the best dates to implant a radio transmitter in a timber rattlesnake; whether a snake with a transmitter feels bloated, causing it to bask longer in the fall; the anatomy and evolution of the rattle; whether a rattlesnake transmits sound to the inner ear through its rib cage, draped over the ground like a series of tuning forks; whether a shedding female releases pheromones through her pores or cloaca or both. George Pisani told the story of an esteemed elderly Kansas herpetologist who used hamburger tongs to catch copperheads; he'd slide the spatula underneath the snake and clamp the gripper portion just behind the neck. The herpetologist's wife forbade him to study the larger, feistier timber rattlesnake because the handles were too short.

Joe Racette is forty-something, tall and rawboned, and not innately drawn to rattlesnakes. He is, however, the biologist in New York's DEC vested with their recovery and has visited many of the state's dens, often in the company of Bill Brown. Racette regularly attends Bashers meetings to learn more about the snake and to seek advice about the state's ambitious Timber Rattlesnake Recovery Plan. His Adirondacks office is close to Brown's research sites, one of the most stunningly beautiful rattlesnake neighborhoods in New York State. For the snake it is all about location, location, location. As you may have already gathered, timber rattlesnakes have an eye for real estate. Indeed, they're landscape connoisseurs: rising above lakes and rivers and green sprawling, stream-threaded valleys like so many solar panels, their dens face the sun and hold heat on chilly October afternoons. They thrive where the human population is sparse — land that is wide open, windswept, and remote. And like Beethoven, who couldn't hear the sound of the very music he composed, timber rattlesnakes can't see the view from where they live. They're as nearsighted as Mr. Magoo.

In 1983, nine years after it discontinued the bounty on timber rattlesnakes, New York State classified the timber rattlesnake as threatened, according it legal protection. The recovery plan, which set an ambitious, if not unattainable, goal, calls for a statewide population of twelve thousand (excluding neonates), two hundred viable dens of which at least half have reproductive intermingling (a.k.a. genetic connectivity) with nearby dens, and the number of breeding females, the benchmark for demographic stability, should account for fifteen percent of the total population. Racette says, New York wants to keep the timber rattlesnake *on* the "threatened species" list so the snakes will continue to be an effective bargaining chip to mitigate future development projects. He estimates the statewide population approximately seven to eight thousand breeding adults, and says the snakes are up for review and their protected status will be in jeopardy if bureaucrats delist them.

Although the endangered species team wants the timber rattle-snake to remain threatened, New York State is *not* interested in either head-starting neonates or repatriating extirpated or diminished dens. That decision, driven by fear—the fear of liability if someone is bitten—was an executive decision. Attorneys versus biologists, said Racette. "As biologists we know that if we really want the species to recover we might have to actively intervene at some locations. If someone else [a university biologist, for instance] came to us for a permit to augment or repatriate a den, we'd evaluate the project on its biological merits. If we granted a permit, the licensee would assume the liability." There are several isolated populations in the state, including a lonely outpost along the Mohawk River, where the drone of the New York State Thruway long ago swallowed the sound of rattles (as well as the drums along the Mohawk). Recently, the state denied a permit to replenish a dwindling population in Region 8 near Rochester, lamented Racette. "It's all about avoiding lawsuits."

Across the dining room table, a Basher articulated what everybody else had thought. "Lawyers function with a whole different lexicon and they bring to the table the same apprehensions and stupidity shared by the general public. I think everyone in this room knows the danger [of getting bitten by a timber rattlesnake] is *not* high." "Our university lawyer," responded another Basher, with whom he admitted to having had a few run-ins over indigenous venomous snakes that prowled rural campus property, "is erudite and proficient in all aspects of law, but he's basically scared shitless of snakes." According to New York's environmental lawyers, when you enter wild country you accept some level of risk. The difference between conserving what is already there and repatriating, said Racette, is "if *we* introduced the snakes, we would increase the public's risk." Racette then told the story of four camp counselors who drowned at the base of an Adirondack waterfall. Their parents sued the state. "Had New York lost the lawsuit the state would have been forced to close down vast areas of public land." It is for the very same reason that New York is unwilling to reintroduce moose (or anything else that might bite you or kick you or run into your car). "If we fostered the development of a moose population and someone died in an accident, the state of New York would be liable." If a moose wanders over the state line from Vermont, however, which is how they returned themselves to Vermont in the first place—having come in from New Hampshire in the eighties—that's another story. In the eighties, Randy Stechert identified two hundred thirty-five dens in New York State (likely there are more), of which one hundred ninety still support rattlesnakes. "We're not looking to increase the range of the timber rattlesnake in New York," repeated Racette. "We don't want them anywhere where they don't currently exist." If an itinerant moose or a rattlesnake were to reappear at an extirpated site, that would be OK.

Bashers W. H. Marty Martin and Al Breisch, the retired New York State herpetologist, coedited with Earl Possardt, W. S. Brown, and John Sealy *The Timber Rattlesnake: Life History, Distribution, Status, and Conservation Action Plan*, which divides the eastern United States and southeastern Canada into fifty-square-mile quadrants. For security reasons, rather than placing a symbol directly on a record, if an active den occurs anywhere within a quadrant, a solid black circle punctuates the middle of the quad. No further information on den location is given. In New York, if a developer proposes a project close to a den, a consultant (often Randy Stechert) is hired by the state to determine the possible impact and to help negotiate a solution. Because collection pressure on the timber rattlesnake makes it vulnerable to the illicit pet trade, public information in the state recovery plan has been censored. It's basically "to hell with the freedom of information act": the New York State Timber Rattlesnake Recovery Plan does not, will not, and cannot divulge den locations. To put this mandate in perspective: if requested, New York State will release information on the location of a bald eagle nest, but *not* on a timber rattlesnake den, the simple reason being that you can't take an eagle home with you.

One facet of the New York State Timber Rattlesnake Recovery Plan that Bashers find particularly annoying, ludicrous really, is the biosafety protocols the state prescribes to field biologists. In order to prevent the spread of disease, Racette reported researchers must disinfect their snake hooks, scale-clipping scissors, and snake bags, basically anything that touches a rattlesnake between snakes processed and site visits, all the while precariously perched on loose talus. Remember that timber rattlesnakes are social serpents. They stack together during hibernation. They entwine in piles when they're babies and when basking. A courting male may accompany a female for several days, all the while caressing her flanks with the side of his head and copulation may last for hours. Also, male-to-male combat dance is a vital part of social hierarchy. Amid beer-enhanced laughter, one Basher suggested that the state insist that field biologists wear HAZMAT suits.

Or, blurted another, "Just spray the damn snake down with Lysol."

Said Bill Brown: "I've got two or three snakes [pinned] under my left foot all nipping away at my leggings and I'm rigging up my snake-bag. It is impossible for me to sterilize my equipment between each snake."

"The point is," replied Stechert in a nanosecond, "you shouldn't have two or three snakes under your foot, Bill."

"I'm not allowing any of them to escape. I'm maximizing my effort."

Perhaps New York State ought to insist that Brown disinfect his boots between snakes.

Racette, who signs off on all New York's timber rattlesnake research proposals and commissions Stechert to monitor snakes at mitigation sites in the Hudson Highlands and the foothills of the Catskill Mountains, admitted that fishery biologists wrote the rattlesnake protocols because both bats and rattlesnakes winter in dark and dank retreats, and they feared the spread among snakes of a disease like white-nosed syndrome, which has already decimated wintering bat populations in the Northeast. Despite the proximity of Racette, the beer-happy Bashers are unanimous: let the state write whatever biosafety protocols it wants, and then let the document idle on the shelf like some outdated textbook.

According to *Webster's Third New International Dictionary*, the verb *spook* means "to stir up or excite (as a horse or steer) esp. by frightening (the entire herd got spooked and stampeded into the mountains)." *Webster's* does not use the timber rattlesnake as an example of a "spooked" animal, but based on what I heard at Bashers, perhaps it should have.

In 1981, Randy Stechert, who has worked with timber rattlesnakes for nearly half a century, coined the term "spook factor," which refers to a quantifiable and qualifiable change in a rattlesnake's behavior toward people, "a distinct handling effect," particularly toward biologists who stuff them in nylon bags, clip their ventral scales, glue transmitters on their back, open them up and poke transponders beneath their skin, weigh them, measure them, paint their rattles, palpate their ovaries or squeeze out their hemipenes as though checking the ripeness of a melon, and otherwise restrain snakes (beneath a boot, for instance) for field or laboratory processing. The pedagogic homolog of spook factor is "intimidation effect," the version more likely to be found in the scientific literature. Brown began his Adirondack fieldwork in 1978. By the early eighties, he noticed that to keep from being caught again rattlesnakes would flush quickly, avoid loitering around their dens, and abandon summer basking sites or bask beneath rock awnings instead of entwining on the rock surface, as they had before the biologists began pestering them. A snake that had been handled, said Brown with an air of resignation, was less likely to be seen. On several occasions, during a fifteen-year period, marked snakes avoided the biologist even though he had visited the research sites twice yearly during that time. Apparently, snakes don't want to be manipulated again. Ever.

On May 11, 2011, I was with Brown on a south-southwest-facing slope above an Adirondack river, at a "transient slide," his term for a long, narrow jumble of rocks several hundred yards from a snake den, where timber rattlesnakes pause for hours or days or weeks to shed or bask en route to the awakening woods and a summer circuit of feeding and mating. At the site, Brown quickly caught a big adult yellow-phase male that was luxuriating in the delicate spring sunlight. The snake had been already marked, ventral scale-clip number

4796, but had worn the paint dab off his rattle. Brown processed the snake on the spot—fifty inches long and slightly more than two pounds. Before he released the snake, he painted the base of its rattle bright yellow so it could be spotted from a distance. A week later, I received e-mail from Brown: snake 4796 was *thirty-two* years old, at the time a longevity record for a wild male timber rattlesnake. (Brown has since recaptured a number of snakes older than forty, including one pregnant forty-five year old.) The snake had been first caught as a five-year-old in 1983 and then recaptured each of the succeeding three years. Although Brown found 4796's shed skin at a basking site in 2001, he hadn't seen the rattlesnake since 1993; for eighteen years the snake had eluded the snake tongs.

Brown began the discussion as though he was directing a high school play. He printed out four quotes, three attributed to Basher stalwarts, the fourth to Howard Reinert, an esteemed timber rattle-snake biologist from New Jersey College, who during many years of study in the New Jersey Pine Barrens and in the mountains of Pennsylvania has contributed a wealth of information on the nature of the snake, perhaps second only to the discoveries of Brown himself (and a handful of others). For reasons unknown to me, Reinert was never invited to be part of the beer-swilling Bashers. Martin, Stechert, and Brown read their own statements, while Stechert, the Jimmy Fallon of the group, triumphantly read Reinert's, effecting the absent biologist's slow, nasal accent. (In the hills above West Point, Stechert endearingly satirized a *National Geographic* photographer blending her speech impediment and Dutch accent to portray an intellectually inquisitive version of Elmer Fudd.)

On cue, the participants read their lines as their text is projected on the screen.

BILL BROWN: The intimidation effect applies to snakes that have experienced a capture event involving sudden restraint, handling, and marking, causing mild physical trauma and stress, and usually evoking defensive behaviors (rattling, striking, spraying cloacal musk). In contrast to capture-recapture methods, studies using ra-diotelemetry seem to have difficulty recognizing the intimidation

effect, apparently because the snakes being tracked are not disturbed
on the surface while under observation ...

RANDY STECHERT (as himself): I have observed Timber Rattlesnakes
that have been exposed to repeated human harassment over a number of years develop a tendency toward reluctance to expose themselves at their primary basking areas. I have noted this phenomenon
(which I nicknamed the "spook factor") ...

RANDY STECHERT in a nasal rendition of Howard Reinert (with
pompous overtones): After years of radiotracking this species ...
timber rattlesnakes allow themselves to be closely approached
while they are engaged in almost any activity. The presence of a non-
threatening human observer usually elicits only brief curiosity on the
part of the snake and practically never a fright or defense response ...
even capture and release or constant approach and observation have
little impact upon behavior patterns.

MARTY MARTIN, stretching his northern Virginia Piedmont drawl to
the limit: Grabbing them with tongs is somewhat of a dramatic experience for the snake. From the snake's perception it has escaped from
a dangerous predator that tried unsuccessfully to eat it. The snake
says to itself, "Don't let that happen again." It probably won't mean a
change in home range but it will mean a change in basking behavior.
It remains alert and ready to pull back at the first sign of a "steppe
ape" [Martin's term for a human, which has been unanimously taken
up by the Bashers]. Timber rattlesnakes will remain partially exposed
under a rock rather than lie on top of it. Let's consider radiotracking.
The snake is caught, brought to the lab, operated on, and brought
back to where it came from and is released. The researcher returns
and the snake freezes and hopes it hasn't been seen. After repeated
visits the snake becomes habituated to the researcher and goes about
its business.

Rhetorically, Martin asked the group whether they would alter
their behavior if, while picking blueberries, they spotted a grizzly
bear. "Are you going to be more alert?" From my own experience
came a more mundane analogy, which I kept to myself: whenever
I pass the spot on a rural New England road where I've previously

been pulled over for speeding, I slow up and look around. A timber rattlesnake grabbed and restrained struggles to free itself with violent lateral thrashes, attempts to bite, and sprays a fine shower of cloacal musk, earthy smelling and long lasting. I've never been in the company of a rattlesnake biologist who exclaimed, "Oh, shit! I've been musked," and then dropped the snake. Everyone carries on with whatever they're doing and tidies up later. And the snake keeps spraying. Conspecific tongue-flicking rattlesnakes, however, read the chemical fine print that adheres to hands and clothes, rocks and tree trunks, and then alter their behavior accordingly; it's like being born again and abandoning a lifelong activity, perhaps forever. Even if only a few snakes are manhandled during repeated visits, "word" gets around quickly and most of the rattlesnakes become skittish. "Sometimes," said Brown, "it is the frequency of visits and the attendant human smell rather than the number of snakes actually captured that makes the difference. You have to assume they can smell you." If a rattlesnake associates a person (the predator) with a bad experience, it avoids people, even if it means forgoing exposed basking or shedding at preferred sites. Dens, of course, are another story; a snake in the Northeast can't survive without one. During ingress, timber rattlesnakes harassed by researchers (or poachers— for the snake there's no distinction between the two) head directly down the portal: no den rock lounging or leisure afternoons basking in the October sunshine.

An old girlfriend of Stechert, a second-grade teacher from Brooklyn, always knew when Randy had been sprayed. How long the lipid-based musk sticks around in the environment was any Basher's guess. A day? A week? A year if protected from the elements? Like the schoolteacher, timber rattlesnakes have deep chemical memories; juveniles learn the woodland neighborhood by scent-trailing adults. If generations of timber rattlesnakes avoid a perfectly good basking rock because sometime in the forgotten past a snake had a bad experience there, young snakes may never find their way to that site, no matter how accommodating it might be. Today, it's rare for Brown to find rattlesnakes unspooled in the sun or entwined in small pods in the open. A few years after he began his long-term

study, group basking, the benchmark of an undisturbed population, ceased and has not resumed for more than thirty years at dens that he still visits on a biyearly basis. Taking a cue from Brown's field-work, Alcott Smith believes the uninterrupted prosperity of rattle-snakes in southwestern Vermont have enjoyed since 1971, when Vermont ended bounty payments, is in jeopardy now that the state began the two-year telemetry and PIT-tagging project. Like Brown's nearby Adirondack snakes, Vermont rattlesnakes, formerly tolerant of voyeuristic naturalists, would become skittish.

To date, one hundred five Vermont timber rattlesnakes have been PIT-tagged (injected subcutaneously with a rice-grain-sized passive integrated transponder) and six males carry radios—by the end of egress 2012, twenty males and two females will be radioed and one hundred forty-four will have been tagged. Smith's twenty-four years of noninvasive research—stealthy observations and record keeping

and then slowly backing off, essentially leaving the snakes alone—
has seldom bothered the rattlesnakes. They have remained unfettered
and unmolested. In Smith's company, I've seen an entanglement of
eight or nine (it's difficult to get a precise head count), sighing and
heaving beneath a rock marquee. I've watched snakes going to and
from the den, mating, basking, in both coiled and vertical-tree
ambush postures, and rubbery, gray-pink babies piled in the open,
under their mothers, and in sunbeams alone on the November forest
floor. On several occasions, Smith himself has watched birthing, and
he once counted eleven snakes in a sentient knot. I was with herpetoculturist Kevin McCurley when, with exquisite nonchalance, he
picked up a *big* basking rattlesnake barehanded, one hand beneath the
vent and the other under the mid-point, turning slowly in the same
direction and at the speed that the snake turned his head, which elicited from the snake no more than a casual tongue flick. There was no
thrashing...no rattling...no striking. When McCurley unhurriedly
poured the snake back onto the basking rock the snake remained
stretched in the sunshine as though nothing untoward had taken
place. (I spied the rattlesnake an hour later; it had moved about six
inches.) On another occasion, a post-partum female leisurely passed
over my boot on her way into the afternoon woods. "The snakes I
process," countered Brown, "behave differently; they thrash and release musk. If Vermont catches every snake they see this year, they'll
be awfully unhappy next year—the snakes are going be spooked
big time." Whenever a snake delivers musk, other snakes are put on
notice.

Contrary to folklore, timber rattlesnakes *don't* smell like cucumbers. Their musk is, well, musky, viscous, off-yellow oil, with
an earthy smell like moldering leaves or garden compost. You can't
handle one without getting sprayed. In fact, you can't watch someone
else handle one without getting sprayed (sooner or later). And once
you've been sprayed, you broadcast a chemical missive everywhere
you go—like eating a fresh garlic bagel; an odiferous plume spreads
in your wake. Brown, who prefers scale clipping to PIT tagging, has
paid the price of cutting snake flesh; his subjects are riled. But as he
pointed out, transponders and batteries don't last forever. Scars do.

"No way are you going to get a PIT-tag return in twenty-five or thirty years." Unfortunately, manipulating a timber rattlesnake provokes the snake, which may intimidate the entire population, a regrettable by-product of hands-on fieldwork. Neither Smith nor Brown believes that marking rattlesnakes in Vermont's nearby Taconic Mountains is entirely necessary.

After Brown finished his introduction to the spook factor, George Pisani reported that a Kansas colleague's time-honored collecting technique was to "pin the snake down and then choke the hell out of it." When he measured the snake, he stretched the animal out so much, said Pisani, straight-faced as Bob Newhart, "you could see daylight between the vertebrae. The guy *never* caught a snake twice." What if an alien captured *you*, reasoned Pisani, "pinned you down, stretched you out, slapped you around, and then cut some part of your anatomy? 'Shit, I'm out of here.'" Bill Brown, who has been accused by some amateur herpetologists of mishandling rattlesnakes, has encountered languorous snakes in the Adirondack foothills (though I've never seen one when I've been with him). In 1984, he first noticed what he has referred to in the scientific literature as the "listless syndrome," lifeless rattlesnakes dishrag-limp that don't rattle, don't strike, and barely tongue flick. He's noted degrees of listlessness: mild, moderate, and severe. The syndrome is unknown anywhere else in the timber rattlesnake's range. "He hasn't cleaned those grungy scale-clipping scissors for years," confided one Basher, with a droll smile, who preferred to remain anonymous, which only reinforces New York's DEC handling protocol for timber rattlesnake researchers. When Brown catches more snakes than he can handle in the field he brings the remainder to his lab to idle in damp shade for a week or more until he processes them and returns them to the ledge drop-off site. No wonder some of the Adirondack rattlesnakes are out of sorts.

To execute research at the same level as William S. Brown, however, to get the results he has gotten for as many years as he's worked in the Adirondack foothills, requires persistence and patience and sacrifice on the part of both the researcher and the snake. Compared to the work of Lawrence Klauber, the most notable of historic rattlesnake researchers, who sacrificed thousands, if not tens of thousands, of individual crotalines over the course of his long career learning, among many, many other things, how long a rattlesnake's body will twitch after its head has been lopped off or when a disembodied head will stop striking, Bill Brown's work is benign. His animals are *not* purposely abused. But of the many, many timber rattlesnakes he's processed since 1978, a few do, unfortunately and in rare instances, get the short end of the modus vivendi. Stechert, who's Brown's herpetological Siamese twin—that is they're welded at the hip through mutual research, life experiences, and thirty years of collaboration and rattlesnake-related litigation and conservation—watched Brown yank an escaping snake out from under a rock. Blocks of ventral scales came off and viscera spilled out. "I could see the snake's heart beating." The rattlesnake died in rehabilitation a few days later. When you handle that many wild animals, regardless the species, the loss of one or two or three is regrettable but unavoidable. In the mid-nineties, when I covered the Florida panther recovery project for *Audubon*, state and federal biologists were embroiled in controversy over the deaths of two cats that had been overanesthetized. At the time, the total population of Florida panthers was less than seventy.

Stechert believed Brown's first concern has always been research results. Hence the spook factor and the listless syndrome on the rocky, western shelves above an Adirondack river. But without William S. Brown's unflagging devotion to research *our* knowledge of timber rattlesnakes would still be in its infancy; anyone who studies the timber rattlesnake stands on his broad shoulders.

The Bashers, however, were *not* a sympathetic audience. As an anecdote to his talk on the spook factor, Brown recounted how a timber rattlesnake kept reappearing every summer along a driveway at an Adirondack camp. Brown, who had organized a compassionate rattlesnake-removal team for the state's DEC, had returned the snake to the hillside every summer for many years. What about the spook factor? I thought.

"Maybe you needed to slap it around a little bit more," someone shouted from across the room. "Or cut off a few scutes."

To illustrate a final point about the spook factor, Brown projected a picture of a timber rattlesnake, which he called 741—the number, like that of 4796, the thirty-two-year-old Brown caught last May, correlated with the array of clipped belly scales. Immediately, Stechert, who recognized the snake, as only a rattlesnake savant could, interrupted Brown in midsentence and corrected his identification. The snake on the screen was not 741; it was a resident of Eagle Ledge Den in the Hudson Highlands, a hundred miles south of Brown's Adirondack study area. The correction hung in the air for a moment.

And then Brown responded. "OK, OK, Randy. It was a good stand-in for 741."

3

Quasimodo in the Blue Hills

These eight or ten rocky knuckles are known as the Blue Hills. I live in a town of twenty-five thousand at their northern base. I can leave; the snakes can't.

THOMAS PALMER

"Hello, my name is James, and I'm obsessed with timber rattlesnakes. I need the twelve-step program," James Condon announces to the wind and trees and corrugations of the southeastern Blue Hills, a wry smile creeping across his face. For the past six years, Condon, who is fifty-six, an unemployed pastry chef with a graduate degree in photography from Cal Arts and a ponytail long enough to be bound in three places, has been combing the Blue Hills for rattlesnakes, unquestionably one of Massachusetts's most enigmatic and most endangered reptiles.

No one knows the rattlesnakes in this place better than Condon. He's discovered where they den, shed, bask, and birth, where they move out, hang out, and dry out. Other critical activities, like mating and feeding, are not as predictable (unless, of course, you're a rattlesnake or Rulon Clark armed with a collapsible aluminum antenna). "Be alert," he warns me, as we bushwhack over a rocky summit stiff with scrub oak. "They could be anywhere." My steps are measured.

Condon visits the Blue Hills four or five times a week during "snake season," which used to be mid-April to mid-October but more recently has expanded into the winter months, ice and snow notwithstanding. "Emergence *never* quits up here. It's one big emergence." He visually examines all the snakes he encounters, photographs them, and catalogs the digital images so that he can recognize individuals by their unique banding patterns, the way a cetacean biologist identifies a humpback whale by the distinctive scallop edging of its flippers. With the permission of Tom French, director of the state's Natural Heritage and Endangered Species Program, Condon carefully and skillfully weighs and measures a select few snakes and marks their rattles with a dot of nail polish. He also speculates on their fitness, laments about their health, and sometimes rants and raves about the object of his obsession. Condon's a whistle-blower who acts on behalf of timber snakes, with—and without—the blessing of the state of Massachusetts. The rattlesnakes in the Blue Hills are in trouble, and Condon knows it.

It's a sunny Saturday morning in early April, well before the full awakening of spring. Condon emerges from the woods on Pig Rock Path and meets me at a narrow pullout in Randolph, along Route 28, a busy two-lane road that cleaves the Blue Hills. Condon calls Route 28 "Death Highway," a reference to the number of animals that get hit here every summer, a roster that includes rattlesnakes and people. Both lanes support a fierce procession of cars, fast-traveling and close together, and are edged with long stretches of dull, dented guardrails, punctuated here and there by replacement sections that are shiny as a new dime. I need to move my car to the pullout on the other side of the road.

"Were they honking at you?" Condon asks when he finds me waiting, incredulous that I executed a U-turn.

This is the largest open space within thirty-five miles of Boston (besides the ocean), a chain of twenty-two rocky, east-west running knobs whose four-hundred-and-fifty-million-year-old bluish bedrock geologists call Quincy granite. The Blue Hills's glacier-scrubbed summits are the highest ground along the Atlantic Coast between Maine and Mexico. When you're flying into Boston's Logan Airport from the west, the Blue Hills stand out, a narrow, emerald island anchored in an expanding suburban slipstream, garroted and Balkanized by highways and roads called Blue Hills, Chickatawbut, Wampatuck, Neponset Valley, Green, Wood, Willard, Ricciuti, and

Old Route 28. If you're driving south on Interstate 93, which curls around the southern end of the hills before merging into Interstate 95, they do not stand out—at least not until you're on top of them, which makes turning onto either the Route 138 or Route 28 cloverleafs a challenge.

More than one hundred and-twenty-five miles of hiking trails weave "snakelike" through the Blue Hills—over balds of blueberry and huckleberry, past crooked pitch pine and towering white pine, beneath canopies of oak and hickory, sugar maple and black birch, and formerly American chestnut, which succumbed to the blight in the forties but persists today as recurring saplings that die back before they bear viable fruit. Trails circumvent wetlands, cross granite outcrops, ridges, and parallel a stairway of sharp-edged boulders, ice-fractured pieces of bedrock that tumbled downhill. A landscape architect named Charles Eliot was the force behind the creation of the Blue Hills Reservation in 1893. "For crowded populations to live in health and happiness," wrote Eliot, "they must have space for air, for light, for exercise, for rest, and for the enjoyment of that peaceful beauty of nature," a prescription that, with the addition of small rodents, describes generally the needs of timber rattlesnakes.

The Blue Hills Reservation is managed by the Massachusetts Department of Conservation and Recreation (DCR), which is subdivided into three divisions: watershed management like Quabbin Reservoir, the sprawling hydroproject west of Holyoke, where the three branches of the Swift River were impounded to provide metropolitan Boston with water; state parks; and urban parks. Because of its proximity to Boston, the Blue Hills Reservation is considered an urban park, regardless of the rattlesnakes and copperheads that haunt the gnarled uplands. Protection of a DCR property, I was told by a DCR resource manager, is based on cultural and natural resources and recreation. Although timber rattlesnakes are an endangered species and Natural Heritage has the mandate to protect them in the Blue Hills, it is the DCR that has to validate their request. "It's political," the resource manager told me. "Natural Heritage can't go against any other state agency—even if we [DCR] violate the rules, they can only say 'Hey guys, let's sit down and talk about this.'"

The Blue Hills have been logged, farmed, pastured, burned, and mined. In 1753, Quincy granite was shipped to Boston to build the King's Chapel. The first commercial quarry appeared in 1825. A year later, the first commercial railroad in America hauled Blue Hills granite three miles to the Neponset River in Milton; from there the stone was freighted to Charlestown to build Bunker Hill Monument. Quincy granite was used in the construction of the Parker House in Chicago and the customhouses of New Orleans, Galveston, Savannah, and San Francisco. On August 8, 1938, Quincy granite made the cover of *Life* as a black-and-white photograph of fourteen teenagers in various positions of readiness as they prepared to dive into an abandoned, water-filled quarry thirty-eight feet below. The caption read: "To young people of Norfolk County, Quincy quarry is famed as the most thrilling and soul-satisfying swimming hole in New England." The last quarry closed in 1963.

I once hiked the summit of Great Blue Hill and climbed Eliot Tower, named in honor of the reservation's founding father, and gazed north at the Boston skyline, which appeared much closer than I had expected. That rattlesnakes still live here is a small miracle of nature and a testament to the inoffensive nature of the snake.

Today, the trees are leafless; the buds, a promise. Skunk cabbage is barely up, and spicebush is in flower, small, thin-petaled, and yellow, like dollhouse confetti. Juncos and robins are everywhere, and busy. Condon leads me off trail. Everywhere we go, I hear traffic, which may be the heart of the rattlesnake's problem. They can't leave. They're housebound, so to speak. It's as though the snakes live in a forested zoo, a seemingly spacious seven-thousand-acre zoo, granted, but restricted to less than a quarter of that total by the roads that section the park. Unfortunately, that may not be large enough to prevent an isolated rattlesnake population from inbreeding, with the consequent loss of genetic diversity, expressed ubiquitously in the snakes' compromised immune systems. "The boys may be mating with their aunties," Condon says, jabbing his walking stick at a boulder. In the case of the Blue Hills timber rattlesnakes, we are talking about an immune system that cannot fend off a skin-eating fungal pathogen called *Ophidiomyces ophiodiicola*. Using a Rube Goldberg snake hook—the end of a wire coat hanger duct-taped to a walking stick—Condon fished his first sick rattlesnake out from under a rock in 2006 and brought it to Tufts School of Veterinary Medicine in a laundry bag. The snake was treated for a fungal infection and then housed by a reptile rehabilitator over the winter. When Tom French of Natural Heritage insisted that the snake be returned to the hills the following spring, Condon kept track of it. He found it four times that summer. It never left the den. Then, on March 3, 2008, with the thermometer reading forty-five degrees, Condon found the rattle-snake in the den crevice, dead. "The animal was desperate," Condon noted. "No vet ever looked at that snake. It was never necropsied."

Until very recently, *Ophidiomyces* was known only as a plant pathogen, but apparently it has found snakeskin to its liking. The outer layer of skin, the dead skin snakes periodically shed, sleeve-like, is made of the tough protein keratin—as are hair, fur, claws, fingernails, and feathers. *Ophidiomyces* loves serpent keratin.

Matthew Allender, the wildlife veterinarian from the University of Illinois who first identified *Ophidiomyces* as a potentially fatal snake pathogen, told me that the infection "looks like the beginning stages of white-nosed syndrome in bats." To date, Allender has necropsied four fungus-ravaged massasaugas, an endangered Illinois rattlesnake. He's not sure whether the snakes' immune systems are suppressed due to inbreeding or environmental toxins, but it is clear that the fungus has taken advantage of the situation. *Ophidiomyces* also has been identified in pine snakes, king snakes, and rat snakes from New Jersey, New York, and Georgia, and as well as New Hampshire where it's believed to have devastated most of the state's last remaining timber rattlesnakes. "We've seen it in captive animals worldwide, but we don't typically find it in free-ranging animals," Allender said. *Ophidiomyces* thrives in wet weather, of which there were excessive amounts in 2010, 2011, 2012, and 2015, but no one has done a health survey on timber rattlesnakes—or any other species for that matter. "It's hard to get people excited about a snake disease."

There are skeptics.

Basher in good standing, Tom Tyning, the leading timber rattle-snake authority in Massachusetts and a former master naturalist with the Massachusetts Audubon Society, takes the opposing view. Tyning, who has monitored the state's rattlesnakes for forty years, teaches environmental science at Berkshire Community College. "Show me the carcasses," he says, whenever the Bashers' conversation turns to *Ophidiomyces*. Tyning believes the sores on the snakes' heads are hibernation blisters, seasonal blemishes that shed off in early summer, nothing more. He and Ann Stengle, who studies the DNA of timber rattlesnakes for her doctorate at the University of Massachusetts, Amherst, track an apparently healthier, outbreeding population in the southwestern corner of the state and adjacent New York.

When I mentioned Tyning, Condon goes ballistic. At the first meeting Tom French convened to discuss the possible fungus issue, Tyning and two of his associates, Brian Butler and Alan Richmond, their arms tightly folded, sat opposite Condon and Kevin McCurley. "These guys sat there nay-saying the whole thing, not only with body language but in discussion challenging us like we were trying to defend our thesis. They treated us like supplicating graduate students. I need a nitro pill." In a recent e-mail Condon continued: "Throwing a blanket of hibernation blisters over the [fungus] issue is IMHO very unscientific." I'm left with the impression that comparing the health of Blue Hills rattlesnakes to the health of Berkshires snakes may be analogous to comparing the health of members of the New York Athletic Club to that of the members of a leper colony.

Condon lobbies for the rattlesnakes day and night, sending salvos of electronic diatribes to anyone in authority who'll read them. His zeal makes him a pest to many conservation biologists and wildlife veterinarians—some of them treat him like a pariah—and a hero among the snake-loving public, a small but perfervid crowd. Because of Condon's badgering, though, Tom French, charged with preserving biodiversity in the state, commissioned the Roger Williams Park Zoo to study the issue in 2011. The jury is still deliberating as to whether the problem is genetically compromised snakes, or a super fungus, or environmental toxins, or climate change, or Condon himself. Everyone in the closeted world of Massachusetts rattlesnakes has an enlightened opinion.

According to French, Tyning considers Condon an uninformed alarmist, the Chicken Little of herpetology, who is convinced that the sky is falling. For Tyning, *Ophidiomyces* is a common, native soil fungus, not a threat to rattlesnakes. Once they leave the den, the sun dries the lesions up. West of the Housatonic River, in rural southwestern Massachusetts, where Stengle and Tyning run their study, there may be almost as many rattlesnakes as there are people. Several animals fitted with radio transmitters have been observed for multiple years. Of these, two had severe fungal infections. Last October, at the Bashers gathering, I viewed photographs of one of the radioed snakes on a poster that Tyning and Stengle displayed on a billiard table. In the lead picture, the snake appeared blind in one eye: the second and third pictures, taken after successive sheds showed the eye clearing up, the fungus disappearing.

When I described to Condon the images that I saw on the poster, he was quick to point out how energy intensive shedding is. On average, a healthy timber rattlesnake sheds once, maybe twice, each year. In preparation for shedding, a rattlesnake may bask and fast for two weeks. If an infected snake sheds three times during a growing season to rid itself of lesions, it's off the meal line and expending energy without replacing it for approximately a month and a half. At best, the rattlesnake's growing season in Massachusetts is five months long so if a snake skips feeding for six weeks during the growing season, it has lost thirty percent of its potential feeding opportunities and also burned off more than the normal amount of caloric reserves. "How much time does that snake hole up basking?" Condon replied, barely containing his agitation. A rattlesnake's breeding and reproduction are directly correlated with its body weight. In New England, where timber rattlesnakes play natural history roulette just to stay suspended on the threshold of survival, additional shedding could delay reproduction for a year or two. And that may be too long. "If, on average, a female timber breeds successfully twice in her lifetime and you take away one of those efforts, you just cut her reproduction [progenative recruitment] in half," says Condon, his voice rising an octave in frustration and disgust.

Rattlesnakes in the Berkshires outbreed. Rattlesnakes in the Blue

Hills don't. Has chronic inbreeding diminished the snakes' genetic diversity and crippled their immune systems? Or, as Allender alluded, could *Ophidiomyces* perhaps be the tip of a much larger toxic iceberg? Has heavy-metal poisoning lowered the efficiency of their immune systems? Or pesticides? Or plastics? Maybe climate change has triggered the rampant growth of a ubiquitous, once innocuous soil fungus at the expense of the isolated, inbred rattlesnakes. There are no answers, yet.

French is caught in the middle: between Tyning, who thinks nothing's wrong—that facial lesions are just hibernation blisters caused by moisture trapped against the snake's skin while it sleeps through the winter, blisters that simply dry out and shed off during the summer, as demonstrated in his poster—and Condon, who thinks that every scabby-looking snake should be taken into captivity and given heat and antifungal medication. "It is James's belief that no one else is taking this seriously enough or quickly enough," French told me. Condon dislikes, disrespects, and mistrusts the researchers and their motives. In a heartbeat, he tells you that they're doing this for selfish reasons. "He says that in e-mails and he says that face to face with no qualms," French said. "James doesn't hold much back. It's not an easy thing to be in this dichotomy, to be between the different kinds of people involved."

To learn more about *Ophidiomyces* I drove to Providence, Rhode Island, to visit the Roger Williams Park Zoo. It was a gorgeous wall-to-wall blue sky, temperature in the mid-sixties, a perfect spring snake day. I was an hour early for my appointment with Dr. Michael McBride, the director of veterinary services, and Lou Perrotti, conservation programs coordinator. I parked in the employee lot, passed muster at the security checkpoint, and clipped on a media badge. Inside the entrance, at the flamingo pavilion, half a dozen birds sat on the ground, legs folded like card tables. With their necks stretched over their backs, heads hidden under their wings, they resembled a flock of broken lawn ornaments. To pass time before my meeting,

I checked out the red wolf exhibit. The twitching ears of a wolf—the only red wolf body part I'd ever seen—poked up from behind a log. Next, I walked to the Conservation Cabin, where a reptile display was advertised. I had hoped to see a timber rattlesnake but instead found a box turtle, a desert rosy boa, a spiny-tailed iguana, a Standing's day gecko, and a black and white tegu, a hulk of a lizard that ambled around its terrarium like a windup toy. No timber rattlesnakes here.

The Roger Williams Park Zoo occupies forty acres of the four-hundred-thirty-five-acre Roger Williams Park, a gift to Providence from the founder of Rhode Island's granddaughter six times removed. One of the oldest zoos in the country, the zoological park opened in 1872 with a less-than-robust collection of small animals: gray squirrels, raccoons, guinea pigs, white mice, peacocks, hawks, and an anteater. By 1890, a pair of lions, a tiger, a leopard, and an elephant named Roger gave the zoo a more traditional look. On exhibit since then have been a parade of elephants named Alice and an evolving menagerie that includes rarely exhibited animals like the red wolf, African wild dog, and Chinese alligator. Today, the Roger Williams Park Zoo has become one of the most progressive zoos in the country, an institution dedicated to education and conservation, as well as to the health and dignity of its captives. In 2011, the zoo unveiled the John J. Palumbo Veterinary Hospital, a larger, better-equipped version of its predecessor. This is where the sickest of Condon's rattlesnakes end up.

Perrotti, who is dressed more for the Blue Hills than the zoo—khakis, a short-sleeved zoo shirt, and field boots—meets me in the Administration Building. He's lean and thin-haired. A Vandyke surrounds his mouth and a nickel-sized tattoo of an American burying beetle (his favorite insect and one he's spent years trying to conserve) decorates the inside of his right arm. Perrotti is animated. Invisible tendrils of cigarette smoke waft from his clothes. His counterpart, Mike McBride, is huge, well over six feet tall and close to three hundred pounds, a Caucasian version of Yankees pitcher C. C. Sabathia. Both men are thoughtful, articulate, and committed; neither one wanted to guess the extent of the timber rattlesnake problem in the

Blue Hills. Sitting on opposite sides of Perrotti's desk they looked like Mutt and Jeff.

To open the conversation, I asked McBride how *Ophidiomyces* affects timber rattlesnakes. He responded, "As with most things in biology it is not cut and dried. We're not really getting a sense of the population as a whole. We see [only] animals that show the worst symptoms." Fungal dermatitis in a small percent of a wild snake population is not alarming, but if *Ophidiomyces* is as widespread as it appears to be in the Blue Hills, it could indicate a significant stressor on the population, a concern for Illinois veterinarian Matt Allender as well. "Our goal [at the zoo]," McBride continued, "is to be a technical resource for them [the Natural Heritage and Endangered Species Program] and to house snakes as they go through the recovery period. It is really the Massachusetts biologists [Tom French, with advice from Tom Tyning] that determine how this process goes. They have to prioritize with all the other projects they have for all the other endangered species and figure out where rattlesnakes fit into the mix."

French, the biologist who got the zoo involved, makes the ultimate decision about the welfare of Massachusetts's rattlesnakes. His plate is beyond full, and projects are in danger of spilling off an overcrowded countertop. French manages a staff of twenty-six people, who collectively monitor, study, protect, and restore the habitat of four hundred thirty-two endangered and threatened plants and animals. "The timber rattlesnake is a special case," he told me. And he brought McBride and Perrotti on board to deal with just one of the snake's most potentially pressing issues: *Ophidiomyces*.

"Is this a widespread problem?" McBride asked rhetorically. Biologist Tyning wants to see carcasses before he'll concede that *Ophidiomyces* is a threat to timber rattlesnakes, and Perrotti and McBride need carcasses to necropsy. Thus far, sick snakes are all they've gotten. The dead ones apparently die in the den, and healthy ones sleep through the winter.

To find answers, McBride needs more samples—to date, he's looked at only thirteen infected snakes. With the blessings of French and French's counterparts in Connecticut, Vermont, and New Hampshire, where lesions on timber rattlesnakes have also been detected, the zoo has applied for a regional grant from the U.S. Fish and Wildlife Service to expand the pilot study, funded by the Disney Corporation, into a regional survey of the health of timber rattlesnakes in the Northeast. (You may recall that timber rattlesnakes are already gone from Rhode Island, Maine, Ontario, and Quebec.)

McBride feels the crunch of limited funding, limited time, and limited space. Because the rehab facility is small and not equipped for visitors, I was denied access. "It's completely unimpressive: a sparse, clean room, just rows of aquariums," Perrotti said, painting an austere picture so I wouldn't feel as though I'd missed out.

The zoo has limitations: money and time. In a city as financially strapped as Providence, the idea of hiring extra keepers to care for timber rattlesnakes was never discussed. "They're figuring out how to pay current employees," said McBride. "We do the best with the resources we have."

After McBride performs surgery and collects samples to culture, Perrotti husbands the rattlesnakes, which means keeping them warm and dry and well fed and applying antifungal medication to open wounds. Eventually, each rattlesnake is returned to the wild. One of these snakes, which Condon calls Stitch, had a suture poking out of its torso after its return from the zoo, and several others were sluggish and became reinfected with fungus. I joined Condon one September afternoon to release a zoo-treated snake that he had collected earlier in the summer. Once liberated, the rattlesnake stretched out in the sun, threaded itself through tufts of grasses, and then remained stone still. An ant picked around the edge of a dried lesion. "He's back in the woods. He's happy," said Condon, giving expression to whatever ophidian thinking might be taking place. "He's trying to figure out what's going on. I don't know if it should be called thinking, maybe processing. His brain is the size of a micron or whatever it is—*very* small."

To be able to make a strong recommendation to French about the direction of timber rattlesnake conservation in Massachusetts, McBride would like to have field samples from two hundred rattlesnakes, healthy and diseased, from throughout New England. To that end, Perrotti wrote the regional Fish and Wildlife grant proposal on behalf of the Roger Williams Park Zoo and the participating states.

"Based on your limited sample of thirteen rattlesnakes, would you call these hibernation blisters?" I asked.

According to McBride, the lesions he sees are on the snakes' faces and heads. He sees very few on the bottom of the snake.

"Are hibernation blisters more or less like bedsores?"

"They're different. Hibernation blisters are not pressure induced. They're from moisture trapped against the snake's body, which is believed to be from snakes resting in the same spot too long."

McBride pointed out that lesions from different sources often look the same: bacterial, viral, fungal, even lesions from chemical burns. His mantra, repeated several times during my visit, was to make sure to identify the source of the lesion. "You must go beyond a visual examination." Perhaps this is why so many of the digital photographs of ill-looking timber rattlesnakes that Condon routinely blitzes to French have been ignored. The snakes just didn't appear sick enough. "When I collect samples, I'm purposely *not* trying to find out where the snakes are from. I don't want to bias myself and say, 'Oh, this one is from such and such an area, so I'm going to look harder for it [the fungus] here.' I try to keep myself unknowledgeable until after we've collected the samples. It's premature to say if it's a greater problem in some areas."

Perrotti, however, apparently peeks at the location data. "Some of the worst snakes we've seen came [from the Blue Hills] in January and February," he told me. I asked if the den could be either the source of *Ophidiomyces* or the incubator that permits it to flourish. "Nobody knows," said McBride. "Anyone who claims to have knowledge of that is making a *big* stretch." Finding the source and the nature of the fungal disease may be like solving the Lindbergh kidnapping. Perrotti and McBride need soil samples from dens (if, in fact, there is soil in the dens) to compare temperature and moisture levels and infection rates between the various bedrooms of afflicted and healthy snakes. Of course, you'd need dynamite and a jackhammer to expose a den deep in Quincy granite.

If the issue is the den, then the problem may be irresolvable. "We've talked about trying to get some fiber optics that will allow us to actually see into the hibernaculum," French told me over the phone. "It has never been done in New England. But it's not as simple as it sounds to stick a fiber optic cable down into a hibernaculum to see what's down there." Condon suspects he knows what they'd find: A bunch of dead rattlesnakes.

At the zoo, I asked McBride if there is any indication *O. ophiodiicola* is an invasive, pathogenic soil organism. "Or is this a native soil microbe that has been liberated from natural checks and balances all of a sudden?"

"I don't think anyone knows the answer to that. People have done seed banks and blood banks, but I don't know of any soil banks, where we could pull samples from twenty years ago and test them [for *Ophidiomyces*]."

Stretching away from his desk and leaning back in his chair, Lou Perrotti said he wanted the zoo "to lend a hand and try to answer the questions in a scientific way. We want data to backup whatever it is that we find. We just can't go out and say, 'Oh my God, the sky is falling.'"

"On the other hand," added McBride, "we can't say there's nothing wrong because we can't prove anything. We may be seeing something like white-nosed syndrome, and there will be a continuous downward slope [in the timber rattlesnake population]. A lot of people assume that *Ophidiomyces* is the problem. It is possible that *Ophidiomyces* is the symptom, and the problem might be heavy-metal toxicity or a virus that we haven't identified yet that has lowered their immune system and made [the snakes] susceptible to all kinds of disease. I'm sorry we just don't know enough to give recommendations. As a medical person, my first goal is to do no harm. That's what I'm doing by *not* making any recommendations. This is a big, big decision because there is no going back."

Lou: "Something's going on. Something's going on. Yeah."

"What if you were told that less than seventy timber rattlesnakes survive in the Blue Hills and if you wait several years to gather and analyze data you might lose the population?"

McBride: "Umm. We can make a very uneducated guess."

"Would you care to make that very uneducated guess?"

"No."

"Can you venture a guess what percentage of timber rattlesnakes die from the fungus?" I asked McBride.

"A very low mortality rate. We're seeing less than ten percent, which is different than Dr. Allender is seeing. I think the difference is in the sampling; we're bringing them in when their lesions are not life threatening." If one in thirteen snakes died at the zoo, that would be about seven percent mortality. This, I thought, is where fiber optics might be helpful. Maybe the corpses are piled in the den.

"Do you know James Condon?"

"I've heard of him."

McBride also believes that the wide range of opinions from snake enthusiasts—from professional to rank amateur—makes it hard to differentiate between story-telling and reality. "We want to approach this in a very standardized and scientific way," he told me. I assume he's referring to James Condon as one of the "rank amateurs." When I mention Condon, French is more direct. "[Dealing with] James is like herding cats," French says. "He's got passion. He's got time, apparently. He feels so strongly, he drives everybody nuts, because he's got the expertise on the one hand with just enough information to be dangerous on the other hand. He really, really needs to leave the science to the people who've got degrees."

For his part, Condon feels the engines of authority are at low throttle. About the Roger Williams Park Zoo, he says, "They're telling us the animal's not sick. I don't think they know what they're doing. Tom French has assured me [McBride and Perrotti] are venomous [snake] experts. They [the zoo-treated snakes Condon has released back into the Blue Hills] didn't look much better than when they were brought in. All three snakes I put back in this spring have lost weight." Throughout the winter of 2012, he saw distorted-faced rattlesnakes that looked like Quasimodo basking in shards of cool sunlight outside their dens, and this persisted in the ensuing spring, when he spent twice the time searching for snakes as he had the spring before and found half the number.

"The time for discussing the fungus has *long* passed. And if they wait much longer, the time for *action* might be past."

§

Condon and I climb up and down the west-facing ridgelines of the Blue Hills through woods of maple, oak, hemlock, and pine, past an old beech with the weathered letters RLA carved in the trunk (Condon knows the carver), across a plateau and a rock-studded valley, around a swamp green with Atlantic white cedar, up a dry hillside through a fortress of naked mountain laurel, over a summit dense with blueberry and scrub oak, past innumerable granite outcrops, blue-gray protrusions from the soft, leaf-littered earth. Up here, the sun heats the rocks that warm the rattlesnakes. In fact, sun time is crucial to a timber rattlesnake's survival in Massachusetts; basking on exposed rock is as nourishing and life supporting as eating a chipmunk or a white-footed mouse. Several ridges run through Blue Hills like a long, linear staircase: rattlesnakes den along a lower ridgeline and bask along the upper. Warm rocks sustain the snakes, particularly in spring and fall, when the air is stiff with chill. Females swollen with embryos lounge all summer on the rocks like sun-seeking co-eds. Lately, the snakes are also sunning on the rocks in winter, when frost can penetrate deep into the ground; snakes plagued by *Ophidiomyces* sunbathe even when it snows. Sadly, when rattlesnakes squeeze down fissures in the stress-fractured bedrock to pass the winter in subterranean retreats—the proverbial snake dens—they could be cozying up to the very source of the fungus: a cool, moist crevice.

Every October in the Blue Hills, rattlesnakes enter vertical cracks in the bedrock, what geologists refer to as "joints." During the geologic time period known as the Silurian, more than four hundred million years ago, when fish first exploited freshwater and coral engineered the planet's first reefs—long before reptiles made an appearance—Quincy granite was a molten intrusion into an overlying mass of sedimentary rock, two miles thick. The granite cooled. Slowly, as the weaker sedimentary rock eroded away and released pressure on the underlying granite, joints radiated through the bedrock. Quincy granite eventually reached the surface. Over time since the end of the last Ice Age, the freezing and thawing of ten thousand winters expanded the cracks until a handful of cracks became wide enough and deep enough for the passage of fat-bodied rattlesnakes. These portals, the keyholes to the rattlesnakes' bedchamber, are critical to the snakes' survival in the Northeast. Since they are not geochemists, rattlesnakes in the Blue Hills choose a den — to which they remain more or less faithful — based on the cleavage and thermal quality of the bedrock. If they achieve frost-free chambers, they survive. If not, they die. For Blue Hills rattlesnakes, survival has always been about ancient rock and climate ... but more recently it appears to be about *Ophidiomyces*.

On an afternoon commute home from Boston in 2003, Condon saw his first rattlesnake — a big, yellow carcass in the breakdown lane of Interstate 93. "It took me four days [of passing the snake] before I realized what the hell I was seeing." Then, a few years later, under the weight of the July sun, he stumbled upon an itinerant male in search of a mate. "I followed that snake for four or five hours, and it was OK as long as I didn't get too close." He became hooked. He started to read extensively about timber rattlesnakes and to correspond with experts. With his newfound knowledge, Condon appointed himself the snakes' caretaker in the Blue Hills. He's chased poachers (a long-standing problem); educated hikers, bikers, and state employees; supplied shed snakeskins for Ann Stengle's DNA research project; and, calling himself a "tree fairy," along with several friends, cut down trees to open basking sites, with and without the permission of the DCR, which he refers to as "a big, disorganized bureaucracy." Condon becomes apoplectic when he describes an illegal mountain bike trail he found in snake country. "The fucking trail went right over a snake den, literally! They pulled rocks from the den opening to improve their descent off the main bedrock outcrop." In response, he returned the rocks and built a beaver-esque dam across the trail to block access, which bikers immediately torched, taking out much of the surrounding underbrush in the process. Looking to be more scientific, he has begun to take copious notes.

I ask Condon to read from his journal.

"First year [snake], nose infected. Twenty-six grams. Black phase blisters. Visited all dens no snakes. Collected for the Zoo, yellow phase, shedding male infected both sides of head.

"Four hundred sixty-gram yellow male, face and lower body region [have] lesions. Collected, rejected by the Zoo.

"January 7, sick timber rattlesnake. 55°. Tom French wants to see pictures first."

Condon closes his journal, looks at me and says, "January 7, that's funny. Yeah, the snake woke up and had his coffee and watched the news." It was fifty-five degrees and unseasonably warm that day that Condon headed out for the dens. "No, [the snakes] don't know it's fifty-five degrees out. That animal was lurking near the entrance, going down [in the den] just enough not to get frostbite," Condon editorializes.

"I am confident in saying that I have learned a lot about timber rattlesnakes in a short few years," he e-mailed me. "Unfortunately, for the timbers, my advancement has been fueled by their very precarious situation, into which I find myself ushered." James Condon is six feet tall, thickset, and wears beige pants and a long-sleeved green shirt, a typical forest ensemble, loose fitting and not bright-colored. When he returned a snake that had been treated at the zoo, he carried the snake in a camouflage-patterned pillowcase. Three years ago, when he first chaperoned me around the Blue Hills, he wore brown canvass chaps to protect against snakebite. Today, the chaps are gone, a statement of Condon's evolving ease in the presence of these venomous snakes. The other day, though, as he hurriedly passed through snake country, he startled a rattlesnake coiled behind the trunk of a fallen tree. As Condon stepped over the tree, he noticed the rattlesnake and, executing a Fred Astaire dance move, he changed direction in mid-stride, just escaping the reach of the startled snake, which struck at the empty space where his foot had been.

Condon once tried to resuscitate a twenty-nine-inch-long copperhead (the other venomous snake of Massachusetts) that a black racer had swallowed then regurgitated. Condon performed ophidian CPR, intuitively puffing air gently down the copperhead's throat,

although the animal died anyway. For many snake lovers, particularly lovers of timber rattlesnakes, black racers are an anathema, the blue jays of the snake world. Racers eat snakes. After chewing their heads, "they suck them down like spaghetti," Condon said. One September, as Condon and I were getting ready to release a rattlesnake that had been treated for *Ophidiomyces* and discharged from the zoo, we startled a five-foot racer asleep in the sun. Like the cracking of a dark, leather bullwhip, the racer turned completely around in a nanosecond and disappeared beneath a rock slab. Not wanting to tempt fate, we released the much smaller rattlesnake a good distance away.

A persistent rumor swirls around the Blue Hills: to control pit-vipers colonists released barrels of black racers into the hills. The lore has been passed down for generations—both Condon and herpetoculturist Kevin McCurley, who grew up in central Massachusetts, heard it from their fathers. Even the *Patriot Ledger*, the daily newspaper of Quincy, once published a story about the importation of black racers into the Blue Hills. From my boyhood on Long Island I know racers are long, thick, muscular, fast, big of teeth, and *very* willing to use them, creatures of perpetual motion, like weasels. It's hard to imagine confining one black racer, let alone a barrel full of them. I imagine they'd burst over the rim like spring-loaded cables.

Condon self-funds his work: originally from his job as a pastry chef in a Boston gourmet restaurant, then from unemployment checks, and various short-term rattlesnake surveys for the DCR, Natural Heritage, and the town of Milton; more recently he's gone into debt. For Condon, who lives in Waltham, twenty-five miles from the hills, gasoline is the biggest expense; his credit cards grow heavier by the day. In the heart of the hills, in an open, rock valley above one of the dens, gay couples cruise and bask like the snakes themselves, oblivious to their reptilian neighbors. When Condon noticed a hiker repeatedly sunbathing fifty yards from a popular snake site, he tucked a large snakeskin and an admonishing note in with the sunblock and towel the guy stored in a ziplock plastic bag behind a log. "I never saw him again." Last summer, when a mountain biker reported a pile of newborn rattlesnakes next to a popular trail, Condon confirmed the sighting for Tom French, who notified the DCR. The trail was closed the next day.

Not everything Condon does is so enthusiastically received. Without divulging the location of the snake, several years ago, he e-mailed Tom French a photograph of a sick animal outside a den in the wilder, southwestern corner of Massachusetts, Tom Tyning's home ground. Thinking the rattlesnake was from Blue Hills, French gave Condon permission to collect it and deliver it to the wildlife veterinary clinic at Tufts School of Veterinary Medicine, then the recipient of compromised rattlesnakes. Condon bagged the snake, while Tyning, who monitors the Berkshire rattlesnakes, happened to be leaning against a tree, thirty feet below the den, watching. Tyning threatened to sue Natural Heritage if the snake died in captivity, claiming the car ride to Tufts was going to traumatize that snake. Said Condon, "What the fuck was he talking about? He sews radio transmitters in rattlesnakes!" Condon has also taken ill snakes off several mountains in the Connecticut River valley, to the consternation of both Tyning and French. "This [the fungal outbreak] is a spur of the moment development [in the Connecticut River valley]. It's all about snake conservation. French should make [Tyning] participate."

In October 2010, James Condon was arrested on Mount Tom, caught with the only surviving adult female on the mountain. The snake had a radio transmitter implanted in her, and a seasonal biologist paid to monitor the snake's movements picked up the signal of the snake in Condon's pillowcase. Authorities were called. Fearing the loss of another population of timber rattlesnakes, Condon had taken it upon himself to rescue the gene pool. "He lost a lot of respect from the research community," French told me. "I suggested to the arresting officer that James could be more use to us and use to the snakes if he did not prosecute him to the full extent of the law." Condon was fined fifty dollars as if it was a citation for a speeding and told not to stray from the Blue Hills. French said, "James has a passion and his passion is true, but he lets his passion rule what he does. I do not keep James in the loop very much anymore like I used to."

"I'm a little bit jealous. I don't hear all the thoughts on this [*Ophidiomyces*]. I just hold my end of the stick as best I can," says Condon, speaking of his excommunication.

Rattlesnake aficionados are fond of pointing out rocks and logs where they've found snakes in the past, and they recall even the most minute detail of each encounter as though thumbing through a photo album. At the rocky summit of a hill, on our early April hike, Condon tells me that last September a female birthed under a particular rock slab and then abandoned the snakelets before they shed, a very unusual behavior for a mother rattlesnake, who traditionally protects her neonates until their first shed, about ten days to two weeks after birth. "Apparently, all mothers are not cut from the same cloth," says Condon, laughing. Farther along the ridgeline, he shows me another granite slab where he once saw two rattlesnakes luxuriating on warm rocks.

"Were they basking?"

"No, they were studying Japanese."

Rattlesnakes have been in decline in Massachusetts since colonial times, when communities such as Medfield, Dedham, North Brookfield, and Canton paid a bounty for every snake tail brought to the town clerk. In 1680, Westborough organized a serpent posse of thirteen to kill snakes, and in 1740, Arlington appointed a day for a "general snake hunt and extermination." In Milford, rattlesnakes were the "terror" of haying season. In 1844, in Manchester, one hunter, John D. Hildreth, exterminated a den of rattlesnakes for their oil, which was considered an antidote to rheumatism. That winter, Hildreth smoked them out of hiding and then flung them in the snow. Historically, according to French, there may have been fifteen rattlesnake dens in the Blue Hills. There are far fewer now, all apparently underpopulated, and many, maybe all, affected by *Ophidiomyces*. One den was obliterated in the sixties during the building of Route 128. In Sharon, west of the Blue Hills, rattlesnakes vanished in the seventies. A rattlesnake-loving friend of Condon's contacted a psychic in Salem, who for a significant fee fell into a trance and reported "seeing" a large snake with several smaller snakes below the summit of a hill once known to have had snakes. Condon and his friend searched for the den but never found a rattlesnake. Today, there may be fewer than one hundred snakes left in the Blue Hills. Unfortunately, the killing of rattlesnakes by humans has long been a problem. "We have so many people in the Boston metro area that hate snakes," French told me one afternoon. People of authority—a Milton police officer, a state police officer, a military employee at the old Nike missile site on Chickatawbut Hill, a Boy Scout leader, all people in positions of responsibility to "protect" the public—have killed rattlesnakes in the Blue Hills. In 2001, an Arlington policeman was bitten on the hand by a pet timber rattlesnake. He claimed he was bitten in the Blue Hills Reservation, but his colleagues didn't believe him. After obtaining a warrant to search his residence, Arlington police found twenty-one nonvenomous snakes in terrariums in the basement and two timber rattlesnake sheds. Assuming the rattlesnake had escaped, police went door to door notifying neighbors that a venomous snake might be loose in the neighborhood, and the school bus began to drop children off at their doors. "I'm shaking," a neighbor told a reporter, "You like to think you're safe because you live across the street from a police officer." Now that they're endangered, it's illegal to kill timber rattlesnakes in Massachusetts, unless done in self-defense.

As I've pointed out (several times), timber rattlesnakes are *not* aggressive and pose very little risk, far less, in fact, than the cloverleafs and traffic on routes 28 and 138. French says there hasn't been a fatality from a rattlesnake bite since colonial times. When Condon finds a rattlesnake sunning itself outside a den, I sense in that instant the wildness of this place. Within moments, however, the sound of a far-off siren interrupts the silence.

Condon knows this snake. He's been watching him bask for a few weeks. Together, we circle the snake, studying his every breath. He neither rattles nor attempts to leave. Northern timber rattlesnakes come in two color phases: black and yellow. This one is a yellow, thirty-two-inch-long male, mustard-colored with chocolate bands, each band set off from the yellow by an off-white border. The tail is coal black, and the hallmark of the species, the hollow rattle, has nine segments that taper toward the tip. As you have already read, every time a rattlesnake sheds, it adds a new segment to its rattle, a self-replicating instrument. Because most rattlesnakes shed once or twice a year, this snake might be five or six years old. The telltale sign of the fungus, skin that looks wet and raw, can now be seen on the left side of his face, along both jaws back to the neck, behind the right eye, and inside the left nostril.

Condon pulls out a laser thermometer. The air is fifty-five degrees, the rocks sixty-eight degrees, and the heat-seeking snake seventy-nine degrees. When the rattlesnake exhales, his crest of vertebrae protrudes like an archipelago, and his loose skin drapes over a rack of ribs. "No one wants to check this one out," Condon sighs. Diamond-back Dave, majordomo of a popular field herping website, believes that shaded-over basking sites, coupled with too much handling, even too much visiting, may act as an immunosuppressant for timber rattlesnakes, making them susceptible to *Ophidiomyces*.

I ask Condon, "Do you have a name for him?"

"No. I'm afraid he'll die. I can't take it."

Not ten feet from the basking rattlesnake, Condon finds a snakelet, barely ten inches long, born late last summer—evident because of its size and its rattle has only one segment. The young snake is dead. Red ants swarm its face but haven't broken through the skin. The snake is supple and odorless and must have died *very* recently. Condon had seen the baby basking just three days before. Its nose is a crusty red. Festering lesions stand out on its face, and the inside of its mouth is the color of fruit punch, raw and ugly.

Amid the chatter of towhees and pine warblers, Condon calls Alexandra Echandi, a natural resource specialist for the DCR, who's working in her office in the hills, on his cellphone.

"Bad news," he reports to a sympathetic Echandi. "I found a dead snake, actually. A little one, it's a baby.... Do you think the zoo might want him? All right, he's all yours. I'm sad.... There was another sick snake there, too, that I've seen a few times. He was up again today ... the one that nobody wants to look at."

We walk the little rattlesnake out to the pullout, where Echandi trades Condon a box of granola bars for the specimen. They mention that there will be a meeting at the zoo in several weeks to discuss *Ophidiomyces* and the direction of the regional fungal research.

"Are you invited?" I ask.

"Are you kidding? What kind of question is that?"

4

Live Free or Die

Apparently wherever they were obvious enough to become eponymous, they were too obvious to survive.

THOMAS PALMER

11 a.m., June 9, 2012. Somewhere in New Hampshire's Merrimack River valley. It's hot and buggy under a cerulean sky. I can't find Kevin McCurley, who has disappeared behind one of the seemingly endless bedrock outcrops that emerge from the woods in a series of terrestrial swells. I'm east of the river and south of the White Mountains, disoriented in a Seussian world, adrift in a run of xerographic outcrops, iteration in stone, like ripples in a pond. While I look for Kevin, Kevin looks for rattlesnakes with blithe disregard for local geology.

As recently as the seventies, this swath of southern New Hampshire was the heart of a metapopulation of timber rattlesnakes, big snakes brick thick and mostly ink-black (yellow morphs were rare then; they're gone now). Snakes would gather in autumn at one of at least ten different outcrop dens within a hundred-square-mile interplay between the rock-ribbed woodlands and a mosaic of fertile wetlands—lakes, ponds, marshes, saturated scrubland, vernal pools—descendants of the prehistoric and meandering Merrimack. There were so many snakes in this corner of the Granite State that four different hills bear the name Rattlesnake, of which all are now sans rattlesnakes. A century of slaughter, both organized and random, reduced the population to a single den that has crashed from a high of perhaps sixty snakes in 2005 to a mere two or three adults, possibly fewer, in 2012. In fact, there are very likely more New Hampshire rattlesnakes in captivity than in the wild. McCurley, who's so incandescent one almost expects to see his clothes smoldering, rarely idles, which is part of my orienteering challenge; I lag behind listening to scarlet tanagers, wood thrushes, and towhees while he vanishes on the far side of an outcrop, a fifty- or sixty-foot-tall bedrock knob capped by boulders and slabs, shaded by white pines, and cut by the geometry of stress fractures. The outcrops look the same to me, though presumably not to the rattlesnakes, which chose traditional sites for basking, shedding, birthing, and denning, while assiduously avoiding others. Apparently, the landscape confuses Kevin as well. During the several trips I've made with him here, whenever we've attempted to regroup and find the den, he broadcasts comments like, "I can't believe I walked right past it," or "Let me get my bearings." Ever since McCurley ran off the end of a ledge several years ago and lost his GPS on the way down, he's reverted to aboriginal navigation.

"What the fuck? This looks familiar. Doesn't it?"

McCurley is forty-seven, and fit. His long, straight, sandy blond hair is typically swept back under either a baseball cap worn backward or, as today, a Caribbean bandana ending in a braid that extends nearly to his beltline, making him look like an extra from the set of *Braveheart*. He wears earrings, and sunglasses perch on his head. A tiny tattoo on his shoulder looks like Pac Man when viewed close up, but from a distance more like a smudge. His metronomic arms keep cadence with his speech, and when the conversation pivots on the welfare of New Hampshire's rattlesnakes, McCurley becomes volcanic and risks serious rotator-cuff injury. During lulls in an oratory, he stands like a heron on one foot, the instep of his left foot pressed against the side of his right knee, pausing for air, collecting thoughts. Although he has a degree in computer programming and worked for some time in the high-tech field, he left the grind twenty years ago and parlayed a lifelong passion for reptiles into a career breeding corn snakes, pine snakes, and rat snakes in his living room. He sold the babies, reinvesting the profits in more snakes. McCurley became a herpetoculturist, defined as someone who keeps either reptiles or amphibians (or both) in captivity, often (but not exclusively) for the purpose of breeding. His career switch was driven by emotion rather than logic. His new company, called New England Reptile Distributors (NERD), has evolved into one the world's foremost sources of designer pythons. Today, NERD is housed in a state-of-the-art fourteen-thousand-square-foot facility on the main street in Plaistow, New Hampshire, just north of the Massachusetts border. "The laws in New Hampshire are a little bit more lax. In Massachusetts, you could keep an emerald tree boa, but you couldn't keep a green tree python," he once told me. "New Hampshire didn't care, and there was this gray area on venomous snakes," which has proven to be both a problem and a blessing.

When he was a young boy in Lexington, Massachusetts, to keep him in sight Kevin's parents would harness him to a dog run in the backyard, where he built up an exponentially expanding reservoir of energy, which still drives his fervid cravings. After his parents divorced, McCurley discovered the joys of reptiles through an unusual channel. "My mother worked. If I was around [the house], my brothers saw me as something to do." His older brother and stepbrother would either wrap him in Saran Wrap and pelt him with rotten apples or handcuff him to a green chair and shock him with a generator. To avoid further torture, Kevin would hide in the woods behind his home, where he discovered spiders and toads, garter snakes, and a smooth green snake in a gray rock pile.

Today, McCurley supervises seventeen full-time employees and a kaleidoscopic array of drop-in friends and acquaintances (including James Condon), who work for an afternoon, a week, or a month. Like Gregor Mendel experimenting with garden pea genetics, McCurley experiments with reptile genetics, creating python, boa, and water monitor (a lizard) color morphs by crossbreeding for various species-specific colors and patterns. His facility annually produces more than fifty unique morphs whose names bear such descriptors as lesser platinum, sable, piebald, chocolate, blue-eyed leucistic, ivory, and ghost striped. Suntiger, a study in pastels, is a reticulated python (herpetoculturists shorten this to "retic") with a beige base overlaid by two repeating patterns of minaret-shaped bands that run the length of the snake along both sides of the body and meet on top, turret to turret, each minaret rimmed in black and graded in color from olive to orange toward the tail. Although McCurley doesn't have a shipping department, NERD morphs sell worldwide. Some go for as much as a Lexus LS, and although most are not so expensive, none are cheap. A half-dozen times, NERD employees have been prosecuted for stealing snakes and selling them on the black market, often to competitive breeders. Visiting the NERD website, the reptile version of Amazon.com, is convenient, interesting . . . and educational. There's a thread on the site about the conservation of timber rattlesnakes, in which Kevin offers a reward of five hundred dollars for information leading to the location of any new snake den in New England, a proposition that makes some regional biologists cringe for fear of the small but determined number of zealots that may be inspired to get off their couch and go into the woods. (McCurley is obsessive in his belief that there are still rattlesnakes in southwest Maine and elsewhere in New Hampshire besides the Merrimack valley.)

I drove to NERD one summer afternoon. On my way in, I passed two peacocks roaming the parking lot, one of which had just set off a car alarm. The bottom floor of NERD looks like a traditional pet shop—puppies, kitties, parakeets, macaws, goldfish, and supplies—with an exotic flare: saltwater aquariums with reef fish, a moray eel, setaceous lionfish, red-eyed tree frogs, map turtles stacked like poker chips, pig-nosed turtles, crested and red-legged and gargoyle geckos, pythons, boas, corn snakes, a cobra, and a jowl-less copperhead. Gifts, both the cobra and copperhead had had their venom glands removed (a state referred to as venomoid), something McCurley would never do, and were not for sale. At any time, three to four thousand snakes are here at NERD, necessitating many thousands of feeder rodents too. Every room, terrarium, and aquarium is individually heated. On average, his utility bill runs fifteen thousand dollars a month. Kevin lives in a three-room apartment on the second floor, above the pet store and down the hall from the reptile and rodent emporium. Four electric guitars hang above his bed—McCurley plays lead in the "thrash" heavy metal band Crotalus, whose official three-and-a-half-minute video on YouTube generates comments like "Keep it brutal my metal herpers" and "Fuckin' rad," whatever that means. To relax McCurley picks Led Zeppelin, often while he talks on the telephone. He shuns alcohol, tobacco, and drugs. For him it's all snakes, all the time.

Thirteen rooms in NERD house breeding pythons and boas. One has a king cobra in a terrarium on a shelf with the constrictors; in another room, a pizza-loving, two-hundred-pound alligator named Wally idles in a tank. Kevin hibernates Wally in the basement each winter. Three rooms host lizards. When McCurley showed his friend Stan the first albino water monitor that had ever been bred in captivity, Stan picked up the three-foot-long lizard and kissed it.

The rodent-breeding room at NERD is wall-to-wall shelving: five hundred drawers full of mating and nursing mice and rats and recently weaned juveniles, waiting to be eaten by the next hungry snake. A kookaburra flutters from shelf to shelf, patrolling the room for escaped rodents, which it bludgeons with its massive bill and swallows whole. A sign on the rat side of the door reads:

Do Not, Do Not, Do Not
Let rodents escape!!!!

EVER

They chew through water lines & flood & kill entire racks.
If any get loose,
You must catch every single loose rodent.

Or you will be
Killed!!!

Next door to the rodent room is one of several rooms where Kevin keeps a collection of venomous snakes (the so-called HOT rooms), some of which have been confiscated from their owners by New Hampshire Fish and Game (NHFG) wardens for illegal possession. NERD is the state's repository for illegal reptiles: an albino manacled cobra, a green viper, puff adders, copperheads, a log-thick Gaboon viper, and rattlesnakes—an eastern diamondback, a western diamondback, a massasauga, and many timbers, including two cream-colored, brown-banded Iowans, a mother and five neonates from West Virginia, several from Pennsylvania, and one from New Hampshire. McCurley removed the latter, a sickly, fungus-infected female, a snake as black as the ace of spades, from the Merrimack drainage system (with the blessing of NHFG biologist Mike Marchand) and nursed her back to robustness. The room is small and neat, with clear plastic cages that vary in size from cigar box to shoe box to sweater box, each reflecting the size of its inhabitant, stacked vertically, and lining nearly the entire perimeter of the room. The largest of the snakes live in glass-paneled terrariums. Streaks of dried, yellowish venom stain the fronts of several cages. Whenever I moved, someone rattled.

I met Kevin McCurley for the first time in mid-September 2010, on a ridge in western Vermont. He had come with James Condon to look for snakes. I was with Alcott Smith. It was the last fall you could visit the site legally without either a state permit or the company of a state biologist; everybody who had had a history with Vermont rattlesnakes wanted to enjoy one last untethered tour. While Smith and I counted snakes that had gathered at the main den portal, McCurley and Condon arrived from below, preceded by their voices. Together, the four of us climbed the ledge and walked the rim, searching for basking and gestating rattlesnakes. A four-foot male, a yellow morph, stretched in the sun, fully exposed. This story bears repeating a second time: Kevin McCurley picked up the snake slowly and thoughtfully ... barehanded. No hooks or sticks. No pinning the rattlesnake's head against the ground. He slid one hand under the snake's cloaca and the other under its belly, and gradually rotated in the same direction and at the same speed as the snake moved his head. The rattlesnake never rattled and never struck, never showed the slightest sign of agitation. When McCurley returned the snake to the rock, it remained stretched in the sunshine as though untouched.

"I handle snakes for a living, all day long, seven days a week, fifty-two weeks a year," he announced to no one in particular, building up a head of steam. "I know snakes. I read snakes. I'm a snake whisperer." Which may sound overblown and pompous but is *very* close to the truth. (Four years later, McCurley reversed his position. There's no room for showboating, he told me in a recent phone conversation. "If I ever got distracted and was bitten, everything I'm trying to accomplish would unravel.")

McCurley had introduced Condon to the world of timber rattlesnakes. When James sent Kevin a topographical map that marked where he had seen his first Blue Hills snake, the I-93 DOR (dead on road) in the breakdown lane, Kevin circled a mid-level ridge on the map, close to the interstate, and wrote, "Search here." The next day, Condon found the den. That afternoon in Vermont was the first time I heard mention of the keratin-eating fungus *Ophidiomyces ophiodiicola*, which McCurley had identified on New Hampshire rattlesnakes in the spring of 2005, when emaciated snakes began to emerge from hibernation with scabs on their heads, and later that year on Blue Hills snakes when James began posting photographs on the NERD website. "I went ballistic when I saw those pictures. If these were my breeding pythons I'd say, 'We've got something we've got to address,'" recalled Kevin. Of course, wild animals are *not* breeding stock, and should only be protected with extreme measures like captive breeding and head-starting when their population is in dire straights—the California condor, for instance. At the time, Condon had no idea that many of the snakes he had photographed were diseased or that the population to which he had introduced himself was about to spiral downhill. Long before James Condon began to inveigh to Massachusetts endangered species biologist Tom French about the fate of the Blue Hills rattlesnakes, Kevin McCurley had alienated the crotaline tribunal with his ranting and raving in three, possibly four, states. "I thought that because these snakes are endangered all you'd have to do is show the state where they are and what was wrong with them and they'd be protected. I was in la-la land.

"The biologists talked to me like I was ten years old."

$$\text{🐍}$$

Ophidiomyces is the reason McCurley can't find adult rattlesnakes in the Merrimack valley on June 9, and the reason I am having trouble finding him. In the woods looking for snakes, McCurley canters; when he can't find snakes he believes ought to be found at traditional sites, he gallops. Keeping up with Kevin is like keeping up with Secretariat. He is unbridled energy, all motion all the time, and lately, sadly, every day has become Derby Day in the woods east of the river.

The lone New Hampshire rattlesnake den sits squarely on private property, surrounded by state and municipal land, and is governed by a landowner—a Boston conglomerate with a Granite State name— that is ambivalent (at best) about the presence of rattlesnakes. In fact, just after the millennium, the company dynamited the last intact corner of one of the four Rattlesnake hills to extract gravel, killing more than a dozen timber rattlesnakes and destroying the only other viable den in the state. NHFG looked on, bound by state law, unable to intervene. In 2006 and 2007, whenever NHFG biologist Mike Marchand searched the property for rattlesnakes, the landowner required that a commissioned, independent field biologist tag along. Today, to see a timber rattlesnake in New Hampshire you either have to be *very* lucky or you have to know someone who knows the location of the den, tucked into a single vertical fissure amid an ocean of granite outcrops in the Merrimack valley. And, of course, that person would also have to be willing to take you there and not get you lost in the process. McCurley is just the latest in a short list of New Hampshire timber rattlesnake advocates.

By the time I moved to New Hampshire in the fall of 1975, timber rattlesnakes had already been flushed from eleven of fourteen historic metapopulations in the state. Although they had once occurred sporadically from Jackson and Bartlett on the edge of the White

Mountains, southeast to Raymond and southwest to Hinsdale, they were already gone from an island, a river, four hills, and seven mountains that bear their name. In 1975, rattlesnakes remained only on the slate beds of Wantastiquet Mountain in the lower Connecticut River valley, in the rocks above Dan Hole Pond in Tuftonboro, and in their dens along the Merrimack River. Four years later, during its 1979 legislative session, the state of New Hampshire passed a comprehensive bill to protect endangered and threatened vertebrates whose ranges include all or part of the state. I was among a team of biologists and naturalists invited to recommend which of these animals warranted inclusion on the state's new endangered species list. After six hours of debating the relative merits of the more than forty animals proposed for the list, we voted, using a point system devised by New York State that assigned numbers to various wildlife categories according to their degree of rarity, as well as their scientific and scenic value. The system was designed to determine which animals needed the most help, a form of triage on the battleground of beleaguered species.

Timber rattlesnakes received a rating of twenty-five, two points higher than the arbitrary cutoff for threatened species and six-tenths of a point lower than the rating given to peregrine falcons, the eventual recipient of a multimillion-dollar, nationwide restoration effort. Except for peregrines and Atlantic salmon, absent from the state since 1798, timber rattlesnakes had the third-highest ranking. Notwithstanding, at the insistence of the fish and game commissioner, rattlesnakes were excised from the official endangered species list submitted to the governor. They eventually made the list in 1987, but by that time they were already thought to be extinct in the state.

Andy Soha, sixty-eight, a recently retired heating, ventilation, and air-conditioning technician at a local hospital, rediscovered the *last* den in 1992, and over the next dozen years, he began the Sisyphean task of tracking the fortunes of the den's residents in his spare time. One time for each, Soha brought his three teenaged children to see rattlesnakes. "They were curious about what their father did in the woods." He would drive them deep into the woodlands along an old logging trail ... blindfolded. "I removed the blindfold only after I parked my truck. I didn't want to be the reason for the demise of the last rattlesnake den in New Hampshire," he told me one afternoon.

During the winter of 1991, while browsing through a four-year-old issue of *National Geographic*, Soha kindled his latent interest in rattlesnakes when he read the article written by William S. Brown that mentioned New Hampshire timber rattlesnakes in steep decline. The truth was that rattlesnakes had *not* been reported in the state for a decade and were thought by many authorities to have already been extirpated. Andy Soha, who worked the second shift at the hospital and had mornings and early afternoons free, was determined to find a New Hampshire rattlesnake. After reading archived newspaper articles that alluded to townships where dens had occurred, examining town histories, and querying outdoorsmen who hunted and fished the backwoods, he followed every promising lead into the recesses of southern New Hampshire. Essentially, Soha did what NHFG had been disinclined to do; he searched for rattlesnakes.

And eventually he found them.

University of New Hampshire herpetologist Jim Taylor had also been combing the south-facing, low-elevation ridges of southern New Hampshire for timber rattlesnakes, but to no avail. In late May 1989, in the company of Taylor, I had spent an afternoon in a remote section of Allenstown and found nothing but broken ledge and intact stories passed down from the days of the Great Depression. Between 1933 and 1942, a team from the Civilian Conservation Corps (CCC), the popular public relief program for unemployed, unskilled, unmarried men aged seventeen to twenty-three, developed a ten-thousand-acre parcel of wild New Hampshire. Local folklore claimed that a man named Pigeon Young raised rattlesnakes near the Allenstown Quarry, and that the snakes were released after his death. Apparently, the CCC crew encountered rattlesnakes, because the former Allenstown police chief recalled his childhood experience of seeing ten or fifteen timber rattlesnakes hanging from a clothesline behind the CCC camp and snakeskins tacked on a bunkhouse door. To safeguard his crew, the CCC camp boss had requested a shipment of snakebite kits from the federal government. After being stonewalled for some time, the camp boss wrapped a dead rattlesnake in a brown-paper bag with a note, "The strawberries are wonderful in New Hampshire," and sent it to the commissary, which prompted a large shipment of snakebite kits. Their work completed, the CCC left behind two legacies: Bear Brook State Park and a Mason jar full of rattles, which remained on exhibit through the early seventies.

Soha brought Taylor to see rattlesnakes. Together they pinpointed the den and, on a shoestring, began to unravel the life history of the state's most evanescent snake, documenting the basking and transient sites that Kevin McCurley showed me in the spring of 2012. To safely manipulate the snakes, Soha modified an Agway cultivator, cutting off the end tines and using the middle tine to hook snakes out from their stony recesses. He drilled a hole in the handle, jammed in a paintbrush, and began to mark rattles from a safe distance—various earthy colors so the snakes didn't stand out—to get a handle on population size, frequency of repeated sightings, number of times a snake shed, and year-to-year survival.

Then Andy Soha took homespun research to another level. Clandestinely, he began a radio-tracking program, eavesdropping on a pair of snakes deep within the rocky outcrops of southern New Hampshire. Taylor provided the receivers; Soha bought the transmitters. Since New Hampshire had already classified timber rattlesnakes as endangered, handling them—let alone operating on them—required a state permit. Soha and Taylor applied for and received such permits in 1992, 1993, and 1994, but to avoid a more complicated application process, Soha drove both snakes into New York State (crossing state lines with an endangered species is also frowned upon), where a state-licensed animal rehabber anesthetized them on her kitchen counter, taped each doped rattlesnake to a yardstick—gauze prevented the tape from sticking to the snakes—and then surgically implanted the transmitters just under the skin. Soha documented the procedure, sharing the pictures with me as we worked over the last of the tuna dip.

Except for that solitary den in the Merrimack valley, the one McCurley and I can't find, which I've taken to calling Andy's Den, Soha uncovered nothing more than stories, memories etched in quarries and outcrops of once thriving colonies of venomous serpents, black as night, whose lives and fortunes had been driven by weather for five thousand years, but were fading, or had faded, into an indelicate swelter of human ignorance.

In 1972, the film *The Rattlesnake King* documented the activities of Frank Young, a grandstanding New Hampshire snake hunter. Forty years later, I viewed the film at the administrative offices of NHFG in Concord. Soha had told me that Young, who died in the eighties, kept more than a hundred timber rattlesnakes locked in crates in a dirt-floor shed and lobbied the state to enact a bounty so he could cash them out. In an effort to sway the opinion of the legislators, in Crockett-esque fashion Young poured a knot of rattlesnakes out of a burlap bag onto the statehouse floor. Fortunately, the representatives, lethargic as the snakes themselves, never brought the proposal to a vote. On this afternoon in the backwoods of the Merrimack valley, when I eventually reunite with McCurley, he takes me to a long-defunct den he calls the Frank Young Den. "The bastard wiped them out," he tells me. "He was a monster." In 1963, Young wrote an article for the *New Hampshire Sunday News* entitled "Something Else to Worry You—Rattlers: Getting to Be a Real New Hampshire Menace."

Frank Young was part of the history of an undeclared Granite State war on timber rattlesnakes. In Concord, at the time of the Revolution, the snake-killing weapon of choice was a white oak or white ash cudgel, about eight-feet long and an inch thick that ended in a rounded knob. Like the artist George Catlin in Pennsylvania, one pioneer is reported to have tied a powder horn to a snake's tail, lit the punk fuse, and released the snake into a den crevice. A family of four killed forty-nine rattlesnakes one afternoon with sticks and hooks. And in autumn, farmers grazed snake-eating hogs around the rocky Concord dens, a practice that was also common on the islands in Lake Erie. In 1933, the *Manchester Telegraph* reported that in Hooksett, three boys picking blueberries killed a four-and-a-half-foot-long rattlesnake, tied a shoestring around its neck, dragged it home, and then displayed it at a local garage. (Four decades later, the same garage would exhibit an X-ray of Mohammad Ali's broken jaw.) Unfortunately, Frank Young became an oddball hero, a John

Lithgow look-alike in bib overalls and a powder-blue T-shirt, who trekked barefoot to snake dens and scaled the granite wall of the Allenstown Quarry with the aid of a climbing rope, also barefooted, hooked snakes out of rock crevices and hauled them home in a wooden crate, where he kissed one squarely on the lips.

In 1980, while teaching in the Department of Environmental Studies at New England College, I learned through a fellow faculty member that Henry Laramie, supervisor of game management in NHFG, knew the whereabouts of rattlesnakes in the Granite State. I wrote Laramie, asking if I could join him on a snake reconnaissance. He replied.

As you probably know by now, the rattlesnake is not endangered in New Hampshire. Now, someone should tell the handful of snakes that are left.

We [NHFG] have no written material and I shall continue to do my part to protect these remaining snakes by not divulging the location of their denning area to anyone.

Laramie was aware of the public's antipathy toward rattlesnakes; an antipathy the writer John Hay suggested had "biblical authority behind it." In September 1974, a surveyor discovered basking snakes on Pinkney Hill on the Hooksett-Allenstown border, not far from Bear Brook State Park, and returned with the landowner. Together they killed ten rattlesnakes, all black, and left their carcasses draped over the rocks. A week later, Henry Laramie and his brother-in-law hiked in to Pinkney Hill and found snakes, both dead and alive. Using a long-tined potato rake, Laramie collected eight or ten live snakes and put them in a big wooden crate, which they had lugged into the woods. The snakes were donated to local science centers, where they languished in captivity for a few years before dying. "It was a mission of mercy," Laramie explained to NHFG warden Jeff Gray, who showed me several faded pictures of Laramie posing with a live rattlesnake slung like an inner tube through tines of the potato rake. Gray, who visited Pinkney Hill a half-dozen times in the early years of the millennium, never saw a rattlesnake.

In autumn 2005, Jeff Gray estimated forty snakes (maybe more) hibernated in Andy's Den. By 2009, the number had fallen to fewer than twenty, and a year later, to eight, of which only two—*two*—were males. The following September, Gray, who had recently retired as head of law enforcement with NHFG, found a gravid female at a birthing site, the first pregnant rattlesnake that had been seen in several years. Gray notified NHFG biologist Marchand, who gave Kevin McCurley permission to collect the snake and bring her to NERD, where the rattlesnake delivered five infertile embryos, or slugs as the herpetoculturists call them. McCurley still has the snake. She was sick when caught, her face swollen with fungal granulomas, the outward manifestation of *Ophidiomyces ophiodiicola*, and she weighed only eight hundred eighteen grams. After the stillbirth, her weight fell to six hundred forty grams, on the small side for an adult female. "When I tried to hibernate her she almost died," Kevin said. "Her face swelled up; she drooled blood." McCurley treated the infection with Neosporin, heated the snake up, and fed her through the winter. At home in a NERD terrarium, bulking on domestic rodents, the snake's weight rose to eight hundred forty-five grams. "We'd love to breed her, but we don't have a male. We don't necessarily need a New Hampshire male," Kevin told me, suggesting that the state might boldly do for the timber rattlesnake what Florida did for the critically endangered Florida panther: bolster a *very* limited gene pool by bringing in potential mates from a healthy population outside the state. "A high-altitude West Virginia snake might be ideal," McCurley said to me one afternoon, glancing at his inventory in the HOT room.

In 2010, the prestigious journal *Biological Conservation* published a technical article about New Hampshire rattlesnakes titled "Decline of an Isolated Timber Rattlesnake (*Crotalus horridus*) Population: Interactions between Climate Change, Disease, and Loss of Genetic Diversity." The lead author was none other than Rulon Clark, the postdoctoral student from Cornell who had unveiled several of the sophisticated behaviors and complex relationships of timber rattlesnakes and had introduced me to the snakes Hank and Travis on the slopes of southwestern New York. Now an associate professor at San Diego State University, Clark was a remote collaborator, who had never actually visited the Merrimack valley. For data, he relied on the eyewitness reports of Soha and Taylor and an entourage of coauthors, which included, among others, NHFG biologist Mike Marchand and Basher Randy Stechert, the successful (and funny) itinerant field biologist from Narrowsburg, New York, who had first visited the rattlesnakes of New Hampshire in the company of Andy Soha in the early nineties.

Unless you live cloistered in a subterranean tunnel and thus have missed one of the saddest and most repeated storylines of the last quarter century, it is now common knowledge that humans are accelerating the planet's loss of biodiversity at an alarming rate, a rate that rivals any extinction event in Earth's history, including meteor strikes, worldwide glaciations, rupturing continents, and wide-scale volcanism. Case studies documenting the synergistic effects of multiple negative anthropogenic impacts on a wild population—fragmentation of habitat, pollution, off-road vehicles, overcollecting, bludgeoning, and so forth—however, are quite *rare*. Enter the New Hampshire timber rattlesnake: the test case.

A timber rattlesnake the color of piano keys is a genetic aberration, the product of a vastly depleted gene pool brought on by recurring inbreeding, what geneticists refer to as a "genetic bottleneck," which in the case of New Hampshire timber rattlesnakes dates back to the killing fields of the CCC, Frank Young, gravel mining, and the Pinkney Hill crowd, who collectively eliminated every rattlesnake colony in the Merrimack valley save the one Andy Soha rediscovered. Although Andy's Den had a stable population of forty plus adults between 1992 and 2005, outbreeding had stopped years before, and the seeds had already been sown for the deleterious effects of chronic inbreeding. By default, mothers mated sons, fathers daughters; grandparents and aunts and uncles, brothers and sisters and cousins had become a swirl of genetic sameness, the rattlesnakes of the Merrimack valley having been forced down the same cultural path as the Florida panther.

To help Clark, Mike Marchand and Jeff Gray collected sheds and carcasses, from which Clark extracted the DNA samples that he used to compare the genetic diversity of the Merrimack valley snakes to snakes from Bill Brown's long-term study in the Adirondack foothills, where members of thriving rattlesnake colonies regularly interbreed. The results were predictable: the number of alleles (defined as one of two or more alternative forms of a gene that arose by mutation and occupy the same site on a chromosome) in isolated New Hampshire snakes was far smaller than the number found in the Adirondack snakes.

Genetic drift is the variation in the relative frequency of different genotypes in a small population owing to the chance disappearance of particular genes as individuals die or cease to reproduce. Genetic drift in the New Hampshire rattlesnakes likely triggered a *very* visible occurrence: yellow morphs, common in Vermont, New York, and Massachusetts, had been eliminated from the population. Mutations, the source of novel genes in a population, swept through the inbred rattlesnakes and resulted in an unprecedented geographically restricted morph, the piebald, an ebony rattlesnake with varying amounts of ivory, a sort of self-made version of a McCurley designer python. Over a combined career of fieldwork covering more than one hundred twenty-five years, with visits to over five hundred different dens in fifteen states, Bill Brown, Randy Stechert, and Marty Martin had collectively tallied more than several thousand timber rattlesnakes and *never* seen a piebald. New Hampshire piebalds are still the only known incidence of the color form in a wild population of timber rattlesnakes.

Like fingerprints, no two piebald morphs were alike. Some had pink rather than black (normal-colored) tongues; others had white running up the sides almost to the spine or blotches on the head or flecks on the sides like stars in a dark sky. Several had cream-white rattles. Soha noted piebald snakes in 1992, none longer than twenty-four inches; two adults and an entire clutch of elegantly colored neonates were seen in 2006; and between 2006 and 2009, at least four litters included a few piebalds; none have ever been seen since, however. I saw one once, a snake so beautiful it left me breathless, a serpent of such pure contrast that it embodied extremes of the Ansel Adams zone system. No wonder that Andy Soha had blindfolded his own children on their way to and from the den; the piebald might be worth its weight in gold to an unscrupulous reptile collector.

Rulon Clark and associates, in their 2010 paper, believed that the inbred, genetically depauperate rattlesnakes might have been rendered especially susceptible to a lethal fungus, since identified as *Ophidiomyces ophiodiicola*, a condition that had been promoted in part by a series of wet years, including 2005 and 2006, the two wettest summers on record in New Hampshire. Which brings us to the issue of accelerated climate change. Like James Condon's isolated rattlesnakes in the Blue Hills, the infected Merrimack snakes basked late into the fall, shed more frequently, fed less frequently, bred less frequently, and looked awful. Many simply failed to rise from their dens in the spring, and the population began to spiral down what conservation biologists refer to as an "extinction vortex."

> Thus, it appears that disease susceptibility in this system may result from interacting effects of habitat fragmentation, inbreeding depression, and climate change.
>
> This combination of factors will most likely lead to the extinction of the NH population, unless it is actively managed. Past research on communally hibernating Viperids has indicated that isolated and inbred hibernacula can be "genetically rescued" by the introduction of novel alleles. We recommend ... a focus on restoring genetic variation from nearby non-isolated populations. If this is not done, it is likely that timber rattlesnakes will be extirpated from New Hampshire.
>
> Although forecasts of anthropogenic climate change for the northeastern US do not predict that summertime precipitation patterns will increase on average, climate variability in general is predicted to increase in magnitude and frequency. This study serves as an example of how extreme weather events can interact with other anthropogenic impacts in unexpected ways to cause population declines. Unfortunately, as global change accelerates, this combination of deleterious local and global impacts will become more common. To preserve biodiversity, we will need to take action at both scales.

The only item Rulon Clark and company omitted from the article's conclusion was by far the most direct and telling anthropocentric cause for genetic impoverishment of the Merrimack valley snakes: wanton slaughter by a litany of misguided residents.

To the amazement of everyone involved in the bewildering collapse of the last population of New Hampshire rattlesnakes, on August 22, 2011, Jeff Gray found a gravid female basking at a traditional outcrop not far from Andy's Den. McCurley had seen the snake entwined with one of the last two males the previous summer, but until now she had been missing and feared dead, perhaps the latest victim of *Ophidiomyces*. Gray secured permission to birth the rattlesnake at NERD and head-start the neonates, which was deemed perhaps the last best chance to save the state's critically endangered timber rattlesnake. For fear they might be denied access by the landowner or encounter a groundswell of critical feedback if the public discovered that scarce tax dollars were being spent to preserve a venomous reptile, NHFG decided that the landowner would *not* be notified of the spur-of-the-moment management strategy, a rather unprecedented move by the state that imprints the motto "Live Free or Die" on every license plate. A couple of days later, McCurley retrieved the snake on a cool, rainy afternoon and set her up in the NERD HOT room, a newbie in the serpent dormitory that already housed dozens of venomous snakes seasoned to life on a meal plan.

When I arrived at NERD the following day, the gravid snake had already acquired the name Blue Tail; her rattle had a dab of blue paint from when she had been marked in 2008, the year she last gave birth. (There were four unmarked segments beyond the blue, which meant she had shed four times in three years, normal for a *healthy* wild rattlesnake.) Blue Tail was black, with faint traces of red-brown blotches down her spine. She was slightly over forty inches long, weighed fifteen hundred grams, and displayed a singular independence. Kevin removed her from her terrarium and placed her on the floor. Unwilling to be corralled into the plastic examination tube, she eluded his snake hook and headed for the darkest corner of the room, right where I stood, forcing me to prop my weight against a sink and raise my legs as she passed by. The more McCurley tried to catch Blue Tail, the more she resisted, coiling and striking at the slightest provocation. Four times, he coaxed her into the first six inches of the tube; four times, she wiggled free. Then, the room became abuzz, literally, as a half a dozen other rattlesnakes became agitated by the activity. "She's going to have babies soon, so she's wound-up," McCurley growled in frustration, all pretense of a "snake whisperer" having vanished. Eventually, Blue Tail surrendered (or just got worn out). Her head snugly inside the tube, a NERD employee rubbed her belly with gel. When McCurley ultrasounded the snake, I saw fifteen tightly coiled, immobile embryos on the screen, two rows, one on top of the other, a rather large clutch for a timber rattlesnake. I can see why gravids often fast during the summer of their pregnancy: there would probably be very little room for a mouse or chipmunk inside their jam-packed tubular bodies.

Blessed miracles are not unknown in the constricted world of herpetology. Over the next few weeks, two more gravid snakes and a clutch of premature neonates that had been abandoned by their mother before they shed, a very unusual occurrence, turned up at traditional Merrimack valley basking sites, not far from Andy's Den. Gray and McCurley collected the snakes, and by the time I returned to NERD in October, there were forty-four snakelets, all black morphs—the *very* lifeline of New Hampshire timber rattle-snakes—each in its own small plastic tub piled on shelves in the HOT room.

Snake sheds freed of purpose hung everywhere drying, delicate thumb-width tubes that unfurled like transparent party favors when-ever McCurley blew into the mouth holes. Some snakelets had al-ready eaten eight full-grown mice in the month since their birth, the number they might eat in an entire five-month active season; several had shed three times (twice the average).

The preemies, much smaller, had just started to digest mice. "Maybe their mother abandoned them because they're genetically defective and they ought to be disposed of," I suggested in passing, which incited a small eruption in Kevin. "It was late in the season. She basically birthed and ditched," he said, defending their mother's bad behavior. All six preemies had a herniated umbilicus, a slit on the belly where the yolk tube had nourished the embryo, which should have cinched inside the mother, but remained open, gathering tid-bits of whatever the small snakes had crawled over. During their first

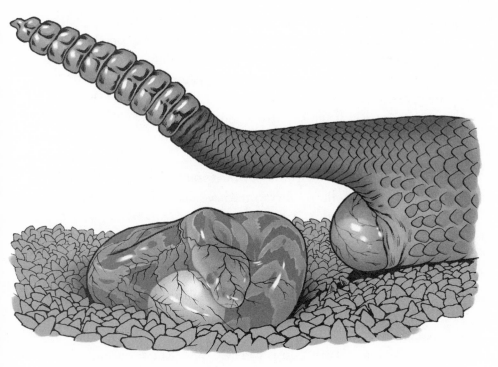

weeks at NERD, whenever they fed, they regurgitated their mice after four or five days. Four died of infections from dragging their vital organs over the ground. The survivors have scabbed-over bellies, about twenty scales anterior of the cloaca, and were a third the size of NERD-born snakes. Time will tell whether they're fit to survive.

Each snake tub has a bed of aspen shavings and a lidded water dish with a silver-dollar-sized hole in the lid so water doesn't spill when the snake drinks or soaks. Tubs sit on electrically heated tape set to a constant eighty-three degrees Fahrenheit, which heats a portion of the tub bottom to 94.9 degrees, an inverted heat source but an ideal temperature for basking. Snakelets regulate their thermal environment by moving on and off the heat-tape portion of the tub. A thermostat on the HOT room wall controls the flow of heat. Daily, with a pair of long-handled tweezers, Kevin (or a NERD employee) removes soiled shavings and dead mice, which were bitten but uneaten; an eastern diamondback grows fat on the discards. To weigh the snakes McCurley weighs the full tub on a digital scale and then takes the snake out and reweighs the tub; the scale flashes the weight of the missing snake. One baby already weighs over one hundred grams. Diligently, McCurley records each feeding, shedding, rate of growth, and in the case of the preemies, patterns of healing. Gently he palpates every snake's cloaca to see if pink hemipenes (boys) or white scent glands (girls) pop out and then he records its sex, which is predominately female. To foster imprinting and bonding, McCurley left the three mothers with their respective clutches until the neonates shed, about ten days, at which point, in the outcrops of the Merrimack valley, the mother would have abandoned her young and headed toward Andy's Den, leaving the babies to follow her scent trail home, their tongues flicking the ground in recognition of the invisible path. Once the little snakes had shed, Kevin returned their mothers to the wild, releasing them close to the den.

McCurley does all this rattlesnake rehabilitation and head-starting for NHFG on his own time and at his own expense, both of which are considerable. "Right now, the state comes to me instead of the Roger Williams Park Zoo [in Providence, Rhode Island]. They made their choice. And I know what I can do is going to be good for these snakes," he tells me. McCurley named one of the three mother rattlesnakes Jefflene in honor of Jeff Gray. "He's my guy. He's got my back." It's deeper than that. If McCurley hadn't had Gray in his corner, NHFG would very likely have gone to the zoo for help.

Gray, then a captain of law enforcement in NHFG, met Kevin McCurley in 2003. Kevin sat at a picnic table in front of NERD eating lunch. "I had just gone over to the salad bar at Market Basket and made an awesome salad, a beautiful salad. It had everything." Then, Gray drove into the parking lot with a pair of backup patrol cars and a search warrant. "I was mortified," Kevin recalls. A disgruntled ex-employee had turned McCurley in for not having a state permit to keep venomous reptiles. "He thinks I'm the outlaw Josie Wales and he's here to serve justice. It was horrible. I never got to finish my salad."

"You certainly looked like you knew what you were doing," Gray had told Kevin in my presence, one afternoon in the woods along the Merrimack. "You had a smaller place then. It would have taken a month of Sundays to go through all the crap you have now." The bust had nothing to do with wild rattlesnakes. After McCurley was fined — not much more than a traffic ticket — Gray walked him through the state's permitting process. "What he did was wrong, but still I recognized his expertise. It wasn't a poaching issue; it was a possession issue." Gray knew that McCurley wasn't removing native rattlesnakes from the wild and selling them; he was keeping venomous snakes without a permit. Through Gray's supervision, Kevin McCurley eventually became a resource for NHFG; whenever they confiscated venomous reptiles from other owners that hadn't acquired a state permit, NERD became the repository for the contraband. "I have to manage where he's putting his efforts. He's very capable, very intelligent. I never met anybody that had that level of snake knowledge, husbandry, handling, and experience. I don't know the people at Roger Williams Park Zoo, or what their level of experience is [with timber rattlesnakes]. I'd say the zoo [staff] has handled a thousand snakes, and that's probably giving them quite a bit. Kevin, on the other hand, has handled thirty or forty thousand snakes at his facility, and that's just in a given year. And multiply that by all his years of experience in genetics producing those designer snakes. He's an intelligent guy who needed to be headed down the right path."

Kevin, being Kevin, can be tempestuous when the welfare of timber rattlesnakes is broached. "I know how to deal with him," Gray continued. If it hadn't been for Gray's guiding intervention on both sides, the fate of the last timber rattlesnakes in New Hampshire might have been sealed years ago. Gray has far more problems with rattlesnake fanciers that have occasionally shown up in the woods. In the days before GPS, one man brought a rod and reel and had planned to tie monofilament around a snake and then follow the line back to locate the den; Jeff Gray disabused him of the scheme. ("Ten yards and that snake would have been tangled hopelessly"); another left a boulevard of hack marks and flagging to the den.

In mid-October, Kevin released eight male rattlesnakelets into the den, representing all three of the recently NERD-birthed clutches, hopefully to join their mothers; the preemies were not yet ready for life along the Merrimack. The remaining thirty-six stayed at

NERD all winter bulking on domestic mice. Kevin's goal was to put as much mass on the young snakes as possible, so they'd be big enough to swallow chipmunks when released in the spring. Many of the month-old snakes were the size of wild-born two-year-olds; a few were bigger.

McCurley lined twelve tubs up on the floor of the HOT room, and dropped a mouse in each. It's feeding time. I settled into a chair and waited ... waited. The snakes were not in a hurry to feed; none were in ambush posture, coiled like a spring ready to explode; instead, they'd cruise around the periphery of their plastic confines, stimulated but distracted—by me, the mice, each other. Several mice ran over the snakes, as though they were nothing more than a tub-side attraction; others burrowed into the shavings; a few preened. One snake struck, more out of annoyance than hunger; I missed the hit, but noticed the mouse convulsing in a corner. Then, the snake struck a second time and the mouse fell on its side, barely breathing. The snake had no interest in feeding. Neither did the four other rattlesnakes that immobilized their prey. When McCurley returned, he removed the dead mice and offered them to a West Virginia timber rattlesnake, which swallowed them with gusto, one after the other.

I returned to NERD in mid-November to check on the snakelets, hoping this time to see a snakelet eat its mouse. Kevin set sixteen tubs on the floor, sixteen potentially hungry snakes, all of which had shed twice. Cicada-like buzzes resonated from their tiny, two-segment rattles, and the eastern diamondback's discordant counterpoint, loud and electric, chimed in every time I changed position, which makes the HOT room rock to the agitation of pitvipers. A snakelet struck a mouse on the hip, apparently delivering a full dose of venom. The mouse fell on its side, rolled over, legs straight skyward like a cartoon casualty. The snake tongue-flicked the mouse's head and moved into position to eat, straight on, and head to head. Downstairs in the pet shop, a puppy barked. The rattlesnake yawned and its jaws spread apart at the corners and in the middle, the bones connected only by expanding ligaments, which transformed the snake into a dark gaping cylinder that slowly, very slowly "walked" its unhinged jaws over the mouse's head and shoulders—the widest part of the body—fangs folded back against the roof of its own mouth. As the mouse slid deeper down the snake, the snake's skin stretched, scales spreading apart revealing thinner whitish skin, which would have remained hidden without the bulging meal. By the time the mouse's head had reached the snake's stomach, a little more than halfway down the tube of the snake, just the tail stuck out of the snake's closed mouth like a flaccid cigar. With the mouse bundled in its stomach, the rattlesnake yawned, realigning its jaws. Ten of the sixteen snakelets struck and killed their mice while I sat there, and five commenced feeding, one while the mouse was still twitching.

As you may have gathered by now, the snake country around the Merrimack supports far more rattlesnake stories than snakes. Reminiscences are *far* more common than sightings. On June 9, 2012, I have driven to Kevin's rattlesnake rendezvous site, an obscure turnoff on a dirt road, not far from the Melville Avenue, which I recognize by an abandoned Sears delivery truck with the word "Fuck" spray-painted across the rear doors. We have to walk along the rim of the mile-long, deeply flooded trail, leaning on stout sticks to keep from falling in. Mosquitoes are everywhere, and hungry. Even under these conditions McCurley is hard to keep up with. Only the wood duckling in front of us moves with more urgency.

Kevin released eight more NERD-born timber rattlesnakes in late May, which joined the eight he had released the previous October, which is why, later in the afternoon, I have lost him as he motors through the woods from one outcrop to another trying to find Andy's Den. Young leaves of hickory, maple, oak, cherry, ash, and birch have begun to coalesce into a translucent canopy that shadows the ground and favors the bloom of shade-tolerant flowers—fringed polygala, pink lady's slipper, Canada mayflower, starflower, blue-bead lily, and, near the water, painted trillium. On the summit of a traditional rattlesnake basking site called Bald Knob a whip-poor-will bolts from a slight depression in the pine needles, leaving two tan, fluffy chicks, huddled against a pinecone. Each chick still sports an egg tooth on the upper bill, tiny, a deciduous calcification as delicate as the chicks themselves. Mother whip-poor-will circles the summit screaming *whip, whip, whip* and then lands on a low pine limb, a soft, brown bird in the soft, green woods.

This is far wilder country than I would have suspected, and it is no wonder timber rattlesnakes have managed to hang on into the second decade of the twenty-first century. Moose graze the wetlands, leaving paths through the emergent vegetation the size of a Boston whaler; bobcats prowl the outcrops; black bears roll boulders looking for hornets; Blanding's turtles lay eggs in sandy soil between the summit rocks and soak in waterlogged swales. In the wetlands, ribbon snakes gorge on tadpoles, and in the woods, hog-nosed snakes devour toads. I could yell Kevin's name, but the sweet chorus of songbirds—wood and hermit thrushes, towhee, pine warbler, and scarlet tanager—stays me from corrupting the stillness of the woods with my voice. Thus far, getting lost has proved an educational inconvenience, not a disaster. Was it the Dali Lama who asked, "How you can you get lost on such a small planet?" The woods fill with the clucking of chipmunks, whose voices commingle with the birds, and whose bodies Kevin hopes will slide down the yawning gullets of the oversized, overwintered, NERD yearlings. On the sunny side of the outcrop, a congestion of milk snakes jam into a horizontal crack, tightly braided and looking oddly out of place without a stonewall or a barn nearby. In the distance, an antiphonal duet of tree frogs and green frogs rises from an unseen shoreline. It's an altogether delightful day in a forgotten corner of New Hampshire, if only Kevin can find timber rattlesnakes and I can find Kevin.

Unexpectedly, I conjure Kevin from the shadows on the north side of Bald Knob, his hair flailing in gusts of his own making, like Jesus with a snake hook. (I'm sure he hasn't noticed my absence.) After fifteen or twenty minutes of traipsing through dwarf valleys and over blueberry-covered summits, we find the Andy's Den outcrop, a fortress of stone that faces southwest. Almost immediately, Kevin finds a snake he released last fall coiled snugly against a gray birch log, a curl of white bark shading it like an awning. The yearling is slightly larger than the two I saw last week in western Vermont, but no monster. Perhaps he's fourteen or fifteen inches long, one of the four (of eight) NERD-born yearlings released in October known to have survived hibernation.

Another yearling, a snake Kevin released two weeks ago, stretches in a bar of sunlight: a charcoal-gray girl with black bands rimmed in white. Then we find a third snake stretched in a grass trough between two small boulders. And another... By day's end, we tally seven yearlings, of which six fed all winter at NERD and were already the size of six- or seven-year-old timber rattlesnakes, three-feet long and plump. "The males are already producing sperm," says Kevin, with an obvious air of fatherly pride. These are oddly behaving snakes. One crawls between my feet, tongue-flicking my boot, oblivious to, or at least unconcerned by, my presence between him and a rock platform near the den portal, where he'll soon luxuriate in the June sunshine. Thus far, like normal yearling timber rattlesnakes, the NERD-enhanced snakes stay within a hundred yards of the den; they look like adults— cold-blooded analogs of the Tom Hanks character in the film *Big*. But while these snakes appear mature, they don't behave like it.

Cottonstone Mountain, in Orford, New Hampshire, across the Connecticut River from where I live, faces Vermont to the southwest, and runs east from the river for two ridgeline-miles before tapering off into the valley of Indian Pond Brook, which drains Indian Pond. A hump on the ridgeline is twelve hundred feet, most of the rest is over a thousand and supports a forest of old red oak and red pine, where itchy black bears scratch their backs and leave wisps of fur stuck in the jigsaw bark. Across the Connecticut in Vermont, the Fairlee Cliffs, a vertical granite precipice, hosts nesting peregrines, turkey vultures, and ravens, and at its base, the windows of a popular breakfast restaurant look across Route 5 toward the rock wall. Lake Morey nestles in a pocket between the cliffs and Bald Mountain to the west.

A friend of Kevin's, a retired high school biology teacher, who taught for several decades at a prestigious New England boarding school, recalls sitting with his grandfather on the warm rocks of Cottonstone, amid basking rattlesnakes, as the sun set over Vermont. It was Labor Day weekend, 1957. "The highlight of a wonderful weekend," he had told me. "I've never forgotten those snakes."

Since Cottonstone Mountain is across the river from my home, I promised them both I'd have a look.

10:00 a.m., August 1. Alcott Smith and I follow a logging road to the Cottonstone summit and then bushwhack west, plodding along the ridgeline, where every so often a break in the forest reveals a sweeping panorama of the Connecticut River valley, south well beyond the arch bridge between Fairlee and Orford. There are plenty of exposed rocks for basking and hiding and shedding, sedges and grasses, and patches of leaf-filtered sunlight spangle the forest floor. It's quality rattlesnake habitat in spades. In an attempt to find a talus slope and a possible den site, we descend off the ridge and backtrack.

About five thousand years ago, thousands of years after the glaciers had given New England a geological facelift, during a worldwide warming period called the Hypsithermal Interval, timber rattlesnakes expanded their range into northeastern North America. From coastal refuges they pushed north along wide, rocky, river valleys—the Delaware, the Susquehanna, the Hudson, the Merrimack, the Connecticut. Wherever they encountered talus slopes or stress-fractured bedrock that permitted access to frost-free winter quarters, colonies formed and began to spread out across a virgin landscape. Eventually, timber rattlesnakes met Woodland Indians, fur traders, and colonists; below riverine basking sites the French and Indians fought the British, and later the Revolution unfolded. Over the past century, they met CCC crews, Henry Laramie, Frank Young, and a host of anonymous zealots, gravel mining, logging, developing, constructing roads, and generally participating in the northward expansion of Boston. Under such circumstances, one wonders that any New Hampshire rattlesnakes survived into the present century.

At the six-hundred-foot contour on the U.S. Topographical Map, running southwest from the base of the shear western edge of Cottonstone is a thirty-degree slope, densely vegetated. Initially, Alcott and I had walked past the rockslide, hidden under a coif of shrubs and vines. Hacking our way slowly upward through sumac, mountain maple, and elderberry, over vines of red-flowering raspberry, grape, and Virginia creeper, we teeter on rocks slippery with wet leaves and moldering vegetation. The overgrown talus might have held snakes in the late afternoon sun more than fifty years ago. It's hard to say, really. I spoke with an Orford historian who claimed that in 1952, a pair of timber rattlesnakes visited a barn on Route 10, a mile south of Cottonstone Mountain, and that many townsfolk stopped by to see them.

No one else seems to know anything about rattlesnakes on Cottonstone Mountain. "Their distribution, *Ophidiomyces ophiodiicola*, this is all preschool," said McCurley; "what we need is a college-level curriculum."

5

Zero at the Bone

They scare me sick.... They always have.

ERNEST HEMINGWAY

Charlie Snyder knew snakes. They were his passion. He hunted
them, four or five times a year; kept them, milked them, displayed
them, and spoke about them, often to large audiences. He was fearless and confident, a world-class herpetologist and a voice of reason
in an age of misinformation. He was also a regional celebrity when
the doings of the Bronx Zoo and its parent organization, the New
York Zoological Society, were more than a curious footnote of city
history. Snyder's public lectures were covered by the *New York Times,*
occasionally up front with the news of the day. While he was still
a teenager, Charlie began working at the zoo, where he served for
fifteen years as a keeper before being promoted to head keeper of
mammals and reptiles, the gold plate of serpentine husbandry. He
worked closely with his mentor and friend Dr. Raymond L. Ditmars, the curator of mammals and reptiles at the Bronx Zoo, and
the first herpetologist to write extensively for the general public, for
whom he popularized snake hunting, fostering the careers of many
naturalists (and poachers), several decades before Staten Island
Zoo director Carl Kauffeld. Ditmars set a high bar recounting the
joys of snaking in remote areas, some surprisingly close to major
metropolitan centers—as you may recall, the last timber rattlesnake
on Long Island was seen in 1962, and the last Bronx copperhead

just after World War II. The *New Yorker* occasionally recounted the quirky activities of Raymond Ditmars, as it would do years later for Kauffeld, who followed in Ditmars's footsteps in both the figurative and literal senses, adding his own contributions to snake lore and literature.

On May 14, 1929, Charlie Snyder and Raymond Ditmars appeared together on the front page of the *New York Times*. The occasion was Snyder's obituary.

The Ramapo Mountains cross the New Jersey border near Suffern, New York, where they form the front range of the Hudson Highlands (the spine of a biogeographic province that includes more than just the mountains), and trend north along the west bank of the Hudson River before fading away in the vastness of West Point. In Lenape, Ramapo means "a river with potholes," and it is also the name given to the river that links the mountains to New York Harbor. More than a billion years old, the Hudson Highlands are a geologic fortress of igneous, sedimentary, and metamorphic rock that has been molten, hardened, eroded, deposited, folded, softened by time and weather, and finally scoured by glaciers, which left behind a chain of rounded hills, none more than sixteen hundred fifty feet high, and a series of rocky, fast-flowing trout streams, ridgelines, and south-facing outcrops. The woods are a deciduous tapestry—oaks (many types), hickories (few types), tulip trees, black birches, black walnuts, and sugar maples. Pitch pines, stunted, pruned, and twisted by unforgiving onshore winds dot the ridges and outcrops, and here and there a few white pines, the very essence of rooted antiquity, rise above the hardwoods.

Sunday, mid-morning, the twelfth of May 1929: somewhere in the Ramapo Mountains, with Manhattan's jagged skyline less than thirty miles away. It's sunny, seventy degrees Fahrenheit. Ideal snake weather. Snyder is on a routine snake hunt, to him a common sport. As he reaches for a basking timber rattlesnake, a second snake, lying in wait, bites the back of his left hand, penetrating a vein. Although rattlesnakes are rarely amorous in early May, and hardly known for altruism, the *New York Times* later claimed that a loyal snake hidden in the grass had "struck for the liberty of its mate." (More likely, after seven months of deep sleep, it had woken up hungry and was aroused by a swiftly approaching warm hand.) Snyder's hunting companion is below the ridge, a hundred yards away. Although they carry a vial of antivenin, Snyder elects not to use it. Instead, he opens the two punctures with a razor and attempts to suck and squeeze out the venom. Then, to slow circulation, with his handkerchief he ties a tourniquet around his arm, and sets out for the hospital in nearby Suffern, both minimizing the gravity of his situation and underestimating the time required to get off the mountain. With the help of a local farmer, Snyder reaches the hospital, five hours later ... exhausted ... venom coursing through his veins, his capillaries leaking, his left arm swelling, his blood pressure falling. The attending physician administers the only three vials of antivenin on hand, and then, at Snyder's insistence, the hospital contacts Dr. Ditmars, who arrives by automobile at eleven o'clock the following morning. Ditmars brings more precious antivenin, but it will not be needed, at least not for Snyder, who has been dead for forty-five minutes. "It was his martyrdom," Ditmars tells the hospital staff.

Monday morning, May 13, 1929. 10:15 a.m.: Charles E. Snyder, aged fifty-seven, died, the last person in New York State known to have been killed by a timber rattlesnake.

Rattlesnakes, as I'm sure you have already gathered, have never had very high approval ratings, particularly after the movie *Snakes on a Plane* rebooted our primal fear. If I told you, "Hey, look, there's a *rattlesnake* under your porch," I would very likely illicit a stronger response than if I said, "Look, there's a *toad.*" First of all, a rattlesnake is a snake (a critical fact already mentioned several times) and snakes make a *very* large segment of the public queasy, merely by being snakes, not to mention the issue of the venom and hypodermic fangs. Make no mistake, a timber rattlesnake's intricate markings and variable base color make the snake (in my opinion) a still-evolving work of art—breathtakingly beautiful, really—but with a decidedly toxic, sometimes deadly, bite. Every few years a timber rattlesnake kills someone somewhere in the eastern half of the United States. In 2012, for instance, a Pentecostal serpent handler named Mack Wolford, whose own father had suffered a fatal bite in 1983, died after being bitten on the thigh during a service in Panther State Forest, West Virginia, the only state that still permits snake handlers to ply their faith. Both Wolfords, it must be said, let God administer their healing.

"Most people," wrote Archie Carr, "have a vague feeling that no reptiles except turtles are to be trusted." Very few images haunt our imagination as much as a chance encounter with a snake, particularly a large, venomous one. For many people the repulsion is primal, untraceable, irreversible, a lesson in demonology, perhaps a remnant of our simian birthright that accompanied us down from the treetops and across the African savanna, as deep and ingrained in our subconscious (and maybe even in our DNA) as fear of the dark. Rattlesnakes are portrayed as sinister, an embodiment of evil, legless, shiftless incarnations of Satan, muscular and venomous cords, pea-brained and deadly. I have known more than a few people who get nauseous when they see a picture of a snake, even the crudest of pictures, say a squiggly line with a forked tongue. One woman I know of wrote a letter to the editor of our local newspaper canceling her subscription after the paper published an illustrated story I had written about Vermont timber rattlesnakes. Was her odium an evolutionary footprint, something coded in her genes? Or did a parent pass along an incorruptible bias? Is there a difference?

Here are a few fast facts about snakebite to keep in mind before we embark on a discussion of the *true* nature and evolution of timber rattlesnake venom. Number one: venom evolved first and foremost as a means for members of several groups of snakes ("families" in taxonomic parlance) to subdue larger and more dangerous or more mobile prey. Number two: for vipers, including all pitvipers, envenomation is synonymous with the onset of digestion. Number three: only secondarily is venom employed in the line of defense. In other words, a timber rattlesnake would rather *not* bite you. *BioMed Central Emergency Medicine* reports that each year in the United States approximately nine thousand patients are treated for snakebite. Pitvipers are responsible for ninety-nine percent of those bites, of which, collectively, the various species of rattlesnakes account for the majority. Over the past several years, on average five people have died of snakebite in the United States in any given year, less than one fatality for every eighteen hundred bites and most of those victims, like Charlie Snyder and the Wolfords, either received little or no first aid, or the treatment was greatly delayed. To put this in proper perspective: the Centers for Disease Control and Prevention tells us that in a given year you are more likely to die from the sting of a bee or a wasp (approximately thirty-three hundred people), a dog mauling (seventeen), a lightning strike (eighty-two), or a recalcitrant farm animal (twenty) than from the bite of a rattlesnake.

For comparison, in Mexico, the epicenter of rattlesnake biodiversity, roughly twenty-seven thousand people are bitten each year, and approximately one hundred twenty die from the bite (one fatality for every two hundred twenty-five bites). If you were to be bitten in Canada, the bite would come from one of three species of rattlesnakes: the massasauga (*Sistrurus catenatus*), the prairie rattlesnake (*Crotalus viridis*), or the northern Pacific (*C. oreganus*). (The timber rattlesnake became extinct in Ontario in the mid-twentieth century; no one knows when it died out in Quebec.) Collectively, these three species bite two to three thousand people a year, killing fewer than one.

In the United States, according to the *New England Journal of Medicine*, twenty-five percent of all pitviper bites are *dry*, which means there is *no* envenomation, not even a vapor. When it comes to venom release, a timber rattlesnake (read any pitviper—copperhead, cottonmouth, bushmaster, diamondback, and so on) that strikes in self-defense, biting with either one or both of its functional fangs (*this is really important*) is parsimonious: it meters the release of venom, often electing to release none. Hence, the dry bite, the timber rattlesnake's choice to conserve venom for a more important quest…*food*.

Here are a few other snakebite facts that bear mentioning: ninety-eight percent of all rattlesnake bites occur on the extremities (you won't believe where some of the others occur, but more on that later), about sixty percent on the hands or arms, and are usually the result of deliberate handling—as in the case of Charlie Snyder or the Wolfords, for example—or an attempt to harm the snake. Herpetologists, who presumably entered the profession well aware of the risks involved, are recipients of many of these bites (six out of sixteen Bashers who attended the 2011 gathering, you may recall, had been bitten in the line of research, several more than once). In 1999, the *New York Times* reported that Howard Reinert, the tenured biologist at the College of New Jersey, while working alone in his lab had been bitten by a timber rattlesnake and nearly died. Reinert told the *Times* that his days of solitary snake work were over, noting that thirty years of his life teaching people the value of timber rattlesnakes "gets undone very quickly when this happens."

Bites provoked by deliberately handling the snake are considered *illegitimate* (really *not* the snake's fault); the typical victim is a white male between the ages of seventeen and twenty-seven, and of these twenty-eight percent had consumed alcohol before messing with the snake. The remaining snakebites, about forty percent, are *legitimate*, characterized by inadvertent encounter between an unsuspecting victim and a snake that presumably either reacted out of its own deep-seated fear or mistook a warm hand for a small rodent. Legitimate bites most often occur on the feet or legs; the victim is bitten while bushwhacking, picking blueberries, moving firewood, baling hay, cleaning the garage, stepping off the porch, working in the garden, peeing along the warm shoulder of a road after dark, that sort of random activity. (To prevent untoward encounters with timber rattlesnakes during Basher field trips in the Berkshires in 2012, twelve of twenty participants [sixty percent] wore snake-proof chaps; although I wasn't one of them, a month later—after telling a friend about the prevalence of chaps—I received a pair for my birthday.) Based on data gathered by the American Association of Poison Control Centers, if you walk off trail in the Southwest or Southeast (and daydream while you walk), you should wear chaps.

It is also worth mentioning that not every timber rattlesnake you

step on bites in retaliation. One Basher (to whom I promised ano-
nymity), while rushing across a huckleberry flat toward the sound of
a rattling snake, stepped on a timber rattlesnake and never knew it;
although he vehemently denies it, a few minutes later, his colleagues
noticed a Vibram-soled print on the snake's loose, pre-shed skin.
And, as I've mentioned earlier, Alcott Smith stood on a Vermont
rattlesnake long enough for the snake to bang its head against the
side of Smith's calf . . . twice, as if to say, "Excuse me."

Here's an exam question: "talons and a hooked beak are to a hawk as [blank] is to a timber rattlesnake." The correct answer, as I'm sure you guessed, would be *venom*, the snake's means of subduing dangerous prey. Without venom, a four-foot timber rattlesnake could not kill and eat a gray squirrel or a long-tailed weasel. In fact, without venom, the weasel might make a meal out of the snake. Even a white-footed mouse, the rattlesnake's principal source of food, if held too long in the snake's mouth might inflict serious damage to the snake's delicate facial organs. To capture prey, a black racer employs long teeth, powerful jaws, and speed (although a four-foot racer could not overpower a gray squirrel); a black rat snake constricts in ever-tightening coils that squeeze the last volume of air from the rat's lungs. Garter snakes eat soft-bodied, less-threatening fare: earthworms, tadpoles, frogs, and salamanders. For a hungry timber rattlesnake, catching food is about both surprise (the ambush) and speed (of execution)—collectively, strike, envenomation, and recoil occur in a fraction of a second (unless the rattlesnake strikes a bird; then, for obvious reasons, it holds on). A snake-struck mouse jumps straight up, runs off, slows down, twitches, and dies. After a few minutes, the rattlesnake unspools and then, head to the ground, tongue flicking, begins to gather and process odiferous clues. (The

BBC, as I've previously mentioned, captured this sequence twice on video with one of Rulon Clark's radioed snakes in western New York. You can watch the action in slow motion on disc two of *Life in Cold Blood* while being lulled by the mellifluous voice of narrator David Attenborough.) The timber rattlesnake sticks to the trail of the dying mouse like a bloodhound—no other rodent path holds any interest. In other words, a timber rattlesnake can recognize the odor of its own venom, now being distributed through the mouse's circulatory system by fading heartbeats.

⌇

Early one morning, a number of years ago, under a warm, clear, tangerine sunrise, I came upon a full-grown pigmy rattlesnake (*Sistrurus miliarius*) coiled in the open by a trailhead in Everglades National Park, not more than six feet from where I had just parked my car. I planned to trek through the snake's primary habitat, a tangle of sawgrass and bushes, to an observation tower to watch wading birds gather in marsh. But not wanting to miss an opportunity to photograph a rattlesnake in the soft light of sunrise, I delayed the walk, put down my pack, and then hastily set my camera on a tripod, splaying all three legs to lower my perspective to snake level. As comfortably as possible, I sprawled in the dewy grass, glued to the image in the viewfinder with the anticipation of opening an unexpected gift. The rattlesnake didn't share my enthusiasm.

Far more temperamental than timbers, pigmy rattlesnakes are pugnacious and prone to bite defensively. For one thing, adults are much smaller, only rarely exceeding two feet in length, which places them on the *carte du jour* for a broader constituency of birds and mammals—crows, grackles, shrikes, opossums, and raccoons to mention a few—as well as snake-eating snakes like racers, indigos, coachwhips, king snakes, milk snakes, coral snakes, and water moccasins. Even a pig frog is capable of swallowing a neonate pigmy rattlesnake. For another thing, the tiny rattle of an aroused pigmy rattlesnake sounds like the buzz of a wasp or the squelch of a cellphone rather than the warning of a bellicose reptile.

While I composed the picture, the snake lashed out sideways at the closest tripod leg, fangs striking metal. Two lines of transparent venom, the color and consistency of stale skim milk trickled down the leg. Alarmed by the sudden turn of events, I hastily got to my feet, which triggered the snake again... *ping*... two more venomous driblets oozed down the metal legs. Faint traces of the dried venom remained on the tripod for months.

If the timber rattlesnake were as mercurial as a pigmy, I might not be writing this book.

Although western Europe harbored a number of species of vipers — some of them *still* remain quite common — the European colonists in the New World had some peculiar notions about rattlesnakes, fanciful ideas forged in the furnace of fear and trepidation, based on something other than direct observation. In 1602, Samuel de Champlain was quick to recognize that a timber rattlesnake posed a threat; he just wasn't sure which end to avoid. "They are very dangerous with their teeth, and with their tail." Several other correspondents from the American wilderness claimed that when a timber rattlesnake fixed its gaze on the eyes of intended prey the stupefied animal would fall out of a tree or run directly into the snake's open mouth. Rattlesnake mojo in today's lingo. In 1794, Samuel Williams reported on the hypnotic effect of a rattlesnake's eyes, which "play with an uncommon brilliancy, and fire, and are steadily fixed on the enchanted animal." The unfortunate bird is intensely distressed and issues "the most mournful accents," while circling aimlessly. If you distracted the snake, the spell would break, and then "the bird recovering his liberty, immediately flies off," claimed Williams. A German Company surgeon by the name of Wasmas reported that a man bitten by a timber rattlesnake fell into a coma within an hour, "never to wake up again." Then (this is the interesting part), within a day, the victim's body mirrors the color of the snake, which from my own experience would be a shade of yellow, black, gray, or brown with up to thirty-six transverse bands (or blotches) that would have faded away near the victim's buttocks. And if the departed lived in the coastal plain of Georgia, he would have a rust-colored stripe running down the middle of his back. Another Revolution-era scribe named Carver agreed with the surgeon, reporting that rattlesnake venom, a "green poisonous liquid," entered the wound and tainted the victim's blood, and this liquid "produces on every part of the skin the variegated hue of the snake." Still another account, written more than fifty years later (certainly enough time for the pioneers to have sorted out fact from fiction), claimed that the victim turned the color of the rattlesnake—blue, white, and green (which sounds more like a description of the flag of Sierra Leone than any timber rattlesnake I've ever seen).

Back on the battlefront in 1777, a lieutenant colonel named St. Leger, who presumably knew something of war strategy but apparently very little about medicine and herpetology, noticed that a soldier bitten by a timber rattlesnake "quickly swelled in a most dreadful manner." St. Leger described the scene as if the solider were devolving before his eyes: skin spotting up, eyes filling with madness, tongue thrusting in and out ("as the snakes do"). The transmogrification was complete when the afflicted began hissing through his teeth with inconceivable strength, after which "death relieved the poor individual from his struggles, and the spectators from their apprehensions." The truth about snake venom, however, is *far* more fascinating (and complicated) than these tales passed down from the American wilderness.

What Venom Is and Where It Comes From

Over the past two decades, sophisticated genetic research has changed the way biologists look at both the phylogeny of species and larger taxa (the who begot whom) and the evolutionary origins of complex organs. The knowledge of genetic inheritance expanded exponentially after 1990 with the completion of the Human Genome Project (HGP), an international, thirteen-year collaboration that unraveled the human genetic code by sequencing our cellular DNA.

The unraveling of our DNA code has been called "One of the great feats of exploration in history ... an inward voyage of discovery."

DNA, you may recall from high school biology, is scientific shorthand for *deoxyribonucleic acid*, the self-replicating material present in cell nuclei (and in the cell organelle mitochondria) of nearly all living organisms and the foundation of genes and chromosomes. Geneticists refer to DNA in the cell's nucleus as nuclear DNA. (DNA found in the mitochondria is, not surprisingly, called mitochondrial DNA, or mtDNA.) An organism's complete set of nuclear DNA is called the *genome*, and it encompasses every genetic possibility of a particular species. The genome of the timber rattlesnake, then, is the entire genetic makeup of the species, the potential unique to the timber rattlesnake. The instructions encoded in cellular DNA— the information that enables a timber rattlesnake to give birth to a timber rattlesnake and *not* to an eastern diamondback—are needed for the snake to develop, survive, and reproduce. To implement these functions, DNA sequences are converted into messages that can be used to produce proteins, which determine *everything* about the timber rattlesnake (or any other species): how it looks, how it catches and processes food, fights fungal infection, sheds, mates, basks, how and when it hibernates, and so on. Until *very* recently, DNA sequences were unknown languages "written" in proteins. The timber rattlesnake genome, then, is the species blueprint—an ever-evolving blueprint, as traits that have been and are being culled by years of trial and error—and the development and changing nature of venom, like every other aspect of the rattlesnake's being, is embraced in the genome, encoded in the DNA, translated into the language of proteins, and then acted upon by natural selection. Coining the phrase "descent with modification," Darwin understood the *law* of evolution; he just didn't have access to the genetic underpinnings. The proteinaceous idiom of life, the mystery unraveled by the HGP, would further prove his point.

It is interesting to contemplate a tangled bank, clothed with many plants of many kinds, with birds ... with various insects ... and with worms ... and to reflect that these elaborately constructed forms, so different from each other, and dependent upon each other in so complex a manner, have all been produced by laws acting around us. These laws, taken in the largest sense, being Growth with Reproduction; Inheritance which is almost implied by reproduction; Variability from indirect and direct action of the conditions of life, and from use and disuse: a Ratio of Increase so high as to lead to a Struggle for Life, and as a consequence to Natural Selection, entailing Divergence of Character and the Extinction of less-improved forms. Thus, from the war of nature, from famine and death, the most exalted object which we are capable of conceiving, namely, the production of the higher animals, directly follows. There is grandeur in this view of life.

And this grandeur begins with *nucleotides*—four nitrogenous bases coupled with a sugar molecule (*deoxyribose*) and one or more phosphate molecules—the building blocks of DNA. Two strands of DNA coil around each other to form the famous double helix, which looks like a spiral ladder (remember those models in biology class?). Each rung of the ladder is built of paired bases—adenine, cytosine, guanine, and thymine, abbreviated A, C, G, and T, referred to by David Quammen in his macabre book *Spillover* as "little pieces in the great game of genetic Scrabble." A, C, G, and T combine in such a way that the sequence on one strand of the helix complements the sequence on the other strand. By reading our letter-by-letter genetic code (written in nucleotides), the HGP identified more than three *billion* pairs of bases. The particular order of A, C, G, and T on the chromosome (the threadlike structure in the nucleus that carries the genes) determines what biological instructions are contained in a strand of DNA. The possibilities are almost infinite. For example, the sequence ATCGTT might code for a timber rattlesnake to be black phased, while ATCGCT might dictate yellow. To grasp the enormity of the human genetic potential, the HGP made this analogy: if the DNA sequence of the human genome were compiled in books, it would fill the equivalent of two hundred volumes each the size of the Manhattan telephone book—collectively, that's more than two

hundred thousand pages; for the computer savvy reader three giga-
bytes of space would be needed to store all the data.

Because cells are very small, and because organisms have many
DNA molecules per cell, each molecule is tightly packaged. The packaged DNA is the threadlike chromosome, of which the timber rattlesnake has thirty-six pairs (we have twenty-two). The *gene* is the DNA
sequence on the chromosome, different numbers and arrangements
of A, C, G, and T that contain instructions to make a protein. The
size of a gene varies greatly, ranging from about a thousand bases to
more than a million in humans. During DNA replication the strand
unwinds so that it can be copied. At other times in the cell cycle,
DNA unwinds so that its instructions can be used to make proteins.
As I've already mentioned, complex organisms also have a small
amount of DNA in their mitochondria, the tiny organelles that generate cellular energy. Sexual reproduction enables a neonate timber
rattlesnake to inherit half of its nuclear DNA from its father and half
from its mother. Because egg cells, but not sperm cells, keep their
mitochondria during fertilization, a timber rattlesnake (and every
other organism) inherits its mitochondrial DNA (mtDNA) from
its mother. Through the study of mtDNA light has been shed on
previously cryptic evolutionary relationships, revealing more precise
phylogenetic kinships. For instance, based on supposed common
ancestry, evolutionary biologists once placed the timber rattlesnake in the tropical rattlesnake (*C. durissus*) group and thought it
most closely related to the black-tailed rattlesnake (*C. molossus*) of
the Sonoran Desert. Technology developed during the HGP has
changed that phylogenetic view. mtDNA evidence places the timber
rattlesnake in the prairie rattlesnake (*C. viridis*) group, which also
includes, among three other species, the Mojave rattler (*C. scutulatus*), which of all snakes in the United States has the most dangerous
venom.

Let me clear up a common misconception: venom may be poisonous,
but poison and venom are *not* synonymous. Poison is ingested (think

arsenic), whereas venom is injected. The digestive system absorbs poison, which is then distributed by the circulatory system; the lymphatic and the circulatory systems receive venom directly, either by bite or sting. More than one hundred thousand species have evolved venom and venom-storage and -delivery systems; biters include certain snakes, lizards, insects, spiders, mites, and the short-tailed shrew; stingers include bees, wasps, ants, scorpions, centipedes, jellyfish, cone snails, anemones, fish, the duck-billed platypus, and the slow loris, the world's only venomous primate. To make the point that most reptile venom is *very* potent, the husband and wife team Sherman and Madge Minton, authors of *Venomous Reptiles* (published in 1969 and still widely cited in technical journals) used a fitting analogy: weapons. They called tetanus, the potentially deadly neurological infection caused by the bacteria *Clostridium tetani*, a "biochemical bullet" that snipes a single molecular target; reptile venom, on the other hand, a toxic cocktail of up to fifty complex molecules, is a hand grenade whose lethal fragments hit multiple targets—red blood cells, white blood cells, skin, nerves, heart, muscles, lymph, lungs, kidneys, gonads, and so forth. I have been told that timber rattlesnake venom tastes weakly bitter and sweetish and leaves a slight tingle on the lips. Why, you may ask yourself, would anyone want to taste venom? I don't have an answer to that question. But the curiosity of my informant corroborates the understanding that venom is *not* as toxic when swallowed; stomach acids such as hydrolases and esterases may actually denature some types of venom. (If you crammed a mouse's stomach full of rattlesnake venom, the mouse would eventually succumb, but the dosage required to kill the mouse, wrote the Mintons, would be seven hundred fifty thousand times greater than if the venom had been delivered by syringe.)

When I was a Boy Scout, in the sixties, busy working on a "nature merit badge," one of the few I was able to complete before being expelled from scouting for climbing high up a beech tree and refusing to get down, I acquired several educational fliers from Ross Allen's Reptile Institute in Silver Springs, Florida, the popular, year-round attraction that catered to southbound tourists (until Disney World siphoned them away) and to snake enthusiasts like me. A self-taught naturalist and movie stunt man, who transformed a hobby into a forty-six-year-long enterprise, Allen exhibited reptiles, birds, and mammals (including a Seminole village, replete with costumed Indians), milked venom from *large* rattlesnakes (for entertainment and medical research), wrestled alligators underwater, published popular and scholarly articles, and sold souvenirs, including snakes and other transiently appealing creatures (some to me) that usually died after a month or two in a suburban home. One of the fliers I ordered from the Reptile Institute was titled *Venomous Snakes of the United States* (or something like that), in which Allen wrote that the venom of pitvipers, at the time classified in their own family, Crotalidae (i.e., not lumped with the vipers as they are now), was *hemotoxic*, a toxic mix that dissolved the walls of capillaries and other circulatory vessels, causing blood to pool in body cavities, the damage radiating outward from the site of the bite, the bitten limb swelling as though inflated. Coral snakes, in the family Elapidae, which also includes cobras, kraits, mambas, and most of the snake species of Australia, deliver *neurotoxic* venom, which distresses the nervous system, shutting down nerve impulse transmission, leading to descending paralysis often very quickly.

You see where I'm going with this? Today, molecular biology, genetic decoding, and sophisticated herpetological research have shown that the nature of a particular snake's venom is *not* so simple and easily classified. Far from it. Many venoms contain mixtures of hemotoxic and neurotoxic compounds (the hand grenade), as well as other elements unknown to Ross Allen, and the composition and toxicity of the venom of a timber rattlesnake, for example, may vary according to geographic location, age of the snake, and possibly even the snake's diet. I mentioned earlier that the bite of a Mojave rattlesnake is not to be trifled with; two of the three discrete metapopulations of Mojaves have a *very* potent mixture of neurotoxic and hemotoxic venom (the venom of the third is mostly hemotoxic). Back to the Mintons: a mere 4.2 milligrams of venom would be lethal to a mouse; by comparison, a timber rattlesnake would need to deliver more than fifty milligrams; for a person, the fatal dose of venom would be a minimum of ten milligrams for a Mojave, seventy-five milligrams for a timber, and one hundred milligrams for a western diamondback, the snake responsible for the most envenomations and the most deaths in the United States.

One of the many specializations for life without limbs was the evolution of nonpareil oral glands to lubricate prey, often large and sometimes dry and bristly. Remember, a snake swallows food whole. No ripping. No shredding. No chewing into bite-sized pieces. Pretend you don't have a gag reflex and then try to imagine yourself swallowing a chicken, unplucked and unchewed. Without lubrication, no way could this be done. Molecular biologists have shown that snake venom began evolving before the Cenozoic era—known in *very* small circles as "the age of snakes"—before snakes and lizards ever went their separate evolutionary ways. In the virtual absence of a fossil record, the origin and subsequent evolution of snake venom had been inferred from the study of biogeography, anatomy, and, most recently and of most importance from techniques developed during the HGP.

"Snake venom is incredibly dangerous, unbelievably complex, medically promising—and the ultimate adrenaline rush. What's not to love," wrote Bryan Fry, head of the Venom Evolution Laboratory at the University of Queensland, a forty-two-year-old Oregon transplant whose shaved head, numerous tattoos, including the chemical formula for adrenalin on the back of his neck, a Komodo dragon skull on each shoulder blade, and the symbol for biohazard on each shoulder, and a penchant for living on the edge—racing motorcycles, skydiving, rock climbing, and big-wave surfing—may obscure the fact that he's a world authority on snake venom. "While my research has significant medical implications, at the end of the day, it is all just a grand excuse to keep playing with these magnificent creatures." Fry's website features a photograph of him sitting on the ground and petting a Komodo dragon, as if the big lizard were a Pembroke Welsh corgi. During the course of fieldwork, Fry has been bitten by several venomous snakes, stung by a centipede and a scorpion, almost fatally, broken twenty-three bones, including his back, when he fell off a termite mound, and received four hundred stitches.

Nothing seems to slow him down, and his website, Venomdoc
.com, is a source of knowledge and amusement. Bryan Fry is an in-
novative, productive herpetologist and molecular biologist whose
groundbreaking (and controversial) research appears in august scientific journals such as *Nature* and *Molecular Biology* bearing titles
like "Early Evolution of the Venom System in Lizards and Snakes"
and "Assembling an Arsenal: Origin and Evolution of the Snake
Venom Proteome Inferred from Phylogenetic Analysis of Toxin Sequences," as well as in more popular venues like *National Geographic,*
Smithsonian, the *New York Times,* and the television network Animal
Planet, where he often appeared opposite the late Steve Erwin. Fry's
the adrenalin junkie who unraveled the mystery of venomous snakes.
Take the Burmese population of the Russell's viper (a distant relative
of the timber rattlesnake). A bite from "the little bastards cause[s]
uncontrollable hemorrhaging of your pituitary gland, where all
your sex hormones are controlled," wrote Fry. So if you are unlucky
enough to survive the bite, you can be rendered permanently impo-
tent and sterile. "You have 40-year old men who have lost all their
pubic hair, and they're experiencing a fate worse than death. It's the
anti-Viagra."

Using evidence found in nucleotide sequencing, Bryan Fry wanted
to see how far back in the geologic record he could trace snake
venom, whether the genes that produce venom were a gift passed
down from an earlier ancestor or whether venom developed independently in several lineages of snakes, after snakes had split off from
their lizard ancestors. Based on significant differences in the anatomy
of venom delivery systems and the *supposedly* distant phylogenetic
relationship, it had been assumed that the venom systems of lizards
and snakes had independent origins. Fry's quest was the ophidian
version of "which came first the chicken or the egg?" Was it possible
that lizards, the closest relative of snakes, still carry those ancient
venom genes? When Fry isolated rattlesnake toxins in the mouth
secretions of bearded dragons and found that proteins encoded in
those genes had the same effect as snake venom, he believed that
scientists might have long overlooked lizard venom because it is not

fatal. Fry discovered that four different, but closely related, lizard families—iguanas, beaded lizards, alligator lizards, and monitors—all have venom genes, many of the same ones possessed by snakes. The venom of lizards, said Fry "knock[s] their prey out so it will struggle less. And while that may not kill the prey, it will still give these animals the chance to tear its head off." Bryan Fry's research in collaboration with colleagues worldwide (some of his technical papers have fifteen or more coauthors) revealed that venom came first, snakes later, about a hundred million years later.

According to Fry, the immediate ancestor of snakes, which roamed the dinosaur tropics one hundred seventy million years ago, already had an arsenal of at least nine different (and sophisticated) venoms that were stored in both mandibular (lower jaw) and maxillary (upper jaw) glands. Fry also discovered that Gila monsters and beaded lizards, which were formerly considered the *only* species of venomous lizard, did *not* evolve venom independently from snakes, as everyone (including Ross Allen) had long thought. A common ancestor to snakes and beaded lizards bequeathed venom to both lineages, and these heirs later evolved their own more potent brands of venom and apparatuses for its delivery; in mandibular glands for Gila monsters and beaded lizards and in maxillary glands for snakes. Fry believes that that common ancestor of snakes and all four families of venomous lizards, which group neatly together into a new clade called *Toxicofera*, retained venom in both the upper and lower jaws (iguanas still do, though their venom is lightweight) and that lizards (with the exception of iguanas) lost their maxillary venom glands and snakes lost their mandibular venom glands.

Because of the technology developed in the HGP, Fry has been able to identify and read the DNA sequence in the suites of genes in venom gland cells. "Instead of spending a couple of months and getting two or three protein sequences done, in a month I can get up to two thousand sequences done," Fry said. "It's an amazing increase in efficiency." What else has he learned (or confirmed)? For one thing, many species of supposedly nonvenomous snakes like my backyard garter snakes, the most common serpent in North America, actually produce tiny amounts of venom and deliver mildly toxic bites, unloading just enough venom to make a squirmy frog quiescent, but not enough to be dangerous to us. For another, long before rat snakes turned to constricting rodents, they had developed venom, which they still retain as a mild vestige, although the glands have atrophied. In fact, Bryan Fry's research, based on his group's (or his) interpretation of experimental data, has raised the known number of venomous snakes from two hundred to around two thousand (over sixty percent of all known species). Of course, not every herpetologist or toxicologist is in agreement.

After Fry published his results, an eight-page letter that cited nearly seventy-five sources appeared in the distinguished journal *Toxicon*, refuting his premise. The authors, two prominent toxicologists, claimed that Fry, in a slight of hand more typical of Penn and Teller, simply changed the *definition* of venom and venom gland to include a host of other reptiles. With an air of concern, the authors argued if an animal is prematurely labeled venomous, society and some academicians might view it differently. If it's common, it might be persecuted. If it's already rare, dollars for conservation may not be forthcoming. Without terminological discretion, declared the authors, a human might be defined as "venomous."

In my in-box was a recent e-mail from Dan Keyler, an original Basher and one of the *Toxicon* letter's authors. "Just because genes are present that may express a toxin, their presence does not confirm functionality of an expressed toxin or venom." In other words, venom genes do *not* necessarily mean the presence of venom; or to put it another way, people that carry genes for blue eyes are *not* always blue-eyed. Harry Greene, professor of ecology and evolutionary biology at Cornell University, the herpetologist who mentored Rulon Clark's graduate and postgraduate research, thinks "an arguably bigger deal in snake evolution" was that the inner two rows of teeth in the upper jaw (left and right palatine-pterygoid bones) began to do all the work of swallowing, which freed the outer tooth rows (the maxillary bones) for all sorts of queer innovations, including but hardly restricted to the evolution of fangs.

Undeterred, Fry continues. "It's evolution on acid." Venom evolved *once* in snakes, he proclaims. It began, perhaps, as a mildly toxic oral secretion, as he assumes it remains today in garter snakes. Later, some lineages (like the pitvipers) evolved a more deadly bite. Fry calls the development of venom "the most important adaptation of snakes." The point being that snakes no longer had to be just subterranean burrowers or dreadfully slow-moving, heavy-bodied constrictors; they could branch out into niches from the roof of the forest to the floor of the ocean. And they did.

No matter what your point of view, venom didn't just happen. It evolved. If you thought that venom glands are modified saliva glands and venom is modified saliva, as many scientists believed before the HGP, your thought process would be logical, but you'd be *wrong*. The origin of reptile venom turns out to be different from that of mammal saliva, *very* different; they are far from homologous. Fry again. The sequences of DNA shed as much light on the origin and evolution of venom molecules, once thought of as seemingly ungraspable as a Zen master's riddles, as it does on the ancestry of snake venom. Fry constructed an evolutionary tree for each of twenty-four different venom genes, searched online databases for the closest relatives among nonvenom genes, and then mapped out and analyzed derivations. Of the twenty-four venom genes he tested, only two appeared to have evolved from saliva genes; most of the rest evolved from genes outside the venom gland—in the brain, for example, as well as in the blood, the pancreas, the liver, the lungs, and the gonads to mention just a few locations. Fry suggests that most of these venom genes whose genesis were in organs located somewhere other than a snake's mouth were accidentally duplicated and translocated, a process known as *gene recruitment*, and that one or more copies then began to mutate into a protein-producing gene in the venom gland. If one of the *new* proteins should prove toxic when injected into prey, then natural selection (Darwinian evolution) would favor those proteins that are most lethal. "Like with rattlesnake bites, which are notorious for turning your arm into goo. They've recruited a very powerful digestive enzyme but they've changed it [so that] in some versions it's more powerful in terms of chewing up the flesh, but in other versions it chews up the blood," Fry points out. Like shuffling and reshuffling a deck of cards, various combinations of venom genes duplicate, mutate, evolve into a dizzying variety of molecules—molecular biodiversity. As eons passed, new lineages of snakes advanced and continued to *inadvertently* borrow new genes from different organs. The point of all this: venom has evolved, and keeps on evolving, more quickly, in fact, than the snake species itself.

Timber rattlesnake venom (and that of all other vipers) is produced in a pair of glands located between the eye and the upper jaw, squeezed out by two special muscles, the *compressor glandulae* and the *pterygoideus glandulae* (*p. glandulae* is found in all *Crotalus* species, but not in *Sistrurus*), and carried by ducts into a pair of hollow, deciduous fangs. The ducts are loosely connected to the fangs, which are periodically shed. Bryan Fry's first envenomation was from a timber rattlesnake, whose venom was a mix of neurotoxic and hemotoxic molecules—remember that DNA research led to reclassifying timbers as part of the prairie rattlesnake group, where some (but not all) populations of the five member species have varying amounts of neurotoxic venom, as well as the more commonly associated hemotoxic. The snake that bit Fry was a southern Piedmont variant of timber, commonly called *canebrake*, the lineage with the bright, rust-colored dorsal stripe. Once considered a separate species and later a separate subspecies, neither classification is now recognized as valid. Fry had no local swelling. No discolored arm or intense pain as though his skin were burning. But ten minutes after the bite, he lost the ability to see all derivations of red and green and blue. His world was shades of yellow and white and black like a daguerreotype photograph. "I got the strangest metallic taste in my mouth like I was chewing aluminum foil. Then I collapsed."

For snakes, venom has been (and continues to be) a key evolutionary innovation, a trigger for advancement and diversification, part of a great arms race: predator versus prey, timber rattlesnake versus white-footed mouse. Keep this in mind: all these elaborate yet random events that have unfolded over two hundred million years, all the chemical warfare that has taken place within the oral glands of snakes are driven by hunger. As Harry Greene reminds us in his best-selling book, *Snakes: The Evolution of Mystery in Nature*, "Humans are largely irrelevant to the evolution of differences among venomous snakes."

Anatomy of Venom Delivery

In order to be operable, snake venom needs a delivery system, an efficient way to get out of the snake's mouth and into the prey. Such a system would include ducts, accessory glands, muscles, and, of course, teeth. And some delivery systems, like those of the viperids, are more sophisticated than others. Historically, herpetologists sorted snakes into four categories based on the design and location of their teeth, which does *not* (as it turns out) suggest evolutionary kinship. Once again, advanced technologies developed in the HGP have shown that serpents did *not* progress step by step across geologic time from a primitive dentition to a more advanced dentition (let's say, from the ancestors of a Burmese python to the ancestors of a timber rattlesnake), like scaling an evolutionary hill. To understand how the venom-delivery system of a timber rattlesnake works will require a brief (and *very* general) explanation of snake cranial anatomy and the origin and evolution of fangs.

First of all, there is a difference between a fang and a tooth. A snake fang is a tooth, an elongated, specialized tooth associated with the transport of venom, a snake's only *tool*. Everything else is just a plain homodontic tooth, simple and small, and recurved (curved backward or, as the herpetologist calls them, *agkistrodontic*). One of the distinctive features of a snake is the loose connection of the bones on the floor of the braincase (cranium) both to each other and to the jawbones. A snake's head is capable of spectacular distortion in almost any direction as elastic ligaments, which connect one bone to another, stretch and stretch and stretch to accommodate the large, bulbous prey the snake swallows without chewing. The two lower jawbones (the mandibles) are also capable of independent motion. Each mandible is made up of two separate fused bones, the dentary and articular, and they don't merge into a solid chin (the mandibular symphysis), as is the case with mammals, but are loosely connected by ligaments. The lower jaws of a feeding timber rattlesnake spread apart, a condition known as *mandibular liberation*, sometimes spectacularly so, perhaps to make room for the passage of a gray squirrel or a half-grown rabbit. A hinge connects the posterior end of each mandible to the strut-like quadrate bone, which swings down from each side of the skull, and connects to the jawbone. Elasticity in the middle of each mandible enables the jaw to flex laterally; collectively, the hinges and struts and elasticity allow for an unbelievably and grotesquely spreading maw. If you have never seen this, you owe it to yourself to watch a snake swallow (engorge is more accurate) something bigger than its own head; it's part of that fabled realm of natural history that triggers unforgettable awe, like seeing a comet blaze across the night sky or molten lava spill into the Pacific . . . or a peregrine plunge into a flock of shorebirds.

In the simplest, most primitive arrangement snakes have two rows of teeth on the lower jaw, one on each mandible, and four rows on the upper jaw; one along each maxilla and one along each pterygoid-palatine. Each tooth is small, sharp, solid, and ungrooved, shaped like a thorn, and set superficially in a socket, an arrangement that facilitates constant loss and replacement. The germ of a new tooth rests in the gum just beneath its functional counterpart, ready to break the surface and become one of many tiny scimitars that fix themselves more deeply, as does a fishhook, when the bitten prey struggles and

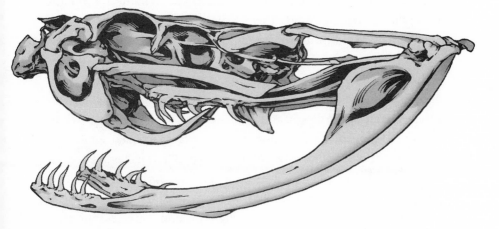

attempts to pull away. These undifferentiated teeth also help to draw food down the snake's gullet, as each quadrant of the toothed jawbones "walks" forward independently, first on one side, then on the other, top then bottom. These teeth are called *aglyphous*—Greek for "without a groove." Aglyphous teeth do not conduct venom, and most important, they all look alike (homodontic). Boas and pythons have a mouth full of them, and rattlesnakes have aglyphous teeth along both mandibles and on the roof of their mouths.

Fangs occupy various positions on the upper jaw. Neither the mandible nor any other tooth-bearing bones have fangs. (As I've already mentioned, the venom glands of snakes reside in the upper jaw, and those of lizards, except for iguanas, in the lower jaw; and the fangs are located accordingly.) Snakes with fangs in the posterior end of each maxilla are referred to as rear-fanged snakes, and those with fangs in anterior end are front-fanged snakes. (Easy enough, yes?) The enlarged, immovable rear fangs, one or two pairs of them, are either solid or slightly or deeply grooved, never hollow-tubular. Biologists call these snakes *opisthoglyphous*, from the Greek word meaning "with a groove at the back." As you may have guessed, the prey must be moved toward the back of the mouth and held there, while the snake tenaciously hangs on, venom dribbling down the enlarged (and sometimes grooved) rear teeth and into the victim's tissues. There is *no* venom injection. Although the bites of a few species of opisthoglyphous snakes have proven fatal to humans, the venom of most—garter snakes and rat snakes, to name a few of the more than fifteen hundred other species that Bryan Fry determined were actually venomous to varying degrees—has little or no effect on people.

In 2008, a team of sixteen biologists, led by Freek Vonk of the Netherlands and including Fry, published an article in the journal *Nature* titled "Evolutionary Origin and Development of Snake Fangs," which attempted to resolve a fundamental controversy in snake evolution, whether or not the fangs of front- and rear-fanged snakes are homologous; in other words, did they share the same evolutionary and developmental pathways? The authors reconstructed the development of forty-one snake embryos, from eight different species, and they determined that rear fangs, the more primitive fang arrangement, developed from undifferentiated (aglyphous) rear teeth and that front fangs originally developed on the posterior ends of both maxillary bones but then migrated forward. Although structurally different in both location and shape, rear and front fangs are strikingly similar in morphogenesis. In fact, they're homologous, like the left wing of a bat, the left flipper of an orca, and the left arm of Sandy Koufax. According to the authors, "The developmental uncoupling of the posterior from the anterior tooth region could have allowed the posterior teeth to evolve independently and in close association with the venom gland. Subsequently, the posterior teeth and the venom gland could have become modified and formed the fang-gland complex—an event that underlies the massive radiation of snakes during the Cenozoic era."

Front-fanged snakes are sorted into two groups: those with fixed fangs and those with movable fangs. Snakes with fixed front fangs are called *proteroglyphous*, from the Greek for "a groove in the front." Fangs are stationed on the anterior end of the maxillary bone and are either deeply grooved (most) or tubular (a few), the groove having closed into a tube. Behind the fangs are a series of short, aglyphous teeth. Elapids—cobras, mambas, coral snakes, sea snakes, and so forth—and harlequin snakes (a hodgepodge of boldly patterned, venomous African snakes formerly classified as either viperids, elapids, or colubrids) are all proteroglyphous. The most recently evolved family of snakes, elapids, have well-developed venom glands and ducts that deliver venom to each fang, which is short and more or less stationary and remains erect when the snake's mouth is closed. Each fang has an entrance lumen that receives venom and a discharge orifice, where venom runs out (into the puncture wound). When a proteroglyphous snake strikes, it stabs downward, mouth agape, a hole-punching bite that penetrates the tough, scaly integument of snakes and lizards (preferred food), and then, unlike a timber rattlesnake, an Egyptian cobra or asp (for example) seizes its prey or perceived tormentor—perhaps Cleopatra by the breast—injects venom, and hangs on … biting, waiting for immobilization, which is often swift and relatively painless. Although the fangs and injection systems of elapids are not as efficient as those of viperids, the quantity and toxicity of their venom more than compensates for the shortness of their erect fangs. Think king cobra, black mamba, tiger snake, taipan, death adder.

Timber rattlesnakes are *solenoglyphous*, from the Greek, meaning "pipe-grooved," and the venom-delivery system of these snakes shows the highest degree of perfection. Solenoglyphs have a pair of *long*, hollow fangs, one on the anterior end of each short, highly mobile maxillary bone. There are no other teeth on the upper jawbone. The surface of each fang and the surface of its venom-delivery canal are enamel-coated and smooth, and when not in use, each fang is neatly ensconced in the gingival membrane against the roof of the mouth. All vipers and a few species of lamprophiids (stiletto snakes) have a solenoglyphous venom-delivery system, the pinnacle of venom injection, the ophidian analog of the syringe and hypodermic needle. When a timber rattlesnake strikes, a series of small, hinged bones on the floor of the skull work in lockstep to rotate the maxillary bones forward, which will erect either one or both of the fangs (snake's choice); the entire arrangement can be understood as a herpetological version of the old folksong "Dry Bones."

> The prefrontal bone's connected to the maxillary bone,
> The maxillary bone's connected to the ectopterygoid bone,
> The ectopterygoid bone's connected to the pterygoid bone,
> The pterygoid bone's connected to the palatine bone,
> Now shake dem skeleton bones!
> Dem bones, dem bones, dem dry bones.

Each small, mobile maxilla pivots on a hinge with the prefrontal and is also hinged to the ectopterygoid, which itself is connected to both the pterygoid and palatine. Muscles connected to the pterygoids pull the ectopterygoids forward, which pushes the maxillae, erecting the timber rattlesnakes' fangs nearly ninety degrees in relation to the roof of the mouth, at full extensions, facilitating deep penetration, efficient venom injection, and rapid release; the result is a very quick stab. Each maxilla supports one fang (or more in some

species)—there are no other teeth on a solenoglyph's maxilla—a classic example of evolution by reduction ... efficiency by reduction, one of several universal themes in biology. The Vonk article color-coded four illustrations of snake skulls, contrasting both the size of the maxillaries and the dentition of aglyphous, opisthoglyphous, proteroglyphous, and solenoglyphous snakes. Besides the enlargement and placement of the fangs, the most notable and progressive change is the shrinking of the upper jaw from the python to the viper. The upper jaw of the viper is short, *very* short.

To summarize, during the evolution of rear-fanged snakes, posterior maxillary teeth uncouple with the development of all other aglyphous teeth (teeth of similar shape and size) on the upper jawbone. These teeth at the back of the maxilla grew longer and connected (loosely) to incipient venom glands, which were simultaneously developing. In embryos of front-fanged snakes, front fangs recapitulate their own evolutionary history and begin their development on the back of the upper jawbone and then migrate forward. All other maxillary teeth disappeared. Grooves in the fangs sealed and became tubes; the shallow bites of cobras led to deep stabs of vipers as the snakes themselves progressed from stab, chew, and hold to inject, release, and wait. Hence, through evolutionary trial and error, a trajectory was set for the eventual appearance of the timber rattlesnake sometime during the *Pliocene* epoch, perhaps as much as four million year ago. (At about the same time in Africa, upright primates called *hominids* had begun to stroll across the savannah.)

Each timber rattlesnake fang is a curved cone, buttressed at the base and rigidly set in a maxillary socket, which arcs back at an angle of between sixty and seventy degrees and then tapers to a surgically precise point. Despite the power of the venom they deliver, the fangs themselves are fragile. Although they're discarded periodically, contrary to the writings of Lucy Audubon, who like her renowned husband was not a critical observer of snakes, the shedding of skin and the shedding of fangs are not coupled. A timber rattlesnake drops a fang every five or six weeks, first from one maxilla and then from the other. As an immature rattlesnake grows each replacement fang grows along with it. Rarely are both fangs shed simultaneously, unless the snake happens to strike something unforgiving like canvas chaps (thankfully) or a tripod leg. Otherwise, a loose fang may impale prey, and then be swallowed and defecated, or simply fall out like a dangling baby tooth. Together with vertebrae and ribs, fangs contribute to the timber rattlesnake's sparse fossil history.

A functional fang occupies one of two juxtaposed sockets that perforate the front of each short, stubby maxilla: one socket holds a functional fang; the other accepts the fang in waiting when the functional fang is shed. For a brief time, however, each socket may support a fang, an overlap in succession that ensures the snake will not be without a functional fang on one side of its head. The edges of the two sockets touch to form a lopsided figure eight, in which the inner socket is slightly in front of the outer. Immature fangs, embedded in the soft tissue of the fang sheath, line up in a double magazine behind the two sockets, alternating from side to side, with the most mature replacement fang *always* behind the vacant socket. A diagram of the arrangement is reminiscent of a tapered peg board in gym class—first one side, then the other; as you proceed to the rear of the mouth each emergent fang is progressively smaller until the last is no more than a nub. Neonate rattlesnakes are born with functional fangs in the two inner sockets and a complete set of emergent fangs; the asynchronous occupation of the sockets, then, is driven by accident and wear, which begins later in the snake's life. An X-ray of a timber rattlesnake's head shows a disquieting toothy grin: longer, more developed fangs are followed by progressively smaller, less developed fangs, all curved back and lined up against the roof of the mouth, looking like a serious orthodontic problem, the epitome of snaggletoothed. Alcott Smith says that technically speaking the rattlesnake dentist would be called a *pleuro-homo-agkistrosoleno-glypho-polyphyodontist* (a combination of Greek and Latin that roughly translates to "same shaped teeth, hooked and knife-like, that sit on the side of the jawbone, and are frequently shed"). Like a cache of artillery, when a timber rattlesnake feeds or yawns, however, the replacement fangs remain hidden in the fang sheath.

Composed of epithelial tissue, each fang sheath loosely connects the venom duct to the corresponding fang. Loose is important because the impermanent fangs alternate socket alignment as they're replaced. Whenever a timber rattlesnake strikes, let's say, a white-footed mouse, a tendon forces the root of each fang against the opening of its duct, creating a stable seal that permits the flow of venom unimpeded into the hypodermic wound. The long paired ducts, the omphalus of the venom delivery system, run from the fang back into the deep muscles of the upper jaw and widen into the paired, triangular venom glands, one behind each eye. The two muscles attached to each gland contract to squeeze venom into the duct; the snake, which meters venom according to a perceived need, controls the entire system. On average, a timber rattlesnake will release more venom during a predatory strike than during a defensive strike, and larger prey will receive more venom than smaller. And then there are those much appreciated dry bites, when the snake determines, for whatever reason, that no venom will be released. Regardless, the ever-prudent rattlesnake never unloads all its venom in a single strike; the ophidian version of the fable about the ants and the grasshopper, sensing the big picture and saving for an emergency, a gesture worthy of at least a footnote from Aesop.

Here Comes the Venom

On a sunny morning in early February, I drive southeast across New Hampshire to meet Kevin McCurley at NERD headquarters in Plaistow. I'm here to necropsy two adult laboratory rats that will be envenomated (for my benefit) by timber rattlesnakes. I take with me a dissection pan, a dissection kit, pins, a tape recorder, notepaper, pencils, and Alcott Smith, who has joined me on the spur of the moment; his fifty years of necropsies on the veterinary table will be a critical addition to our rat-postmortem team. NERD, as usual, is a diverse panoply of distractions: peacocks squabble in the parking lot, and inside can be found an emerald tree boa; red-eyed tree frogs; a Tokay gecko; pastel-colored ball pythons damp with amniotic fluid peering out of leathery, elliptical eggs, the tops of which have been sawed off by each snakelet's egg tooth (chick-like); an alligator snap-ping turtle as big as a truck tire; a snorkel-nosed Florida soft-shelled turtle; three caimans and Wally the Alligator; a friendly, twenty-foot long, hundred-fifty-pound reticulated python, female; a Pakistani cobra; a rhinoceros viper; and a montage of crustaceans and fish and lizards and frogs and big spiders and snakes ... snakes ... snakes that make the pet center a cold-blooded outlet (there are no puppies or kittens here today, just a world-class gathering of reptiles and fish and two small rooms filled with tropical birds).

We pass up a stairwell, down a hallway, beyond the kookaburra perched atop stacked rodent trays in the rodent room, a squirmy rat pup in its claptrap bill, and join Kevin in the HOT room, where disenfranchised rattlesnakes stir and a humorless puff adder tracks our movements. McCurley tosses a rat by the tail into one cage, and then a second one into the adjacent cage, homes to a pair of rattlesnakes that I first met as grainy images — tightly wound and highly developed embryos — on an ultrasound screen. Both snakes are girls born at NERD on September 13, 2011, offspring of a charcoal-colored mother, named Farina, the last wild-caught gravid female in the state of New Hampshire, a legacy of centuries of abuse and neglect. Both snakes have prospered, first on mice and then on rats. Today, they're more than three feet long; one weighs more than six hundred grams, the other five hundred eighty-four, a size usually not reached in the Northeast until a timber rattlesnake's seventh or eighth year. These sisters will stay at NERD indefinitely, part of New Hampshire's captive breeding stock.

It's hard to tell whether the snakes are hungry or annoyed (they have eaten three days prior to our visit), but after several seconds, both strike, a quick thrust of the head — so quick I'm not sure what I've actually seen — and then they each retreat into the far corner of their own cage. Waiting for a timber rattlesnake to strike is like waiting for a crack of thunder; you know it's coming, but you're never quite prepared. If not for the rats' almost immediate convulsive stagger and their spastic panting, rapid, shallow, and out of control, I would not have been sure the snakes actually made contact. Both strikes, which drew a single spot of blood, appear to be single-fanged. Did the second fang miss? Or was it the snakes' intention to download only one? For a timber rattlesnake (for any venomous snake, really), the metabolic cost of producing venom is astronomically high, at least an order of magnitude higher than the cost of producing an equal mass of muscle and bone. So when a snake delivers either a dry bite or a bite in which only scant amount of venom is released, or strikes with a single fang, metering venom can be interpreted as an intentional act of energy conservation.

Metabolic autonomy appears to be beyond the rats' control. Alcott times one collapsed rat's respiration rate at one hundred twenty breaths per minute. How fast is its pulse? Three hundred beats per minute? Five hundred? More? Are its lungs congesting, swamping with blood, like the tide overwhelming a retaining wall? Is its heart bursting ... or dissolving into a gelatinous slush? The rat lingers on its side, twitching and unsuccessfully trying to hyperventilate, the muscles that control its lungs, which have lost their ability to oxygenate blood, in obvious paralysis; the other, more ambulatory, still wobbles around the cage, eyes glazed as though it had too much to drink. Neither rattlesnake is interested in feeding. In fact, one snake, seemingly vexed, coils up and rattles, while the other wedges herself in the far corner.

Kevin walks by and asks, "Are the rats dead?" No. "Half-ass strikes! When they're hungry they go down right away." One rat, lying on its side, voids. Kevin thinks the snakes will come over and bite the rats again, which doesn't happen. Neither snake appears hungry. Were the strikes defensive?

After twenty minutes, the first rat dies, while the second rat continues to stumble along the edge of its world, a cocktail of digestive and function-disrupting enzymes coursing through its system. Safely, Kevin removes the carcass with long-handled forceps and places it in the dissecting pan. Before unbuttoning the rat, Alcott and I pluck patches of fur as we search for a fang puncture, which proves harder to find than I had thought. Little eddies of rat fur swirl around the dissecting pan whenever I move my hand.

Careful not to cause too much trauma to the rat—we don't want to obscure the effect of the venom—Alcott opens the abdominal and chest cavities and probes out the lungs, which are pink, spotted with pinpoint purplish dots—pulmonary petechiation Alcott calls it, otherwise known as microhemorrhaging—and fluid filled, the result of venom-induced vascular-wall permeability (leaky vessels) and procoagulant enzymes that prevent clotting. (Venom destroys the blood components required for clotting: platelets are clumped, and fibrin digested.) Alcott milks hemorrhagic fluid from the lungs. Although there is no other visible evidence of lesions, the ruptured pulmonary capillaries and the collapsed arteries and veins were enough to both suppress oxygen delivery to vital, energy-starved, high-maintenance organs—lungs, heart, liver, kidneys, brain—and lower blood pressure, which sabotaged the heart, making it progressively weaker and harder to pump the decreasing volume of blood. The result: the rat died of cardiopulmonary shock.

On the macro level, timber rattlesnake venom is an amalgam of tissue-destroying enzymes and other proteins that cause circulatory vessels to hemorrhage and sequentially aggregate platelets, inhibiting clotting (porous vessels are *not* plugged), which collectively diminishes oxygen distribution, initiates fluid shifts, and sets off organ necrosis. The result: shock. On the micro level, more than twenty-five different enzymatic proteins disrupt cell function and cause a collapse of cell metabolism by dissolving lipids and proteins, which disables cell membranes, breaks down connective tissue, and aids the spread of venom throughout the rat's body; enzymes digest DNA, RNA, and cell nuclei, disrupt cytoplasmic homeostasis, warp nerve impulses, contract smooth muscles, and dilate blood vessels, which causes a rapid decrease in arterial pressure. If all that wasn't enough, metallic ions in the venom—copper, aluminum, cobalt, tungsten, and boron, for example—each have a corrosive effect on a particular organ, and serve as cofactors for the function of other venom components.

In the nineties, two studies published in the *Journal of Comparative Biochemistry and Physiology* divided timber rattlesnake venom into four types, labeled A, B, C, and D, which vary geographically from virulent to weak, a distinction likely the result of populations having been isolated for scores of centuries in temperate refuges during the Ice Age. One component, a specific isotope of *phospholipase A2*, a protein also called *canebrake toxin*—a paralytic neurotoxin that blocks muscular nerve impulses in mammals (this was what happened to Brian Fry when he fainted after his timber rattlesnake bite)—has been noted in snakes from two separate regions of the southeastern United States, which may allow southern morphs to prey on cottonrats and wood rats (this is purely speculation on my part), both of which have a high tolerance for or resistance to the hemotoxic components of timber rattlesnake venom.

By the time we finish our first necropsy the second rat has died. If we called the effects of venom on the first rat subtle, the effects on the second were blatant, an all out assault on the circulatory system that left the rat with a massive and debilitating hematoma (or fluid shift), a contained pool of blood the size of a half dollar that contributed to shock from fluid loss in the vessels that lead into the subcutaneous tissue between the dermis and muscle wall. Being clubbed with a blunt instrument might have left a similarly massive (and fatal) bruise, but this was biochemically induced trauma, an inoculation of toxic enzymes, and blood loss alone would have been enough to have killed the rat. The hematoma extends into the back muscle; permeable capillaries, small veins, and arteries leak blood, which is everywhere, and unavoidable as our fingers pull back the rat's skin. The fang had punctured the muscle wall and entered the abdominal cavity, and the lymphatic system did the rest . . . pumping and pumping (circulation is slower through the lymphatic system than through the blood, and small toxic proteins, called peptides, move through the lymphatic system), whisking venom throughout the body.

I hold a flashlight to guide Alcott's necropsy. Tissue around the bite site is dissolving; there's so much bruising the rat looks stomped on. Even the stomach is blood-speckled with lesions larger than the purplish spots on the lungs of the first rat, a condition Smith calls *ecchymosis* (the subcutaneous leaking and spreading of capillary blood), midway in size between a petechiation and a hematoma. One of the core vessels near the stomach has hemorrhaged. Because of necrotizing cellulitis, back muscles have atrophied, died, and liquefied. The excessive internal turmoil of the second rat, which was also the product of a single fang, may have been caused by one of three scenarios (or any of them in combination): 1) the fang made a direct hit to a large vessel; 2) the rat received more venom than the first rat; 3) the venom had a longer time to act on the tissues in the second rat, while we worked on the first rat.

Alcott cuts liver transects, pulls out the spleen and pancreas, squeezes the lungs, examines the heart. All seem good, but as he observes, "No two [envenomations] are going to be identical. Quite frankly, I thought we'd see more traumatic lesions dispersed around the body. The venom damaged the entire circulatory and other systems; but much of that damage is at cellular and microscopic levels and we just can't see it. If we had five rats, we'd probably learn more."

The results of the NERD envenomations were a window into what might happen to our own bodies if we were unlucky enough to be bitten—a careless arm or leg in the wrong place at the wrong time; a momentary distraction; a preoccupation with the wrong activity. Like the real estate mantra—location, location, location—the essential difference between the rattlesnake envenomation of a rat and that of a human is all about location: the snake usually strikes a rat's torso, whereas it is more likely to strike a man's leg or finger (I'll discuss a few bizarre exceptions later), which, unless the venom directly enters a vein or artery, gives the victim a reasonable opportunity (an hour or so) to seek medical attention. So—since neither Alcott nor I have ever been bitten, and wanting to understand the phases of timber rattlesnake envenomation, I asked several Bashers, who *had* been bitten, what the experience was like. Bob Fritsch, a retired police officer, nipped on the finger some years ago by a Connecticut timber rattlesnake—an event covered by *Yankee Magazine* and the *Hartford Courant*—said the initial sensation had felt "like slamming my finger in a car door, hinged side … several times."

Dan Keyler, an authority on snakebite toxicity and treatment, recently retired from the Division of Clinical Pharmacology and Toxicology, Hennepin County Medical Center and the University of Minnesota. In retirement, Keyler continues antivenom research in Sri Lanka and Costa Rica, studies timber rattlesnakes on the rolling bluffs of the upper Mississippi River—the snake's northwestern frontier—and remains an elite snakebite consultant available to medical personnel across the continent, twenty-four/seven, to discuss the appropriate treatment protocol. (I carry his cell number in my wallet.) Keyler, a native of Indianapolis, graduated Broad Ripple High School with David Letterman and, faithful to his Midwest roots, has a droll and provocative sense of humor, even when discussing grim subjects.

Before I visited NERD, I had e-mailed Keyler to ask him what I should look for when dissecting the rats, and his response was that I should not confuse the trauma of envenomation with the trauma of dissection (we were careful). Keyler also sent me a synopsis of the most important features of timber rattlesnake venom and what they do to a mammal's body, a collection of electronic reprints of scientific papers on timber rattlesnake venom, envenomations, and a history of snakebite treatment he had prepared for *Mithridata*, the Toxicological History Society newsletter. One reprint, and several letters to the editor of *Annals of Emergency Medicine*, a publication of the American Colleges of Emergency Physicians, were responses to the previously published article titled "'Kiss and Yell,' a Rattlesnake Bite to the Tongue" that had appeared in an earlier issue and confirmed a long-standing suspicion I have that although medical personnel have a sense of humor, rattlesnakes don't.

> To the Editor:
> We enjoyed Gerkin et al. "kiss and yell case," report [July 1987; 16:813–816] of *Crotalus* envenomation to the tongue. We note that the patient's blood alcohol level was 1.78 g/L. The relationship between the type of alcohol consumed and the site of envenomation is of particular interest to us. In this case, it would be valuable to know if the patient, residing in Arizona, had consumed any distillate from the fermented mash of agave.
> Our interest in this relationship was prompted by a strikingly similar case we recently encountered in Kentucky. A 23-year-old man successfully enticed a 4-foot *Crotalus horridus* (timber rattler) to envenomate his tongue while mimicking the snake's characteristic tongue-flicking activity at close range. Criothyroidotomy relieved his glossal airway obstruction and 41 vials of antivenom were administered. Unfortunately, autologous debridement of the central tongue may result in his only potential sequela, a forked tongue deformity.
> It is clear after reviewing the other reported cases of oral envenomation that glossal stimulation of the heat-sensitive pit organs is provocative to these reptiles. More importantly, aromatic vapor stimulation of Jacobson's organ (in our case by Kentucky bourbon) may influence the site of envenomation.

Hopefully, Gerkins' patient's taste runs to sour mash bourbon and not tequila. If so, we theorize that consumption of aromatic distillates should be avoided prior to osculating venomous snakes to minimize the risk of illegitimate tongue envenomations.

Daniel F. Danzl, MD, FACEP
Gary L. Carter, MD
Department of Emergency Medicine
University of Louisville
Louisville, Kentucky

The authors of the original report respond:

In Reply:

We enjoyed Danzl and Carter's flight of medical fancy, but regret to say that we cannot support their hypothesis as our patient cannot recall what type of ethanol he consumed prior to his "kiss of death," but is sure it was not bourbon. We trust their comments were not written tongue in cheek, yet we doubt that aromatic vapor stimulation influenced the snake; if anything, it influenced the patient.

The question by Sarant [the author of a third letter] about the deposition of the snake allows us to tell another interesting part in the saga of our patient we did not feel would be appropriate as part of our original report. After his hospitalization the patient killed his pet snake and placed it in the freezer. We requested that he send us the head of the snake so we could determine if the snake was *C. atrox* or *C. scutulatus*. For reasons that are still not clear to us, he was compelled to thaw the head before mailing it. Unfortunately, he attempted to do this by placing it in a microwave oven, where the head promptly exploded. Fortunately, by examining the tail stripes, we were able to determine the correct species.

The patient did skin the snake and mounted the skin along with some pictures of himself during his hospitalization to create a graphic display of his misadventures. At this writing, he is attempting to contact the patient described by Danzl and Carter in order to share experiences personally.

Richard Gerkin, MD
Kathy Clem-Sargent, MD
Steven Curry MD, FACEP
Michael Vance, MD, FACEP

Central Arizona Regional Poison Management Center
Department of Medical Toxicology
St. Luke's Medical Center
Department of Medicine
Good Samaritan Medical Center
Phoenix, Arizona

Keyler has consulted on, cataloged, and written about scores of envenomations (he's received several bites himself, including a dry bite by a timber rattlesnake while conducting fieldwork). The following describe only some of his consultations. A nine-year-old boy, while attending church, impales his thumb on the fang of a freeze-dried rattlesnake-head tie tack and requires hospitalization. A guitarist is bitten on the left hand in front of a live audience by a timber rattlesnake prop wrapped around the neck of his guitar. A woman cleansing her liver with diluted bushmaster venom purchased at the local health food store develops cardiac palpitations. A cow attempts to nap on a timber rattlesnake and is bitten. A forty-five-year-old man, hunting golf balls in the rough, is bitten on the ankle. A woman is bitten stepping off her porch, and a neighbor captures the snake in a pillowcase, brings it to the hospital, and then lets the pillowcase buzz over the telephone in the ER to the attending physician in the ICU, who confirms the identity of the snake. A Minnesota man deliberately drives his pickup truck over a timber rattlesnake, kneels down to sever the snake's head, and is bitten on the eyeball. One man is bitten on the penis (don't ask). Another takes a strike to the cheek. A thirty-six-year-old man is envenomated on the thumb by a freshly severed timber rattlesnake head. Another man, preparing a meal of timber rattlesnake, dies from the bite of a severed head. Having read both of Keyler's reprints about envenomations by a severed timber rattlesnake head, I followed up by consulting the bible of rattlesnake biology (including both the abstruse and the practical aspects), Lawrence M. Klauber's two-volume, fifteen-hundred-page *Rattlesnakes: Their Habits, Life Histories, and Influence on Mankind*, first published in 1956. Klauber, who devoted six pages to "Experiments with Decapitated Rattlesnakes," wrote that "the head of a rattler, when separated from the body, is dangerous for at least 20 minutes, and sometimes for almost an hour."

The Biology of Rattlesnakes, published in 2008, a multiauthored compendium, featured Keyler's paper "Timber Rattlesnake (*Crotalus horridus*) Envenomations in the Upper Mississippi River Valley," an analysis of twenty-seven rattlesnake bites in Minnesota, Wisconsin, and Iowa, which confirmed (although it bears repeating) more men are bitten than women (twenty-two of twenty-seven, and nineteen of these male victims were between the ages of twenty and forty). More than half the bites (fourteen) are illegitimate and men were involved in all cases—apparently, women are better behaved around venomous reptiles (or simply choose not to *be* around them). Of the illegitimate bites, four involved the consumption of alcohol. And all five bites that occurred in metropolitan areas (again inflicted on men) involved truculent pets. So the upshot of all this is that the likelihood of being bitten legitimately by a timber rattlesnake is very, very small. But . . . it does happen. And the subtext of Keyler's study? Timber rattlesnakes make poor pets.

Unfortunately, should you get bitten, these may be the symptoms you'd experience (unless your bite is dry): swelling of the bitten limb, pain, weakness, difficulty breathing, giddiness, nausea, vomiting, hemorrhaging and no clotting, weak or racing pulse, heart failure, paralysis, stupor (or coma), sweats, overheating, chills, and general nervousness. (Except for the hemorrhaging and bloated limb, I experienced very similar manifestations the morning of my bar mitzvah.) As Alcott pointed out during our rat necropsies, unless you receive a dry bite, like the unique pattern of our fingerprints no two envenomations are ever quite the same. The severity of a bite will depend on your age, size, and health; the location and nature of the bite (one fang or two, deep puncture or shallow, vein or artery penetration); the amount of venom injected; the conditions of the fangs and venom glands; the victim's sensitivity to the foreign protein in the venom (more on this later); the amount and diversity of pathogens in the snake's mouth; and the toxicity of the venom—remember that venom varies both geographically (there are four distinct timber rattlesnake venom types) and even between individual snakes in the same population.

Proteins account for more than ninety percent of the solids in timber rattlesnake venom: enzymes, polypeptides (a long chain of amino acids), glycoproteins (a combination of proteins and carbohydrates), and low-molecular-weight compounds. And as Alcott and I saw during the rat dissections, many of the proteins work on the circulatory system, disrupting blood flow, digesting the walls of vessels, preventing coagulation, causing blood to pool in body cavities and between tissue layers. The skin around the bite site often bubbles with blood blisters (called blebs), bleeding may occur from any or all body orifices, and the lungs, a particular target, may harbor concentrations of venom ten times those measured in the liver. In the rare occurrences where death results, it will be caused by blood loss, congestion (drowning in your own bodily fluids), or renal failure (or a combination of all three). Again, essentially death from shock. If left untreated, a swollen extremity will test the very tensile strength of skin; a bloated, fluid-filled limb may constrict nerve and blood supply, which may eventually cause toxified muscles to wilt and necrose. The result: an arm (or a finger) or leg may turn black and gangrenous, and amputation would then be required.

What to Do and What Not to Do after a Timber Rattlesnake Bite

There is little agreement on the appropriate treatment for snakebite once someone has been bitten. There *never* has been. So if you do nothing in the way of first aid, you've done nothing wrong. Staying calm and getting to a medical facility as fast as possible should be the first, the most decisive step you take after snakebite—what Charlie Snyder unfortunately failed to do. (Car keys and a cellphone are *far* more useful than a razor and tourniquet.) According to Sean Bush, an envenomation specialist at the Loma Linda Medical University Center, "Time is tissue." In 1988, Findlay Russell, a toxicologist at the University of Southern California School of Medicine at that time, and author of *Snake Venom Poisoning* as well as numerous articles on related subjects published in scientific and medical journals, had this to say about first aid protocol: "I can't tell you the best first aid measure, but I can tell you which is the least worse." When I was a Boy Scout, before a three-day, outdoor adventure at Camp Wauwepex in Wading River, New York, my parents and the parents of several other members of Wantagh Jewish Center Troop 104, having glanced at Ross Allen's pamphlet on venomous snakes and the treatment of snakebite, purchased for their sons a standard snakebite first aid kit, which included a razor, iodine, gauze, nylon strap, and several rubber suction cups—never mind that timber rattlesnakes were nearly (if not) gone from Long Island and that likely none of the scouts and scout masters (and certainly none of our middle-class suburban parents) knew there had ever been rattlesnakes on the South Shore in the first place. To reinforce the kit's treatment regime, the Allen pamphlet contained a set of illustrated directions on how to slice an appropriate "X" through each fang puncture, how to suck out the venom, and how to tie a *loose* tourniquet on the bitten appendage to impede the spread of venom through the body. Dan Keyler and most other medical personnel agree that "cut and suck" often causes more permanent damage—sliced tendons, vessels, nerves— than does the actual snakebite. And unless he happens to have suction cups, there's a good chance that the Good Samaritans who suck venom from someone else's bite may expose themselves to its toxic effects through open sores in their mouths or through the lining of their throat mucosa, potentially leading to life-threatening airway closure. A tourniquet, it turns out, does little to stem the flow of venom but is marvelous at cutting off a limb's supply of blood and oxygen, which may starve and kill muscles. At the conclusion of his essay "Venomous Snake Bites and Remedies over Millennia," Keyler quoted an eighteenth-century scientist named Fontana, who experimented with snakebite treatment and concluded that when a patient recovers from a bite all that has been proved is that the drug administered to the patient did not kill him. With Fontana's statement in mind, Keyler summarized the long (and checkered) history of envenomation treatment with a timeline: chanting, laying on of hands, herbs, cutting, scarification and amputation, alcohol, stones, almost every chemical in the periodic table, ligatures, cut and suck, ice, more alcohol, more cut and suck, surgery/fasciotomy (lengthwise incisions cut into the limb to reduce swelling), electrotherapy, and finally immunotherapy. "Not surprisingly," Keyler concluded, "even the most bizarre remedies appeared to save life and limb."

One set of remedies begs for more detail. A folk treatment common throughout the range of the timber rattlesnake involved applying a poultice made from the split and bleeding carcass of a freshly killed chicken directly to the snakebite. When the chicken's flesh turned green or black and its comb blue, or if its feathers fell out, the venom had been successfully drawn from the victim's limb. Although the practice was called the "split-chicken remedy," in the absence of a chicken, almost any other animal was acceptable: mouse, lamb, kid, toad, frog, rabbit, skunk, etc. A poultice of warm deer liver and heart saved a snakebitten man in Marjorie Kinnan Rawlings's novel *The Yearling*. If a cow happened to be available, it was slaughtered and the victim immersed his bitten limb in the heat and gore of the abdominal cavity. The list of animal products employed for snakebite relief also included turkey vulture gizzard, saliva of a fasting man, eggs, melted cheese, pulverized crocodile teeth, pulverized human molars, and powdered crayfish.

In 1787, the scientist Fontana experimented with static electric discharge as a treatment for viper bites, and two centuries later, a missionary doctor named Guiderian published his version of shock therapy for snakebite; the doctor used the spark-plug cable from an outboard motor running at half-throttle to deliver five one-second high-voltage, low-amp shocks. He claimed it worked, which led to the development of the "Snake Doctor," a stun-gun model for snakebite treatment, which was endorsed by *Outdoor Life* and other sportsman magazines. Wrote Keyler, "The consequences of such treatment were astounding." In a less than Mensa-level display of ingenuity, one victim treating himself for a rattlesnake bite to the *lip*, took shock therapy to a new level. The victim lay down beside his car, clipped a spark-plug wire to his upper lip, and had a friend start the car and repeatedly rev the engine to three thousand revolutions per minute. The man lost consciousness on the first rev, but he survived the treatment.

In 1887, when the split-chicken remedy and all of its derivatives were popular in the Appalachian foothills, Henry Sewell, a young professor of physiology at the University of Michigan, took a more scholarly approach to the prophylactic treatment of snakebite. Sewell wondered, "If immunity from the fatal effects of snake-bite can be secured in an animal by means of repeated inoculation with doses of the poison too small to produce ill effects . . ." To test his hypothesis, Sewell milked venom from three massasaugas (*Sistrurus catenatus*) and then mixed six drops of venom with eighty-eight drops of glycerol, a colorless, sweet, viscous liquid formed as a by-product in soap manufacture. (Sewell force-fed hamburger to the snakes, hoping to increase their venom production, although he found no positive correlation between the input of burgers and the output of venom.) Next, he diluted the venom-glycerol mix with distilled water, and began to inoculate pigeons, which are highly susceptible to rattlesnake venom. During the first trial, all pigeons that were given a fifth of a drop of venom mixture survived, and those given two-fifths of a drop died within three to twenty hours. Sewell gradually increased the survivors' sublethal dosage until one pigeon survived an inoculation of more than four drops, ten times the original lethal dosage. Sewell's experiment kick-started antitoxin therapy and led directly to serum-therapy treatment for diphtheria and tetanus, which remained the basis for treating bacterial infection until antibiotics were developed.

The results of Sewell's research also reached the Pasteur Institute in Paris, where, in 1895, a team of French bacteriologists developed the first antitoxin serum for snakebite therapy from horses immunized against cobra venom. Horse-derived antisnake serum, which the French called *antivenimeux* (*antivenin* in America), was used worldwide until the beginning of the twenty-first century. Following Sewell's immunization procedure with pigeons, the French team administered between a tenth and a hundredth of the lethal dose of venom to each horse, upping the dosage at intervals, until the horses manufactured enough antibodies to neutralize the venom and to confer immunity. Then, the horses were bled, the serum fractionated, and the antibodies separated out, yielding an antivenin that targeted either one species of snake (monovalent) or several species (polyvalent). Over six or seven years, a single healthy horse might provide one hundred eighty gallons of blood. Rattlesnake antivenin was first produced in the United States in 1903 but was not commercially available until 1926. (Charlie Snyder took a single vial on his ill-fated snake hunt but never used it. That one vial, says Dan Keyler, "would not likely have made a difference given the severity of his bite.")

Today, sheep serum has replaced horse serum in the United States, and the name *antivenom* has replaced antivenin. The switch from horses to sheep, which produce antibody fragments with slightly different characteristics than horses, was driven by the occasional allergic reaction some people had to the larger immunoglobulin proteins contained in the horse serum, a reaction that sometimes triggered life-threatening anaphylactic shock. This made it necessary to treat snakebite victims for potential anaphylactic response to the antivenin given to treat the toxic effects of the venom. However, as noted by a doctor in a 1983 issue of *Annals of Emergency Medicine*, "the morbidity and mortality from envenomation usually outweigh any adverse reactions to the antivenin therapy." Notwithstanding, in the early seventies, when I worked at the Bronx Zoo, you couldn't be hired as a keeper in the Reptile House if you were allergic to horse serum—anyone considered for a job had to pass a scratch test—because the zoo relied on its store of antivenins for on-site emergencies, hence Charlie Snyder's request of help from Dr. Ditmars.

In 2001, the British-based pharmaceutical company Protherics PLC launched CroFab, a sheep-derived antivenom that exclusively targeted North American crotalines (rattlesnakes, copperheads, and moccasins). The first new snakebite antivenom made available in fifty years in the United States, CroFab sheep antibodies are enzymatically digested into smaller fragments that still effectively bind venom toxins. With CroFab the identity of the snake (unless it's a coral snake) no longer has a bearing on the treatment.

CroFab, which is short for Crotalidae Polyvalent Immune Fab (Ovine), is a concoction made of antibodies from the blood of healthy Australian-raised sheep that have been immunized against the venom of one of the following four crotalines: western diamondback, eastern diamondback, Mojave rattlesnake, and cottonmouth. The final product, an equal mix of the four monovalent antivenoms (thus, a polyvalent antivenom), neutralizes the venom of American pitvipers, including, of course, the timber rattlesnake. The antibody fragments in CroFab bind to and neutralize the toxic proteins in venom, which are then redistributed away from the target tissues and removed from the body. If excess CroFab is dosed (more than is needed to neutralize the venom components), the excess antibody fragments are eliminated in the urine. Thus far, CroFab has caused a single case of anaphylaxis—a woman that had allergies to wool—although it does on occasion invoke serum sickness, a much milder allergic reaction characterized by skin rashes, joint stiffness, and fever (a very small price to pay for the neutralization of snake venom). Protherics recommends an initial dose of four to six vials of CroFab, infused intravenously over the course of an hour. Before intravenous delivery, each vial is reconstituted with eighteen milliliters of sterile water and gently swirled; the reconstituted vials are then further diluted with two hundred fifty milliliters of normal saline solution. If local clinical manifestations of the snakebite—swelling, edema, local bleeding, and so forth—are not completely arrested or reappear, and systemic signs and coagulation tests have not become normal, an additional four to six vials (reconstituted and delivered as described above) are recommended. Once the symptoms are under control, two vials of CroFab are given at six-hour intervals for up to eighteen hours, for a total of three additional doses. The company recommends a follow-up dose of two vials (again, reconstituted), if necessary. Protherics, the drug company, recommends CroFab for mild to moderate North American pitviper envenomation (I'm not sure what they recommend for severe envenomation—maybe chanting, laying on of hands, a split chicken ...) and emphasizes early treatment, within the first six hours after the snakebite. The dosage of CroFab antivenom is not governed by the age or weight of the victim—a child receives the same dosage as an adult. As a general rule, CroFab is administered until the venom is rendered harmless: the worse the bite—collapsed blood pressure, progressive swelling, racing heart, necrotizing cellulitis—the more antivenom is given. The governing principle here: the amount of venom and its effect on the victim determines the course of therapy.

Snakebite and Medicine Gone Awry

By the time Paul DeGregorio reached the six-wheeler, Joel had collapsed, one hundred eighty pounds of slumping, unresponsive, down-throttled youth, bleeding from both nostrils. The bleeding had been triggered by venom from a timber rattlesnake bite. The fainting and shallow breathing were something else.

There was no cellphone reception in the isolated valley where they hunted snakes. So, under the weight of the midafternoon sun, DeGregorio strapped Joel into the seat, steadied him with his right hand, started the vehicle with his left, and gingerly maneuvered it off the hill, out of the woods, and onto a service trail several hundred yards away. Then, flooring the Gator's accelerator, DeGregorio followed the winding trail across the college's field-station property. He reached the station's headquarters—and its landline—in less than ten minutes, about twenty minutes after the bite. Meanwhile, Joel remained limp and semiconscious, draped across DeGregorio, who parked the Gator in the shade, away from the fierce sun, and then stacked Joel upright and ran into the vacant station to dial 911.

"And that's when things got completely amusing," Paul DeGregorio told me, long after the fact, a grin blooming across his face, pink as a prairie coneflower.

The 911 dispatcher asked a series of prepared questions—as they're required to do when dealing with the general public—that challenged DeGregorio's diagnosis.

"Are you *suuure* it was a timber rattlesnake?"

"Listen, lady, I'm sure it was a timber rattlesnake. I've studied timber rattlesnakes for forty-five years. If you give me an hour, I'll go back and tell you what sex it was, but in an hour the boy will be dead. You have all the information you need to respond. His life's in peril; you have the address, the directions. I'm going out to unlock the front gate."

"Oh, no, you must stay on the line."

"Listen lady, in a moment you'll be talking to the wind; I'll do what I have to do, and I've just told you what you'll have to do. If I don't unlock the gate, the first responders can't get the emergency vehicle here. The gate has to be unlocked."

"Wait, wait ..."

Click.

To open the gate, DeGregorio had to ferry Joel, still unconscious, down the long, stony descent of a driveway, and then back into the diminishing shade. In the distance, the sound of a siren, a rent in an otherwise noiseless afternoon, grew progressively louder as the first-response vehicle approached. Everything seemed to take forever. Then, as DeGregorio watched in abject horror, the vehicle cruised right past the gate. The EMT had not only driven past the driveway; he had driven past the big sign with the boldly printed address that announced "College Field Station." With a sound of braking tires and pebbles spitting, back it came, past the gate *again* in the direction from which it had come. As the driver re-recalibrated, DeGregorio watched the vehicle's ascending line of dust and imagined Abbott and Costello in a film called *The First Responders.*

"Domino's would have found us quicker. These people live up the road, for God's sake."

Finally, having corrected whatever warped navigational system they used, the first-response crew pulled into the driveway, drove up to the field station, and leaped out the doors. They put an oxygen mask on Joel and checked his vital signs. He'd stopped breathing. No heartbeat. Now, in complete panic, two of the men lifted Joel from the six-wheeler, and then accidentally dropped him on the ground. One EMT listened to Joel's heart with a stethoscope and exclaimed, "There's no sign. We're losing him."

They inserted a short, plastic endotracheal tube into Joel's trachea and replaced the oxygen mask, as DeGregorio thought, "I didn't haul this kid's body out of the woods to lose him here."

In the midst of this apparently futile medical intervention, De-Gregorio tossed a Hail Mary. He screamed, "Joel, breathe!" And Joel took a deep breath. The guy giving heart compressions stopped and looked up at Paul, nonplussed.

"Don't look at me. Keep pumping."

While the first-response crew panicked, DeGregorio, an original and stalwart Basher, backed off and called his pal Bill Brown in upstate New York, who quickly gave DeGregorio an overview of the bite (Brown had been bitten three times), and said he'd call Dan Keyler, the Minnesota toxicologist on the Medical Advisory Committee with the Antivenom Index, and ask him to call DeGregorio ASAP.

Less than forty minutes earlier, DeGregorio had instructed his protégé, "Joel, don't try to impress me. If we miss one, who cares? There's always another day. No data is worth getting bitten for, so stay next to me and don't do anything dangerous."

Up there, on the prairie fringe, the hills, populated by dwarfed and thirsty trees—Osage orange and honey locust, stunted oaks— give way to grassy valleys of varying widths, invaded by shrubs— blackberry, coralberry, sumac, dogwood—that are periodically pruned by fire. The limestone outcrops that pock the valley rims, where professor and student had hunted snakes, are layered slabs of water-soluble rock embedded in rich prairie earth, the ideal zone for the western timber rattler. They bask in the sun when they emerge from hibernation and retreat to their dens during cool weather. On that hot late-April afternoon, the snakes were waking up, and Joel had been invited to DeGregorio's study den to watch the spectacle.

Joel Shapiro, a Bronx native, had gone west to college a year before, at nineteen years old, already a reptile aficionado. On this day, he was on an afternoon snake hunt along the Missouri-Kansas border. After the bite, and for the remainder of his undergraduate career, Joel attended college in New York City—well away from venomous snakes.

As a junior high student, Joel had begun studying snapping turtles in the city parks of New York, where his father was employed. The nature and dimension of his research, sophisticated for someone who had not yet begun to shave, so impressed Bronx Science High School that they reconsidered their initial rejection and admitted Joel as a freshman. Joel, who lectured school groups and nature clubs throughout the five boroughs and had raised a menagerie of reptiles in his parents' home, was unrestrainedly devoted and enthusiastic, a goodwill ambassador for snakes. He chose a college west of the Mississippi, where there were decidedly more snakes than in New York City.

As Paul and Joel progressed along the lip of the hill, DeGregorio caught several snakes of various species, but only one rattlesnake—a snake he had marked the previous day. After he released it, Joel, who hadn't caught anything, became increasingly anxious. He wanted to catch something, anything. After they reached the far side of the ledge, he caught an adult copperhead, which he proudly displayed.

Things seemed to be improving.

Then, Joel, fueled by adrenaline, spied a young-of-the-year timber rattlesnake, eighteen inches long.

"I got this one."

With the tongs DeGregorio had loaned him, Joel grabbed the snake by the tail, which, unhappy with its pernicious change of circumstance, yanked forward to escape beneath a large rock slab. Eyeing the situation, DeGregorio instructed Joel to release the snake; he then grabbed the snake with his own tongs, but decided that it would be too difficult to catch the snake without a dramatic (and possibly traumatic) effort; by now, it had nearly vanished under the rock.

"Let's let it go. We'll come back in an hour, and it'll be out again, basking."

While Paul backed off, without warning, Joel, intemperate and swift, had darted forward and grabbed the snake's tail, barehanded.

Before DeGregorio could yell "No!" Joel had been promptly bitten by a larger rattlesnake concealed beneath the same rock. The snake had delivered what DeGregorio had thought was a feeding bite—mouth wide and both fangs extended—to the base of Joel's left thumb, nicking the radial artery. (The bite was eerily similar to Charlie Snyder's fatal bite in 1929: reach for one snake, get bitten by its camouflaged neighbor.)

In retrospect, thought DeGregorio, "The snake must have sensed something small and warm." Food. In the timber rattlesnake's world, which is amoral and free of consequence, an infrared-radiating hand reaching over the rock must have been mistaken for a small rodent. Instantly, the snake had bitten and disengaged, and withdrawn into the darkness beneath the rock, coils piled on coils, forked-tongue reading the wind—not the customary defensive strike and retreat. The fangs had struck five-eighths of an inch apart. (The next day, Sunday, Paul DeGregorio returned to the hill alone and caught a young timber rattlesnake with a tong-bruised tail, the snake that had likely involuntarily instigated the bite.)

Joel stepped back, looking pale. "I got bitten."

"We're not marking that one today. We're outta here, now!"

Joel mumbled something incoherent.

As they began to work their way back to the six-wheeler—off the ledge, up a hill, and across a creek—Joel's eyes started to glaze over and his movements became laborious. Then, DeGregorio saw him stumble and noticed that his breathing had become strained.

"Maybe he's scared."

Ninety seconds after the bite, Joel Shapiro had trouble standing, talking, and breathing, and DeGregorio realized that this immediate and intense reaction to snakebite was more than calcified fear. In fact, it was more than venom. The kid's system had begun to shut down.

"Stay with me, Joel. Stay with me."

Shapiro's legs sagged, and DeGregorio, who didn't want to sling the kid over his shoulder, where his head would hang lower than his heart, draped Joel's left arm over his own shoulder, grabbed him around the waist, and jacked him up by the belt.

"If I never do this again, it will be really good."

As Paul side-carried Joel off the ledge, he kept yelling, "Breathe Joel. Stay with me."

They reached the six-wheeler, blood already dripping out of Joel's nostrils. Paul propped him up and strapped him into the seat, ran around the Gator, and started the engine. The flow of blood had increased.

It took another five minutes to get to the field-station headquarters.

A first responder pumped Joel's chest, up, down, up, down, palms together, compressing and releasing, a hundred compressions a minute. His face pallid and expressionless, he screamed, again "I'm not getting a pulse."

"I didn't drag this kid all the way up here to lose him on the grass," thought DeGregorio again.

Joel stopped breathing, again. So DeGregorio yelled, "Joel breathe, dammit! Bronx kids don't crap out like this in the middle of Kansas ... Breathe!" And he started to breathe. And once again, the guy stopped the heart compressions and looked up, dumbfounded.

"Quit looking at me and do what you're supposed to do."

Then, the county ambulance crew arrived from rural Washington County—without missing the field station sign—and took over from the first responders. Someone on the ambulance crew noticed that Joel had refluxed, plugging the short endotracheal tube. They tried to remove the tube, but Shapiro's steel-trap jaws were locked shut. They began to prepare for an outdoor tracheotomy, right there on the lawn.

Hastily, his confidence bolstered by his previous success, DeGregorio reimplemented plan A: "Joel, open your mouth."

Shapiro no longer had the strength to open his mouth, but he did relax his jaws. Someone pulled out the tube and (again) looked quizzically up at Paul.

"Quit looking at me and get the other tube in."

An EMT inserted a longer tube down Joel's throat and clipped the oxygen mask back on.

The ambulance whisked Joel to the local hospital, and DeGregorio helped the first responders, who still looked "positively freaked," gather their stuff back into the van.

"These guys were used to dragging bodies out of cars," he told me. "For them, snakebites were otherworldly." Before they left, an EMT told DeGregorio, "I'm sure glad you stayed calm during all this."

"Somebody had to."

Joel Shapiro, who was exhibiting symptoms of a massive snakebite with his bleeding nose and organs shutting down, had a more urgent issue to deal with. He was experiencing a life-threatening allergic reaction to the suite of foreign proteins in the venom, and this had caused almost instant anaphylaxis, or shock.

Just after DeGregorio drove home, Dan Keyler called to discuss snakebite basics.

The key, said Keyler, "Make sure the hospital has at least six units of CroFab up front and more available later."

After he spoke with Keyler, DeGregorio called the ER, and the attending physician said, "We're having trouble stabilizing him, and we have to transfer him to the regional trauma center." What she didn't tell DeGregorio was that the hospital only had four units of CroFab on hand, even though rattlesnakes and copperheads range throughout the neighboring woods.

"I mean people do occasionally get bitten."

The ER doctor also neglected to tell DeGregorio that they had tried to treat the anaphylaxis by giving Joel an intracardiac injection of adrenaline. Unlike John Travolta in *Pulp Fiction*, who successfully slammed an intracardiac injection into a coding Uma Thurman, as the ER medical staff aimed for the intraventricular chamber of Joel's heart, they missed, and punctured his left lung. Already, as a result of the bite, Joel's platelets were low, his blood vessels permeable, and his clotting factors not working, causing him to bleed internally.

Now he was running on one lung, and the hospital was out of antivenom. The severity of Joel's symptoms being clearly out of their league, the local hospital transferred Joel.

The story got worse before it got better.

Just after DeGregorio terminated his landline conversation with the ER, his cellphone rang. It was a local off-duty MD with affection for rattlers, who wanted to go snake hunting. As he explained the irony of the doctors' request, Paul's landline rang.

"A tremulous male voice, a voice I've never heard before, asked for an unfamiliar name."

"Sorry, wrong number," DeGregorio told the caller.

And the caller asked again. "And I'm thinking, 'It's a fuckin' telemarketer.'"

DeGregorio again told the caller he had the wrong number, and as he started to hang up, the voice said, "No, wait. Please, can you ..." Click.

"Then I thought, 'Telemarketers don't usually beg.'"

DeGregorio retrieved the number from caller ID: "Robert Shapiro." It was Joel's father in New York City. DeGregorio called back.

"Maybe you want to talk to me."

"Were you with my son?"

"Yeah."

"Well, the hospital gave me this number and another name."

"I don't know who the other random name is, probably the janitor. This is I, Paul DeGregorio."

The father asked if Paul would advocate for his son. "Yes, of course, except I'm not a medical doctor. I'll gladly do what I can do."

Joel had been airlifted to the regional trauma center, but at the time of the phone call, neither DeGregorio nor Joel's father could confirm his admission status. After speaking with DeGregorio, in an attempt to locate his son, Shapiro called the trauma center two more times. Although the center had no record of a Joel Shapiro, the switchboard operator agreed to contact the ICU and track him down.

Joel had been admitted through the heliport late Saturday afternoon (hours after the bite), too late to be logged in. (His name wouldn't appear in the system until Sunday morning.) Once he confirmed that Joel has been admitted, Robert Shapiro reported back to DeGregorio, who, sensing the father's anxiety, called the ICU direct line and spoke with the attending physician. Next, DeGregorio placed another call to Dan Keyler, who was now at a wedding cookout. Although he sounded perfectly lucid, Keyler had already had a couple of beers and he felt as he spoke with DeGregorio that given the extreme circumstances of Joel's symptoms, he wanted to be one hundred percent clearheaded to deal with Joel's case.

"I can't compromise somebody's health."

Sometime later, Keyler wrote to me in an e-mail: "I do have to confess, that it was one of the most unfortunate times in my life to have had a couple of beers on board when [DeGregorio] called. I truly believe even a couple of beers can compromise optimal judgment... I would never forgive myself if something went wrong under those conditions."

Over the phone to DeGregorio, Keyler went over the initial things Paul needed to do to advocate for Joel Shapiro—CroFab, CroFab, CroFab—and then he referred him to two other out-of-state toxicologists on the snakebite advisory committee, both eminently qualified to discuss the proper protocol for timber rattlesnake envenomation and the accompanying anaphylactic shock.

"Keep me in the loop, but let them run with it."

DeGregorio relayed Keyler's suggestion to the weekend ICU doctor, who was both gracious and grateful.

"He knew he needed help. He was trying to do what was the best for Joel."

"I'm going to leave it to you to contact these people [snakebite advisors]," DeGregorio told the doctor. "Go with whomever you get a hold of first."

Cognizant that he was neither an MD nor a toxicologist, DeGregorio advocated for doctors to speak to doctors. He didn't want to risk mispronouncing some five-syllable Latinized medical term and have things get more out of hand than they already had.

So Paul directed the ICU doctor to have the nurse on the floor use the computer to pull up the names and phone numbers of the two experts, and then track them down. He asked Joel's father to make sure the ICU doctor reached one of the toxicologists.

"If you have to fly somebody in to deal with this, do it. Money's no object," Robert Shapiro pleaded. But no one wanted a consultant tied up on an airplane for three to five hours when symptoms and advice could be relayed electronically. The ICU nurse contacted Bob Embry, in Phoenix, and the center e-mailed pictures and the chart signs over the Internet for Embry, who took on the case. In the meantime, Keyler and Embry had spoken about the preliminary details.

"I didn't realize there's a classification of envenomation called *very critical*," DeGregorio told me, "which is apparently one step up from rubber bag."

On Monday afternoon, two days after the bite, Joel's parents, who had flown in the day before, just after he regained consciousness, were sitting by his bedside when the EMT on the helicopter burst into the room.

"You're alive."

Joel looked up weakly.

"I gave you CPR all the way from the hospital to here. None of us [aboard the helicopter] expected you to live. Can I take your picture?"

Joel had regained consciousness a little more than twenty-four hours after the bite.

When I asked Paul DeGregorio how he knew what to do under such dire circumstances, he responded: "I don't panic and go into shock mode." DeGregorio developed a mental checklist, an item-by-item commonsense response to an emergency. First, get Joel off the hill and to the field station; second, get medical assistance; and third, call Dan Keyler. Joel was off the hill and into the ambulance in twenty minutes.

"I didn't have to go to the gym for a week.

"Speed was critical. The Hail Mary play, I don't know where that came from. I had read somewhere that if a person is anaphylactic and fading they can still hear, but they may not be able to execute commands or respond." Apparently, Joel not only heard but managed to respond.

"Shapiro had *no* recollection of anything after he sagged in the woods, except the assertive command, 'Kids from the Bronx don't die in the woods in Kansas.'" His next recollection was the swishing of helicopter blades.

Joel was conscious on Sunday evening when his parents arrived. His father had had a plane warming up at Teterboro, New Jersey, with a case of CroFab (parks department connections), which proved unnecessary because the trauma center had finally located a closer source. Joel's first words to his parents were, "Hi, Mom. Hi, Dad."

And then, "Don't get Paul in trouble for this. It was all my fault."

"Trouble? I want to give the man a hug and take him to dinner," said his father, tearfully.

The snakebite had taken place on Saturday, April 30, 2011, about 2:30 in the afternoon. By Sunday evening, things appeared to be going well; Joel had progressed. Then, Monday morning, two days after the bite, Joel relapsed and became acutely ill, and the weekend ICU doctor didn't know whether the reaction had been caused by severe anaphylaxis or by envenomation.

DeGregorio recalled that "we could have used an epi-pen on site at the field station if we had known what the problem was. The EMT had one in the ambulance. At the time, I wasn't sure this was an allergic reaction to the snakebite. I knew the epinephrine wouldn't have made things worse, but I'm no doctor and events had unfolded *pretty fast*. I didn't want to suggest options, giving that those medical attendants were themselves freaking out. So, I thought, 'Go with something very basic and I fed Dan the information and we tried our best over the phone to figure out what to do.'"

Said Dan Keyler, "When you go into a hospital you're the hospital's patient regardless of what your consultant suggests. The attending physician can always override a consultant. You can give your expert opinion. But the bottom line is it's their patient."

That's where snakebite therapy gets a little bit tricky.

"If a doctor used the epinephrine at the wrong time, you could trigger cardiovascular arrhythmia."

Teasing apart whether the response is venom-induced anaphylaxis or a toxic effect of venom is difficult.

"This is *not* an exact science. Joel Shapiro needed both antivenom and the epinephrine, and when he was given the epinephrine he showed signs of cardio stabilizing."

Then came a scenario to top the worst-case scenario.

That Monday, Keyler called DeGregorio and inquired, "What's going on at the trauma center?" The weekend doctor had called Keyler to say that the weekday ICU staff, which had come on duty, abruptly "booted [the consultants] off the case." The weekend ICU staff is different from the weekday staff, and both are different from the ER staff, which also changes on the weekend.

"We're just consultants," lamented Keyler. "Joel was the regional trauma center's sole responsibility."

By late Sunday afternoon, Joel had been upgraded to critical and by evening to serious. Then, the next morning, as the prairie sun rose above an ocean of sameness, he began to decline. His blood work was abnormal again: platelets and fibrinogen (a blood-clotting protein) were low. Breathing had become labored, which is not uncommon for a victim of a timber rattlesnake bite. The mismanagement of the treatment had created more of a problem, for, as Keyler pointed out, "despite the best of intentions with medical care, things can get really fucked up." Fuck-up number one: the new attending physician, the local toxicologist, decided that he could handle this on his own and gave Joel epinephrine and interthoracic injections. Fuck-up number two: at one point on Monday, the ICU staff administered blood products (fibrinogen and platelets), unaware that the treatment was inappropriate, since the remaining venom would immediately destroy the infusion. Fuck-up number three: by Tuesday morning, the state Poison Control Center hadn't been notified, even though their office was two floors above the ICU and they had a staff doctor specifically trained for envenomation treatment.

Then, Joel's father called DeGregorio. "Our son deteriorated after having started to recover, and we [Joel's parents] were all over them [ICU] to get Embry and Keyler back on the case." The Shapiros had become livid in the ICU, threatening to sue the crap out of the trauma center if the consultants weren't brought back on the case. Acquiescing, the trauma center brought Embry and Keyler back. Joel began to improve.

On May 7, a week after the bite, Paul DeGregorio visited Joel, who was now out of ICU and in a private room. A *new* doctor entered the room and reported that Joel's platelets had dropped from seventeen thousand to fifteen thousand. (One hundred fifty thousand is normal.) Timber rattlesnake venom always drops platelets, which may bottom out at zero.

"CroFab would be preferred," DeGregorio said. Consultant Embry had said as much earlier in the morning.

The attending doctor had been checking Joel's platelet count every four hours, which to DeGregorio seemed excessive. "If the kid's not making platelets, why check them so often?" The doctor agreed. The tests were cut back.

By Saturday morning, one week after the bite, Joel Shapiro had been given fifty-two vials of CroFab, thirty-four vials more than the average envenomation patient. It was time to be weaned.

$$\text{\textsnake}$$

Prior exposure to foreign proteins had apparently caused anaphylactic shock, which is triggered in the brain and cardiovascular system. An anaphylaxis victim needs a breathing tube almost immediately. According to Keyler, the release of histamine by cells in response to an allergic reaction contracts smooth muscles and dilates blood vessels, which initiates a cascade of pharmacological events, including the redistribution of fluids outside the vascular system into other body tissues and cavities.

Bob Fritsch, who equated being bitten by a timber rattlesnake to slamming your finger repeatedly in a car door, had been bitten twice. His second bite dropped his blood pressure to thirty over barely measurable, a severe allergic reaction that had been precipitated by his body's allergic reaction to the proteins in the first, rather uneventful, bite.

But Joel Shapiro had never been bitten before. What could have provoked his life-threatening allergic response to timber rattlesnake venom?

Joel had handled timber rattlesnakes before, many times. People who clean snake cages can get exposed to a variety of allergens, said Keyler. "It's not common, but it happens. You don't have to get bitten to be exposed to the antigens; you can pick them up from snake sticks, snake bags, snakeskins, handling snakes, and so forth."

An average timber rattlesnake bite usually requires eighteen vials of antivenom, which is the standard quantity for the majority of pit-viper envenomations in the United States. With a really severe bite, however, you don't know how much venom has been injected. The sooner you begin CroFab treatment, the sooner the toxic effects of venom can be neutralized. The first four units were administered at the local hospital (that's all they had), and then there was a lag until Joel was transported to the trauma center, where four more vials were given in the ICU. Then another lag until the regional trauma center located more CroFab. Prolonging the treatment in this way allowed many of the venom effects to go unchecked and the symptoms to flare.

According to Keyler, the resident toxicologist at the trauma center, who had returned from a weekend off and intervened in the management of Joel's treatment, ran against the wind. "You're stuck unless you can convince them to listen to you."

Since Joel was a minor, the trauma center also had convinced his parents, who were trying to sort through a barrage of medical advice, much of it wrong, to sign a release form for the doctors to do surgery to relieve the swelling in his left arm.

Said Keyler,

Joel may have had swollen, discolored, waxy appearing fingers, yes. But did he have compartment syndrome, which occurs from pressure buildup beneath the fascia [the sheaths that cover the muscles], compromising circulation and neurotransmission and consequently the viability of the underlying tissues? Before you perform surgery for compartment syndrome [called a fasciotomy], it is important to measure the intracompartmental pressure in the region around the bite. Typically, intracompartmental pressures are measured using either a Doppler ultrasound [a noninvasive procedure that senses blood flow into tissue], or the invasive Stryker Intracompartmental Pressure Monitor [a needle is placed down below the fascia layer that covers the muscles and pressure is measured in millimeters of mercury]. Normal pressure in the compartment below the fascia would be two millimeters of mercury. If Joel had greater than thirty millimeters of pressure, were his hand and fingers still perfusing?

How would you tell?

You can push on the nail beds of the fingers or toes, and if you get capillary refill—blanch white returns to red—and you can feel any pulse in a distal extremity such as a finger or toe; then despite the compartment pressure measured, things are still functioning, tissues are still perfusing and being oxygenated. Then, you usually *don't* need a fasciotomy to relieve the pressure in the swollen limb.

Unfortunately, medicine is not perfect; doctors sometimes panic when they see severe swelling and discoloration, which almost always accompanies a rattlesnake bite. Surgeons, as Keyler points out, are trained to cut, and you *can* treat rattlesnake bite this way, but the question is does it need to be done?

In Joel Shapiro's case we won't know. The surgeons opened up his hand to release the pressure and debrided damaged tissue from the bite site. Having judged from photographs of Joel's hand, the fasciotomy extended from wrist to the ball of his thumb. Skin was grafted back, and during the slow recovery process, said Keyler, "morbidity can be significant."

"Snakebite is biochemically and toxin-induced trauma," said Keyler, "far different from a physically induced trauma. If you give antivenom within a reasonable time period after the bite, the antibodies in the CroFab sheep serum will begin to bind up the toxins, neutralizing the venom. Most of the time, if you let nature take over, the limb comes back a lot better than you'd think without any surgical intervention."

The late Findlay Russell, the Arizona toxicologist, treated over a thousand snakebites, and he never performed a fasciotomy.

"I'm not saying it's always wrong, but it is often an overreaction," said Keyler, who has never recommended cutting to relieve pressure.

"Tissue looks deep purplish blue. You can still debride the dead surface tissue to try to get granulation in the healing process back."

What is the window for the delivery of antivenom?

"The sooner the better."

To reduce pressure and to save tissue, you want to redistribute the fluid out of the swollen limb as soon as possible. To do a fasciotomy, you have to document extremely high fluid pressure within the fascial compartment.

Joel Shapiro remained in the trauma center for thirteen days. Before returning to the Bronx and a life away from the broken limestone hills, he stopped by Paul DeGregorio's home to say good-bye … and thank you. They discussed collaborative research projects (non–pitviper) that Joel could execute in New York and DeGregorio could mentor from the prairie fringe.

The total cost of his snakebite treatment was about three hundred eighty thousand dollars. Both the local hospital and the large regional trauma center, after they had acknowledged their mistakes, wrote off whatever the Shapiros' insurance failed to cover. The out-of-pocket expense for thirteen days in the hospital, the surgery, and fifty-two vials of CroFab—each vial cost more than a thousand dollars—was just three hundred dollars and a signed letter of release to allow the local hospital and the regional trauma center to use Joel's case history as a training experience for their respective ER and ICU staffs.

Joel returned to the Bronx, and six months later, Paul DeGregorio visited him.

"His range of motion is OK. His skin looks good. Like a lot of nineteen-year-olds, he's proud of his scar. It's a conversation piece. There's a lot of turtle research in his future."

One year after the bite, *Reader's Digest* published Shapiro's story "Snakebite," one of ten essays the magazine selected from the nearly seven thousand entries in the contest, which was called "The Best Life Stories." Joel concluded the essay with, "I was blind from hypoxia, but I could see my future clearly. I cannot deny my passion. Though they nearly killed me, I have dedicated my life to the study of snakes."

He was paid twenty-five hundred dollars for the piece.

"Don't be surprised if I appear on *Oprah*," he e-mailed DeGregorio.

"I told him not to mention my name. But if Oprah wanted to give me big bucks and have me on the show, he could whisper my name to her. I'd appear with a mask like if I was in the Witness Protection Program.

"Or is that Witless Protection ..."

6

A Long, Muscular Tube

A serpent's body being long and narrow, its contents are as it were molded into a similar form, and thus come to be themselves elongated.

ARISTOTLE

Pressing the timber rattlesnake to the table, Alcott Smith coaxed the scalpel through nearly two hundred ventral scales. The posterior end of each rectangular scale overlapped the one behind it; so collectively the scales spread away from each other like a long line of roofing shingles. For more than a half-dozen years, these belly scales, also called scutes, had hugged the floor and shelves of western Vermont, rubbing against fallen oaks and hickories, abrading loose configurations of pebbles and stones and rock dust, polishing narrow chambers deep below the surface of the ridge. Respectful of the snake's personal history, Smith carefully scraped away moist bits of meat as he peeled the snake's husk back across its tubular body, the way you'd peel the rind off an orange. In May of 2007, Smith had disengaged the rattlesnake, an eight-button, yellow-morph female approaching the prime of life, from the main den portal, where she had gotten stuck during emergence. Already weakened by having metabolized her fat deposits during hibernation and prevented from crawling backward by her shingled scales, the rattlesnake stayed wedged between stones, a prisoner in her own doorway, and eventually starved. Smith removed the snake, loosely knotted her to the frame of Paul Jardine's backpack, and continued his fieldwork, while the limp snake became a metronome that kept cadence with their pace, slapping against Jardine, who worried all afternoon about being punctured by a lifeless yet potent fang. When Smith returned home, he stuffed the rattlesnake in a plastic ziplock bag, placed her in the refrigerator for two weeks, and then brought her to a lecture he gave to a University of Vermont herpetology class, where, to the horror of the professor, he inadvertently injected himself with residual venom while attempting to show the students the arrangement of the snake's teeth. His index finger remained numb for two weeks, and the snake was repackaged and tossed in the back of the freezer, where she idled for four years amid ice cubes and frozen broccoli, suffering progressive freezer burn.

I've opened up rattlesnakes before. But this was different. This snake was to be dissected, combed apart so I could see, really *see*, the inner workings of this efflorescence of cold-blooded evolution—snake with the toxic bite and the admonishing tail. To prepare for the dissection, I cleared off the dining room table and resurfaced it with an old soap-stained rubber-ducky shower curtain. Our dissecting pan was a cookie sheet paved with wax. I had melted four blocks of paraffin on the stove the night before that slowly coalesced and spread across the sheet like a warm glacier before hardening into a reusable dissecting pan.

Methodically, working his fingers between flesh and hide, Smith continued to shuck the snake, opening her body cavity, carefully pulling away the transparent peritoneum and mesentery, which revealed a run of long, mostly narrow internal organs. We anchored the cutaway skin flaps to the waxed pan with T-shaped pushpins. When Smith withdrew his scalpel, the splayed-out curl of a timber rattlesnake, more than thirty inches worth, filled the cookie sheet like a muscular S.

We were not the first naturalists to unzip a timber rattlesnake. In December 1682, a rattlesnake arrived in London in a barrel having traveled from Virginia with a layover in the West Indies, a four-month trip without food or water. Not surprisingly, the snake died a few weeks later, and its keeper, a prominent London merchant, gave the cadaver to Edward Tyson, a British medical doctor and the country's foremost comparative anatomist, who dissected the specimen and promptly published the results in the *Philosophical Transactions of the Royal Society*—the first scientific journal in the world—under the rather cumbersome title "*Vipera Caudi-Sona Americana,* or The Anatomy of a Rattle-Snake Dissected at the Repository of the Royal Society in January 1683." Tyson thought the snake "so curious an animal" and wished "mightily" that he had had more rattlesnakes to dissect to "provide the most accurate account, and the exactest anatomy." Having read the thirty-three-page manuscript, I can unequivocally say that although Tyson's specimen was drying out and conceivably as fat depleted as our own, he did yeoman service locating organs and interpreting rattlesnake anatomy—particularly because he didn't have access to the color-coded illustrations that Smith and I had spread across the dining room table.

Tyson's biographer, none other than Ashley Montagu, called the doctor's career "modest," but considered Tyson creative with scalpel and scissors, which the doctor employed to dismember strange animals and "monstrous and abnormal births." Besides the timber rattlesnake, Tyson dissected and wrote scientific papers on everything from tapeworms and roundworms to shark embryos and lumpfish to lions, opossums, and Mexican peccaries. Although he thought them a link between fish and terrestrial quadrupeds, he was the first to recognize that a porpoise is more mammal than fish. Tyson was also the first to recognize the similarities between man and chimpanzee. In 1699, with the publication of "Orang-Outang, Sive Homo Sylvestris, or The Anatomy of a Pygmie Compared with that of a Monkey, an Ape, and a Man," Tyson became the "father of primatology." The fact that he actually dissected a young chimpanzee rather than an orangutan and later addressed the question of whether an ape and a human could cross, reporting that an "Orang-Outang" has a taste for "white blonds" hasn't diminished his anthropological standing. Literature, too, owes a debt to Tyson, wrote Montagu; the Yahoos in *Gulliver's Travels* derived from the portrayal of the young chimp in "Orang-Outang." Dr. Tyson is also remembered in medicine as the discoverer of the preputial glands in the foreskin of the penis, which were later named Tyson's glands, in his honor. A proponent of the "great chain of being," which suggests links between major "groups" of animals, one of the early articulations of evolution, Dr. Tyson shared a link of his own with Charles Darwin—his maternal grandfather was Darwin's great-great-great-grandfather.

In the late sixties, when I was an undergraduate studying comparative vertebrate anatomy, a subject that fascinated Dr. Tyson, I learned that all members of our august phylum Chordata, which includes wormlike sea squirts and lancelets as well as our more familiar relatives—sharks, bony fish, amphibians, reptiles, birds, and mammals—possess a notochord, a cartilaginous skeletal rod that supports the body, during at least some stage of their development. In the more highly evolved chordates—fish through mammals—a vertebral column, either cartilaginous or ossified, encases a spinal chord and replaces the notochord as the body's central joist.

I was taught that the traditional classification of the subphylum Vertebrata that backboned chordates is divided into eight classes supposedly based on evolutionary affinity. Four of the classes represent what we call "fish," a loose, ineffectual term analogous to "tree" or "bug" that includes lowly, jawless lampreys and hagfish (Agnatha); armored, extinct marine fish with paired fins, the evolutionary derivative of limbs (Placodermi); cartilaginous sharks and rays (Chondrichthyes); and the pinnacle of "fish" evolution, the statistically abundant and statistically diverse bony fish—tuna, swordfish, goldfish, guppies, and so forth—(Osteichthyes). The remaining four classes of the subphylum, the terrestrial vertebrates—Amphibia, Reptilia, Aves (birds), and Mammalia—are familiar to anyone who grew up watching *Sesame Street*. Like tracing Abraham's descendants in the book of Genesis, the relationship between the classes of vertebrates was reasonably easy to plot: jawless fish begot placoderms, which begot both sharks and bony fish. Amphibians, which tentatively invaded land four hundred million years ago, derived from a group of lung-bearing, bony fish called lobefins, the *Crossopterygii*. And then, fifty million years later, during the steamy Carboniferous period, when the hot breath of Earth welded together all the landmasses of the planet into one gargantuan equatorial welt of profound and uniform lushness, amphibians begot the reptiles, which eventually begot both birds and mammals many more millions of years later, during the legendary Age of Dinosaurs.

My comparative anatomy course subdivided the class Reptilia into approximately sixteen orders (dinosaurs comprised two of them), of which only four are still extant: turtles, crocodiles, Squamata (lizards and snakes), and Rhynchocephalia, a single, lizard-like genus of only two living species, called tuatara, which make their last stand on islands off the New Zealand coast. Of course, that classification was before the advent of sophisticated molecular research, when the phylogeny of vertebrates was more an artifact of the human mind than an actual evolutionary blueprint. That older classification lumped groups of animals into a pyramid of ascending anatomical sophistication, at the apex of which humans stood alone, but it often failed to portray actual evolutionary relationships between groups of animals.

Today, a student of comparative vertebrate anatomy is introduced to a phylogenetic tree, or cladogram, where relationships between groups called clades are based on genetic markers, biogeography, fossil history, and internal anatomy. This system more precisely portrays animal groups as equally indented sister taxa on a horizontally radiating tree that looks like a tipped-over menorah. For instance, all terrestrial vertebrates—amphibians, reptiles, birds, and mammals—are four-limbed and share a common ancestor. Collectively they make up the taxa Tetrapoda. All reptiles, birds, and mammals have three membranes that enclose the developing embryos, and belong to the clade called Amniota. (As the name suggests, the amniotic membrane appears for the first time in this group.) More to our point, crocodiles and birds sprang from a common ancestor;

in fact, an alligator is more closely related to a chickadee then it is to either a turtle or a lizard or … a timber rattlesnake—which is to say, birds are late-blooming reptiles and crocodiles are latent birds.

Let's take a closer look at the phylogeny of snakes. Like all other vertebrates, which include various "fish" and the tetrapods, snakes have a segmented backbone, but unlike either fish or amphibians, to which they are more distantly related, they also have a shelled egg and an amniotic membrane. When I was an undergraduate, squamates were divided into three suborders: Lacertilia (lizards), Amphisbaenia (mostly limbless lizard-like reptiles related to something called a whiptail), and Serpentes. No longer. Squamata now fractures along truer evolutionary lines into Iguania (iguanas and chameleons) and Scleroglossa (all other lizards and snakes). Scleroglossa subdivide into Gekkota (geckos and flap-footed lizards) and Autarchoglossa (monitor lizards, alligator lizards, beaded lizards, amphisbaenians, and snakes). All male squamates have paired copulatory organs, but only members of the Autarchoglossa have tongues that deliver airborne chemicals to an interpretive center on the roof of the mouth. Within the Autarchoglossans, monitor lizards (family Varanidae), which include the massive and well-known Komodo dragon, and alligator lizards (family Anguidae) are more closely related to snakes than to any other family of lizards; in fact, monitors and alligator lizards possess several snakelike traits: a long, forked tongue, a sophisticated Jacobson's organ that discriminates the most subtle chemical message conveyed by the tongue, and a pair of moveable mandibles (or lower jaw bones) that work prey. "Actually," wrote herpetologist Harry Greene, "snakes are lizards in exactly the same sense that humans are primates, primates are mammals, and so forth."

The evolutionary history of snakes is not accommodating. They die in the wrong places and don't fossilize often or well. In addition, there is not a great deal of bone diversity, so fossils that *are* found— ribs, vertebra, occasionally a jaw or skull and later in their evolution fangs—are not as revealing as for many other taxa. The point of origin of snakes has been highly controversial. One hypothesis asserts that snakes evolved from mososaurs, an extinct group of large marine reptiles, and that leglessness was an adaptation to an undulating life in tropical seas. According to this marine theory snakes returned to land only secondarily (like whales but in reverse). A second asserts that snakes evolved from subterranean monitor lizards. Then, during an indeterminate period of eons, to facilitate burrowing in loose sand their legs atrophied and eventually disappeared—along with pectoral and pelvic girdles, sacrum, and eyelids.

The debate may finally be over. In 2003, in a terrestrial deposit in Patagonia, an Argentine paleontologist unearthed a fossil snake with a well-defined sacrum that supported a pelvis and hind legs, which extended beyond the rib cage. The fossil dates back to the Cretaceous, more than ninety million years ago, and is a *true* "missing link," a long, slender snake with hind legs. Blair Hedges, an evolutionary biologist at Pennsylvania State University, claimed: "In one fell swoop, this new fossil kind of casts doubt on the aquatic hypothesis." Hedges's research on DNA sequencing further suggests a terrestrial origin for snakes. "We see many cases where animals that walked on land eventually evolved lineages that invaded the oceans. Almost all of them kept their limbs and turned them into fins or paddles." Think whales and seals, manatees and sea turtles, sea otters, penguins.

In 2010, paleontologists identified the fossil of an eleven-and-a-half-foot snake looped around a broken egg in a dinosaur nest as though waiting for the egg to hatch. Outside the coil were two intact fossil eggs and a hatchling *titanosaur*, a juicy and collapsible baby, about a foot and half long, just bite-size for a snake without the flexible jaw joints that would one day become an adaptive hallmark of serpents, a trait that today permits exotic Burmese pythons to swallow whitetail deer and bobcat in the Florida Everglades.

These prehistoric fossils suggest that from limblessness to vision, everything that makes a snake unique among tetrapods was forged long ago in the darkness of subterranean burrows, and that a snake is a palimpsest of its own long history. Ancestral snakes had little need for eyesight. Invisible odor plumes of their prey drove them forward. Their eyesight degenerated. In all but the most primitive families (like the blind snakes), eyelids fused into a transparent shield called the brille. The outer layer of the brille is shed during ecdysis and appears as twin cups that protrude from the head portion of the shed. As far as snakes go, blind snakes are just weird. If snakes are regarded as degenerate tetrapods, blind snakes, wrote a South African herpetologist who was familiar with them, are "degenerated degenerates ... no reptile approaches so nearly the stage of being sans teeth, sans eyes, sans taste, sans everything." In Texas, blind snakes that escape the grip of hungry adult screech owls have been known to survive in the owl's nests, dining on the maggots of parasitic flies, which would have otherwise plagued the birds—owlets grow faster and healthier in the presence of commensal blind snakes.

When snakes reappeared and colonized Earth's surface, their eyes had to be rebuilt for daylight. All tetrapods except snakes focus by changing the shape of their flexible lens using muscles within the eye. During thousands of generations underground, snakes lost those muscles. Instead, using a divergent set in the iris (the iris ciliary muscles), they focus by moving their nonflexible lens forward and backward from the retina, the same principle that applies when focusing a camera—the lens moves in relation to the stationary plane of the film, a detail missed by Dr. Tyson and one of the many missed by Smith and myself. (In all other Tetrapods, the iris ciliary muscles regulate light rays entering the eye.) Consequently, snakes are nearsighted and respond only to motion. Within the purview of a timber rattlesnake a stationary chipmunk is veiled in a swirl of radiating heat. If it moves, however, it's dead.

To the uninitiated a snake may appear to be a neckless head and a very long tail. The actual tail, however, begins at the vent, the horizontal slit more than three-fourths of the way down the animal's body, where at various times our timber rattlesnake's cloaca had unloaded digestive or urinary wastes and would have birthed snakelets had she lived long enough. Both sexes have paired anal scent glands in their tails, which Dr. Tyson called "scent-baggs." He wrote, "The liquor included in them was something crass, and of a strong very unpleasant smell." To draw our own conclusions, Smith squeezes a couple of drops of musky smelling, brownish oil from each gland that contrary to hill-country folklore smells nothing like cucumbers but *not* nearly as bad as one description in a nineteenth-century text, which characterized the odor as "extremely disagreeable." Although a female snake has larger anal scent glands than a male, males have longer, thicker tails than females because their tails are packed with the forked hemipenes. Dr. Tyson, who had dissected a four-foot male, wasn't able to compare scent "baggs" between the sexes and died without knowing what lurked inside the tail of a female timber rattlesnake.

Looking at the timber rattlesnake laid open in the dissecting pan, I noticed one other obvious distinction between the body and the tail. A pair of mobile ribs was attached flexibly to each trunk vertebra, except the first two neck vertebrae, the atlas and the axis. All internal organs were housed in one long body cavity; no division existed between the cavities of the thorax and abdomen. From the base of the neck to the vent, Alcott Smith's lengthwise incision exposed the tips of more than one hundred fifty of these rib pairs. Beyond the cloaca the short, thick caudal ribs and vertebrae fused into a less distinguishable ossified wedge that supported the energetic shaker muscle, responsible for the vitality of the rattle. The thoracic, cervical, and abdominal ribs, long and curved, never converged ventrally into a rib cage—there was no breastbone; instead, the ribs flared outward into a pliable archway that allowed meals to pass through the snake's body unimpeded. Up to thirty iterating sets of discrete muscles on each side of the rattlesnake laced vertebra to vertebra, ribs to vertebrae, ribs to ribs, and ribs to the inside edges of the ventral scutes. The deepest of these muscles bound adjacent vertebrae to each other and to more distant vertebrae; the middle layer bound ribs to vertebrae and ribs to ribs; and the outer layer connected the rib tips to the lateral edges of the ventral scales. The musculature in its entirety was braided together like a long, cylindrical pot holder. Elastic ligaments yoked bone to bone. And equally elastic tendons connected muscle to bone. Even though Smith's snake had been frozen and defrosted, her body remained limber yet firm, a faint echo of the living serpent.

The snake itself was a link between an opaque, underground

Mesozoic past and the unimagined future of her species, an evolutionary work in progress, and as Smith reverently peeled back the integument and parted the peritoneum, he exposed the culmination of a million generations of procreation, where natural selection had interpreted the environment into timber rattlesnake design. How could anyone not marvel at the evolution of a timber rattlesnake, ongoing and incomplete, sophisticated and splendid, which lay on the dining room table?

"Everything about this snake is an unfinished masterpiece," I blurted out, unable to contain myself and realizing for the first time exactly how far a snake had to go from basic lizardness to become a snake, and how far a rattlesnake had to go from basic snakeness to become a rattlesnake.

"If she was fresher, we'd see a lot more. I had her in Sherry's freezer twice. Jim Andrews had her in his freezer. Last week, Kiley Briggs left her for me in The Nature Conservancy field office. No one told me. So, she stayed in the conservancy's refrigerator until I picked her up a couple of days ago. Then, back to Sherry's refrigerator." This rattlesnake had been frozen and thawed more times than an orthopedic ice pack.

Like all snakes, the opposing ribs pressed out laterally against the body wall and the sharp tips of each pair attached by muscle to the inner edges of a ventral scale, one per side. Although Dr. Tyson counted one hundred sixty-eight ventral scales from the neck to the vent, which equaled the number of vertebrae, he never mentioned the relationship of either to the number of ribs, a relationship described by his French predecessor M. Charas in 1669. After dissecting a European viper (*Vipera aspis*), a species common to England as well as continental Europe, Charas wrote, "there are as many large scales beneath the belly, as there are vertebrae provided with their pair of ribs." Dr. Tyson did note "the scales of the belly were joyn'd to each other by distinct muscles; the lower tendon of each muscle being inserted about the middle of the foregoing scale." Smith and I found it impossible to tease apart the shriveled muscles of the integument of our snake.

With no breastbone and no diaphragm, the timber rattlesnake is simply an exquisitely muscular tube, shaped like a wide, inverted *U*, framed by ribs and supported by backbone. The belly, dull yellow and speckled black, was Great Plains flat. Whenever the rattlesnake eats, contracting and expanding sets of muscles squeeze the meal through the esophagus toward the stomach—a process I recall from high school biology, peristalsis—while the paired ribs spread away from each other like a drawbridge and allow for the passage of meals wider than the snake's own body. One summer, in the woods above West Point, Randy Stechert and I found a number of timber rattlesnake ribs at the entrance of a granite slab den, where Stechert had seen big rattlesnakes basking on a rock shelf beyond the den entrance and where a mother red-tailed hawk had taught her fledglings to hunt lethargic snakes each autumn. A pair of those ribs sits in front of me now, crossed like swords, two and a half inches long and wider at the base where they once interfaced with a vertebra. The ribs taper to a point sharp enough to pick spinach out of my teeth—something I bet Dr. Tyson never did—and I can imagine them having spread apart as a full-grown gray squirrel passed into biological blackness.

For cryptically patterned, heavy-bodied snakes—adult boas, pythons, and vipers, like our timber rattlesnake—there is no need to hurry. Travel for them is leisurely and straightforward. Unlike a skinny black racer, which hunts by sight and chases down prey, a timber rattlesnake smells its way to success, slowly and surely. When this rattlesnake found a potential ambush site, rife with rodent odors, she waited looped into spring-action coils or partially stretched out propped straight up against a tree trunk (vertical ambush). For a timber rattlesnake, movement between life's critical stopovers is more or less in a straight line, a pattern of crawling called "rectilinear" locomotion that employs the backbone and ribs and the weave of trunk muscles transmitting force to the edges of the ventral scales. This timber rattlesnake would have *walked* herself forward by anchoring the edges of regularly spaced groups of ventral scales against small irregularities on the surface and then contracting her torso in bilaterally symmetrical waves. A rattlesnake progresses wave after wave, undulation after undulation, like ripples expanding on the surface of a lake. The ribs don't walk; it's the ventral scales that walk attached as they are by muscles to each other and to the ribs. In 1683, Dr. Tyson had already unmasked the timber rattlesnake's gait, "the scales are as so many feet." When I watch the side of a moving rattlesnake stretched in a nearly straight line, the caterpillar-like rhythm is evident. But when I watch the dorsal surface it isn't; the snake's back appears almost motionless, the scales going passively along for the ride.

To demonstrate the integrity of the vertebrae, Smith tried to clean muscle off a portion of the backbone, which proved impossible without first boiling the snake on the kitchen stove, a project that would not have gone over well with my family. (Somehow, Dr. Tyson succeeded in striping the meat from the vertebra.) Instead, I ran my fingers down the length of the rattlesnake's backbone and felt the lateral processes of the vertebrae, two spikes on each side bumped against my inquiring touch. These winglike projections control the amount of the snake's sideways movement—too much twisting might damage the sensitive spinal column that runs through a canal near the top of each vertebra. (Lateral and ventral movement between adjacent snake vertebrae is no more than twenty-five degrees.) Since I had limited access to the rattlesnake's backbone (short of boiling off the meat), I examined a run of nine green watersnake vertebrae I had collected many years ago on a rural road in central Florida. Even though the discs were long since gone, all nine fit nicely together in a series of balls and shallow sockets, the ball of one vertebra imbedded in the corresponding socket of the one behind it. Or as Dr. Tyson stated, "the vertebrae of the spine seem admirably contrived; there being a round ball in the lower [posterior] part of the upper [anterior] vertebra, which enters a socket of the upper [anterior] part of the lower [posterior] vertebra." Sockets in front; balls behind. Gently, I moved the group sideways until the loosely interlocking, lateral processes prevented further torque. It was a precise arrangement that permitted a measured movement, flexible within the limits proscribed by the vertebral projections. You can't bend a snake in half without crippling it—they're not rubber. I imagined our snake crammed in her doorway, ventral scutes hooking rock, muscles contracting, backbone twisting, going nowhere and unable to back up, wave after wave of unproductive movement, while life went on all around her.

This bears repeating: snake scales are not separate and removable like fish scales, which can be flaked off without damaging the skin. Snake scales are precisely arranged folds and creases within the continuous sheet of the outer skin, or epidermis, and can't be scraped off. Stretching our snakeskin revealed the interstitial skin, thinner than the scales, which is responsible for the living snake's outer elasticity. This portion of the skin would remain hidden unless the rattlesnake was distended with either food or embryos. A free, rear margin of each scale neatly overlapped the scale behind it, an arrangement that was particularly evident in the ventral scales.

This positioning of scales, called squamation, is reasonably uniform for each species and subspecies, and has been of critical importance in the classification of snakes. The skin itself has a rough texture, made rougher on the back by thin, longitudinal ridges or keels that run the length of each rhomboid scale. (Dr. Tyson compared the scales to parsnip seeds.) The keels felt less prominent as I ran my fingers down the side of the rattlesnake, and they eventually disappeared altogether just above the ventral scutes. The living skin is thick and deep and composed of two layers: the dermis and the epidermis. The entire outer layer of the epidermis, known as the stratum corneum, is a thin sleeve of dead cells made of keratin—the same tough protein that makes up fingernails and talons and feathers—one continuous sheet that is sloughed off periodically as the snake grows. The nearly impervious stratum corneum had protected this timber rattlesnake against the abrasion of talus and kept her from drying to a twisty crisp during the dog days of August. (But apparently couldn't protect her from freezer burn.) Her shed skin would have stretched out twenty percent longer than her body, soft and oily at first, but dry within a few hours after shedding. Wind and rain, ultraviolet light, bacteria and fungus would have rendered it unrecognizable in two or three weeks if hungry crickets and sow bugs and nest-building great crested flycatchers didn't get to it first.

Shed skins are inverted, usually translucent, and nearly colorless. The palette of the color patterns that camouflage a timber rattlesnake in a sun-speckled forest comes from pigment cells called chromatophores, which are buried at the juncture of the epidermis and dermis, well below the stratum corneum. Melanin translates into black, brown, dull yellow, and gray; guanine into white; and carotenoid into red, orange, and bright yellow. Collectively, the pigments make the fifteen to thirty-two brown crossbands, chevrons, and blotches edged in white or yellow; the black-speckled, cream-white ventral scales; the gray, brown, deep-ebony, or neon-yellow base color of adults, and the pinkish gray of neonates; and the ink-dark dorsal and lateral scales of all tails. Crossbands and blotches are always brownish and darker than the base color. The jet-black timber rattlesnakes of the Northeast and the higher elevations of the Appalachians are the product of dense concentrations of melanin that overwhelmed the other pigments as green created by chlorophyll overwhelms the other pigments of a leaf. I find vague patterns on timber rattlesnake sheds with hints of banding and barring, but once, in the foothills of the Adirondacks, in the company of Bill Brown, we found a strikingly fresh shed, as black as obsidian, which revealed the faintest suggestion of a pattern. Less than an hour later, we found the undressed snake basking on a slab, so dark and shiny he appeared to ooze melanin.

On the floor of the rattlesnake's mouth was a soft, white-humped encasement, the glottis, gateway to the lung, which concealed the bifurcated tongue that had once flicked out of the mouth through a small notch in the scales at the synthesis of the two mandibular bones of the lower jaw (we know it as the chin) and read the wind with its fabled tines. Although Aristotle thought a snake's glottis was the beginning of the trachea, Dr. Tyson realized that the glottis preceded the trachea. Hidden beneath the glottis was the hyoid apparatus, the thinly forked bone that anchors the tongue and its associated muscles and is so important to modern snake taxonomy that in 1968, the University of Illinois published David Langebartel's best-seller *The Hyoid and Its Associated Muscles in Snakes*, which stated that serpent hyoids take the shape of one of four letters of the alphabet—*M, V, Y,* or *U.* Crotalines like our flayed timber rattlesnake have U-shaped hyoids, the twin tines curving around a short handle, so that in the book's illustration the bone looked like a tiny tuning fork. Whatever letter shape the hyoid takes, when a snake, our rattlesnake, for instance, swallows a chipmunk or white-footed mouse, the glottis extends outside the mouth like a snorkel to facilitate breathing and the tongue remains below the glottis moored by the hyoid.

Although Aristotle had noticed the elongation of a snake's organs more than two thousand years ago, it was still a revelation to me when I saw the rattlesnake's tubular leitmotif, her external anatomy mirrored in the arrangement of internal organs. Smith inserted a probe in the glottis, and we followed it into the trachea, the long conduit between the inside and outside of the rattlesnake. Tough, cartilaginous semicomplete rings, one of the more prominent internal features of our snake, kept the trachea from collapsing while the snake swallowed a meal. A collapsed trachea would cut off oxygen from reaching the lung, a prescription for quick asphyxiation.

The ringed trachea reminded me of the pleated breathing tube on my father's ventilator, and it was tough enough to roll between my fingers. The upper surface of the back of the trachea was lined with respiratory alveoli and functioned as an adjunct lung merging so imperceptibly into the real lung that I couldn't pinpoint the boundary between the two. Dr. Tyson did, however, commenting that this feature is common to the timber rattlesnake and "to the viper-kind," a rather profound glimpse of the close relationship between pitvipers and vipers, which wouldn't be unraveled for several centuries. Like Dr. Tyson and Charles Darwin, the viper and the rattlesnake shared a common ancestor in the not-too-distant past (geologically speaking).

The long, thin right lung picked up where the trachea left off (wherever that was) and extended through the body cavity past the stomach. The left—useless and vestigial anyway, another evolutionary accommodation to the tubular blueprint—had vanished in the course of the snake's many trips in and out of the freezer. The vascular portion of the right lung gave way to an even thinner, nonvascularized bladderlike air sac, a reservoir without a direct connection to the circulatory system. Dr. Tyson noted that the "air cells" of the lung gradually disappeared, eventually leaving "only a large bladder without any cells, composed of thin, but strong transparent membrane," which he compared to the smooth lung of a salamander. Of course, Smith and I saw nothing of this, only the deflated sides of the air bladder stuck together like cellophane. This air bladder, which Dr. Tyson calls the "store house," increases buoyancy and submersion time in aquatic snakes and permits a rattlesnake to float high on the surface like a linear balloon. It may also store air during food swallowing, and it allows a threatened rattlesnake to inflate itself to appear bigger and to prolong its hiss by forcing air across a membrane in the glottis (though I must confess that I've *never* heard a rattlesnake hiss, perhaps because the delirious rattling eclipses any suggestion of breathy sound).

Smith ran the probe through the glottis, the trachea—bumping the probe's round head against the semicircular, cartilaginous rings—and the vascular lung, and out the transparent cellophane of the air sac. The entire respiratory system, of which half—the lung and tracheal lung—was vascular, extended approximately three-quarters of the length of the rattlesnake. It had once inflated at a rate that reflected the ambient temperatures of its natal ridge. Cold-blooded snakes breathe more rapidly when it is warm and more slowly when the weather is colder. Each breath begins with a half-full lung, which is then filled to capacity and immediately deflated back to half capacity, followed by a prolonged breathless wait. Never is the snake's lung completely emptied. Watching a rattlesnake breathe is like watching clothes dry on a laundry line; it happens so slowly it is barely noticeable. If you blink, you won't miss; if you nap, you may not miss either. Absent a diaphragm, there is no negative pressure in the snake's chest cavity to expand the lung; instead, the intercostal muscles woven through the long line of ribs expand the ribs, alternating laterally and medially, pulling air into the lung.

Smith went out to his car and came back wearing a skullcap replete with surgical headlamp and attached magnifying goggles, a device he wore when operating on dogs and cats that now made him look as though he were headed down the shaft of a coal mine rather than down the gullet of a rattlesnake. There's no telling what Dr. Tyson might have discovered during his dissection had he had access to Smith's space-age paraphernalia.

"I can't find the pulmonary vein. There's too much drying and decomposition."

"Where're you looking?" I asked, a small Maglite clamped in my teeth, and not sure where along the protracted lung to begin my search.

"We're fighting long odds. The pulmonary vein leaves the right lung with a load of oxygenated blood and enters the left atrium of the heart," replied Smith, tapping the grape-shaped heart with the probe and strumming the few vessels he found.

As circulatory systems go the pulmonary vein and artery are anomalies. Generally, blood flow through arteries is pressurized, and blood flow through veins is passive; all arteries carry blood rich in oxygen through the body, and all veins bring blood low in oxygen, and loaded with carbon dioxide, back to the heart. In the case of the pulmonary artery and vein, these roles are reversed: the pulmonary artery carries the oxygen-starved blood to the lung for aeration, and the pulmonary vein brings oxygen-rich blood from the lung to the heart, which then pumps it through the bloodstream, distributing oxygen via capillaries to every organ, tissue, and cell. No matter, however. In our snake, we couldn't find either vessel.

"I don't see it in this pasty conglomerate. It's too thin, too short. There's bile leaking out of the liver and gallbladder. Likely I ruptured it making the initial incision. We're lucky we see what we see."

The heart was one of the few organs I was sure I could identify and the only organ that seemed, at least on the surface of things, to be unaffected by the tubular template. The size and color of a red grape, the heart was about a third of the way down the rattlesnake's body, near the vague juncture of the trachea and the lung, opposite ventral scale number seventy-one, said Smith. A timber rattlesnake has between one hundred eighty and two hundred ten ventral scales, a number that does not vary with the age of the snake, but we had no inclination to make a precise count—had we tried, the organs would have spilled out the lengthwise incision. The heart appeared to be a third of the way down the body, and we left it at that. Because tree snakes hang vertically, increasing the need for oxygenated blood to overcome gravity, their hearts are closer to their heads; the heart of aquatic snakes is closer to mid-body.

An amphibian heart has three chambers: two atria and a ventricle, a *big* step up from the two-chambered heart of a fish. Oxygenated blood and deoxygenated blood are kept somewhat separate, largely as a result of the timing of the contractions of the left and right atria. The ventricle pumps blood returning from the right atrium into the pulmonary circulation. After most of the deoxygenated blood has left the ventricle en route to the lung, oxygenated blood enters the ventricle and is pumped through the amphibian body. Because the moist, tissue-thin skin of salamanders and frogs functions as an auxiliary breathing organ, a septum inside the ventricle to separate the oxygen-rich blood from the oxygen-poor blood was *not* an evolutionary priority.

The leathery, almost impervious hide of the reptile triggered a corresponding change in both respiration and circulation. A reptile heart is more efficient than earlier versions of the vertebrate heart. Like that of a bullfrog, a timber rattlesnake heart has three chambers—a right and left atria and a ventricle—but in the rattlesnake's case, an incomplete wall partitions the ventricle, further reducing the mixing of oxygenated and unoxygenated blood. Because there is no diaphragm to hold it in place, a snake's heart is suspended within a fluid-filled sac called the pericardium that is held in place by the mediastinum; whenever a large meal passes through the esophagus and pushes against the pericardium, the fluid-filled sac absorbs the brunt of the displacement. Eventually, in the course of evolution separate pulmonary and systemic (body) systems were perfected by complete separation in the four-chambered hearts of crocodiles, birds, and mammals. Of course, Smith and I saw nothing of all this. Neither had Dr. Tyson, but he was still closer to the mark than William Harvey, the British physiologist who unmasked the circulation system in 1628; Harvey thought reptiles had two-chamber hearts and that the interior of their hearts was smooth and "entirely without fibres or muscular bands." We saw only the reddish grape of a heart hanging in an empty pericardium, adjacent to the collapsed right lung, which was transparent and beyond thin.

I remember studying latex-stained circulatory vessels in comparative vertebrate anatomy the way a motorist would study a highway map. Arteries were pink and veins blue. Everything led somewhere, and if you took time and thoughtfully used your probe you might navigate your way through the circulatory system of a shark, a bony fish, a frog, a turtle, a pigeon, or a cat. Not so with our rattlesnake. Her carcass resembled something from a crime scene—a mass of intact and ruptured organs not decipherable by any atlas of the circulation. Smith did manage to point out the dorsal aorta, the interstate highway of the circulatory system that delivered oxygenated

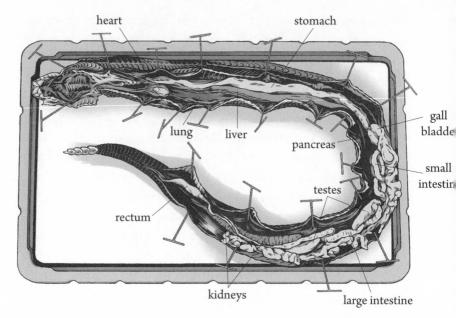

heart

stomach

gall
bladde

small
intestir

lung

liver

pancreas

testes

rectum

kidneys

large intestine

blood from the heart through the snake's body. And near the head, we found the paired carotid arteries, which exited the aorta and had kept the back roads of the snake's brain supplied with oxygen.

The snake's throat was the opening to the digestive system. Unlike the more mysterious circulatory system, Dr. Tyson grasped the digestive system. "The whole length from Throat to the Anus is but a continued Ductus." Any prey unlucky enough to cross this threshold would proceed into the esophagus, the stomach, the small intestine, the large intestine (colon), the cloaca, and ultimately in different form out the vent into daylight. This long, open-ended tube of varying widths was more or less like the rattlesnake herself with no room for loopy hose-like convolutions in the small intestine as in mammals or for a storage pouch and a grinding pouch like the crops and gizzards of birds. Because of their easily digested diet of meat, the attenuate shape of a snake's digestive system is a simple tube within the tube of the body—a straight shot from mouth to vent.

Not wanting to envenomate himself twice on the same dead snake, Smith carefully avoided the sheathed fangs and ran the probe through the mouth and throat into the thin esophagus, which yielded to the pressure of the tiny, round metal head and bulged outward. The esophagus, which runs parallel to the stiff-ringed trachea, is capable of expanding as a large meal lubricated by saliva passes toward the stomach. Esophageal muscles and rib muscles squeeze moistened prey into the stomach, a journey that might take fifteen minutes for a white-footed mouse, but considerably longer for a squirrel. As you've previously read, on several occasions at NERD, I've watched Kevin McCurley's yearlings swallow lab mice, and the process is time consuming as waves of muscular contractions slowly compress and express the mouse headfirst down the gullet. Peristalsis again.

I wasn't sure when the probe arrived in the timber rattlesnake's stomach; the line of demarcation between the two organs was opaque (to say the least). The filiform stomach was thick-walled and long and filled with folds. A surgically precise slice of Smith's scalpel revealed the interior of the stomach as (like our own) a series of muscular corrugations, longitudinal ridges and valleys that enhance the mechanical breakup of food. Smith then passed the probe through the stomach into the small intestine, which, except for a few small curves, ran straight to the shorter, wider, multitasking large intestine.

Venom injected by the hypodermic fangs begins digestion; the stomach concludes it. From the stomach, the chemically and physically pulverized mouse enters the small intestine, where the bulk of the nutrients are assimilated. The large intestine reabsorbs moisture while transporting fecal material—hair, feathers, nails, teeth, scales, but *not* bones, which have been dissolved in the stomach—into the multipurpose cloaca. Here, indigestible excrement mixes with semisolid uric acid from the kidneys. One autumn, in the Blue Hills of Milton, Massachusetts, I saw amorphous swaths of black and white rattlesnake waste that had been voided just before the snakes retreated below the surface for the winter. The chalky layers were uric acid; the dark, gummy smears, crumbled and soiled mouse mostly.

Dr. Tyson, who suspected that the timber rattlesnake ate only one meal a year, believed that their large mouth accommodated more than just a big meal. "And if what is confidently reported by many, be true, that on occasion of danger they receive their young into their Mouths, these are fit places for receiving them." Avoiding the four rows of tiny, sharp fishhook teeth not to mention the brutal fangs would be a feat for a mouthful of neonates.

Between the stomach and the heart Smith and I found the liver, the biggest internal organ in the snake. "How many lobes do you see?" Smith asked, forcing me to finger the hotdog-shaped organ.

"Two of unequal length," I replied, imagining chopped rattlesnake liver on Saltines the way my nana had served chicken liver at every opportunity, which I had avoided almost as much as borscht.

The long, twin-lobed liver, another adaptation to the tubular body, began at the heart and ended near the juncture of the small intestine and the stomach. It stored blood, detoxified the body, and secreted bile, the most well known of digestive enzymes. Wearing his magnifying goggles and headlamp, Smith, after considerable searching, teased the more than an inch-long, deflated balloon of a gallbladder out of hiding on the far end of the stomach near the junction of the small intestine. Aristotle noted that the gallbladder of snakes "was usually beside the gut" not attached to the liver as it is in most other animals. Had the gallbladder been attached to the liver, it would have compressed against a prey-distended stomach, stimulating the premature discharge of bile into the small intestine. Positioned on the stomach, the snake's gallbladder secrets bile only after the small intestine has already filled with digested food.

Grass-green bile, stored in the gallbladder and released into the small intestine, now leaked throughout the snake's mesentery, tingeing everything, including our fingers a bright, translucent green. Although Dr. Tyson didn't taste the bile of his timber rattlesnake, he claimed that the bile of a European viper tasted at first "salt, then a sweet bitter." I'll take his word for it; neither Smith nor I was motivated to make a comparison.

Next to the gallbladder was the spleen and near that the pancreas. The spleen made and removed blood cells, and the pancreas released digestive enzymes into the duodenum, the portion of the small intestine closest to the stomach. As recently as 1900, comparative anatomists and herpetologists have misidentified the two organs. Dr. Tyson called the pancreas the spleen and said it was "about the bigness of a large bean." Oddly, his French contemporary, M. Charas, knew a snake spleen from a pancreas and compared the later to a "good-sized pea." Leguminous analogies aside, Smith and I found neither organ.

Embedded in the pancreas—but invisible to us—were the tropical-sounding islets of Langerhans, the source of the blood hormones insulin and glucagon. In the liver, glucagon transforms the complicated polysaccharide glycogen to the simple sugar glucose, and insulin controls glucose levels in the blood.

"If the snake was fresher, we'd find a lot more," Smith lamented, as he dug deeper for the remote organs of the digestive system.

"If it came from Carolina Biological Supplies, we'd find far more too," I rejoined, remembering my fruitful undergraduate days in comparative vertebrate anatomy. Of course the snake would be bathed in formaldehyde, our fingers would wrinkle and reek, and my dining room table would be forever unfit to eat on.

Dr. Tyson might have found the true spleen and pancreas had he not gotten distracted "by the ravishing beauty of . . . the kidneys." He saw the timber rattlesnake's kidneys as "so very curiously contrived and with so great beauty, that I want Words to express what the Pencil could not imitate, much less be represented in a Print."

Reptiles were the first Amniota to showcase the sophisticated metanephric kidney, which houses millions of capillaries and purifying tubules that remove and concentrate nitrogenous waste from the blood and then transport it via two conduits (the ureters)—one from each kidney—directly to the cloaca. In mammals, the kidneys are paired, compact, and lima bean shaped; in snakes, like almost everything else, they're long and thin. Unlike mammals, snakes do not pee liquid urea (urine). As a water-saving measure, snakes expel nitrogenous waste as a concentrated, semisolid uric acid. Snakes have no urinary bladder, another streamlining modification. Metanephric kidneys also control blood chemistry by regulating the body's fluid and salt levels, a major improvement from the primitive urinary apparatus of fish and amphibians. Like all snakes, our snake had two kidneys, right and left, a detour on the theme of organ condensation we had encountered in the lung and liver. The kidneys were offset. The right was closer to the heart, the left adjacent to the large intestine. Both were long, browner and lighter in color than the nearby liver, and had twenty-five or thirty lobules that looked like a tipped-over stack of pennies. It was the lobules that had driven Tyson to poetry. He recognized that each kidney subdivided yet retained its integrity. "Tho one continued body yet plainly distinguishable into several lesser kidneys." Each lobule was crammed with uncountable capillaries and tubules bound by a tough kidney casement, a network of microscopic vessels that had swept the rattlesnake's blood clean of impurities, allowing the circulatory, respiratory, digestive, and reproductive systems, and the musculature, to function as an integrated whole.

Smith searched in vain for the adrenal glands, moving everything this way and that. Theoretically, the pinkish adrenals are located between the kidneys and ovaries, which in our snake appeared long and flattened and contained very small, hard follicles that would never produce a next generation. Finally, inserting the probe into the cloaca, Smith pinpointed five tubes — one from the large intestine, one from each ovary, and two from the ureters — that had each unloaded freight into one of the cloaca's three chambers (*cloaca* in Latin means "sewer"). Eventually, he found one of the ureters and the tip of the probe reached a kidney, but the adrenal glands must have withered long ago, because we never did find one. The adrenals are the "stress glands," the fight-or-flight glands that once galvanized the timber rattlesnake into action, either to assume the legendary "threaten" posture — stacked coils, head up, tail a noisy blur — or to vanish into the shadows beneath a rock slab, or to freeze in a weft of spotty sunshine on the forest floor, the dynamic mosaic of reptile hidden in full view, an illusion that appears as evanescent as a bubble.

I imagined our yellow morph stuck in her stone doorway, struggling mightily to escape, adrenaline streaming through her circulatory system, her fat reserves metabolizing until the last spec of fat body eventually blinked out like a dying ember, the snake all the while clamped in place by the unyielding grip of stone, but pushing, pushing, pushing day after day, until too much adrenaline and not enough energy brought an end to her struggle. For Alcott Smith and me, the rattlesnake had become the Missouri River of crotaline anatomy and adaptation, as we traveled like Lewis and Clark upstream and downstream in the company of Dr. Tyson and a half-dozen maps of snake anatomy, through a watershed of organ systems, and in the end, though our specimen was less than satisfactory, we reaffirmed, at least for ourselves, life's splendid solutions to complicated problems.

"I'm amazed we've seen as much as we have," I announced, picking bits of rattlesnake meat from under my fingernails and making sure snake juice hadn't seeped through holes in the porous shower curtain onto the dining room table.

"If Jardine only knew what had become of that snake ..."

Interlude

HIGH SUMMER

ON THE TRAIL WITH
TIMBER RATTLESNAKES

7

The Dangers of Being Male

*Being with a woman all night never hurt no professional baseball player. It's
staying up all night looking for a woman that does him in.*

CASEY STENGEL

Testosterone is a hormone produced mainly in the testes, a mind-numbing internal drug that alters consciousness, skews judgment, and urges men into unguarded realms. When testosterone levels are high, behavior considered normal degenerates into a single-minded quest for sex. It should come as no surprise to anyone who has kept a pet that human males are not alone here. We belong to a long roster of evolutionary cohorts—offhand I can't think of a species in the animal kingdom in which the male doesn't abandon all sense of decorum during the breeding cycle. Think stallion, buck, boar, rooster, alligator, and so forth. During an undeviating search to mate, the male lowers his guard—you might say, "behaves inappropriately"—stops eating, stops sleeping, and often engages in ritualistic combat, all of which collectively deplete vital stores of energy. For the human male, carnal behavior peaks in the twenties and then fades with the passage of time; for an old male timber rattlesnake, however, decrepit is not part of the game plan, and only death trumps the libido, which may percolate for more than forty years. And for the rattlesnake the annual arousal may lead to death, particularly when a road is involved.

Shortly after graduating college, I took a job in the education department at the Bronx Zoo. Six months later, I accepted the zoo's generous offer to enroll in a night course in mammal ecology taught at Hofstra University. The course was so much fun that I joined the American Society of Mammalogists and attended their 1972 annual meeting at Busch Gardens, both on the zoo's nickel.

On my way home from Tampa, I detoured around southern Georgia, hoping to see a large snake. Any large species: indigo, eastern diamondback, coachwhip, yellow rat. I wasn't fussy. After several hours of cruising, I spotted an adult diamondback, gargantua of the rattlesnake clan, emerging from a bouquet of palmettos. I pulled over, got out of the car, and watched the big snake move, slowly and deliberately, in a straight line, like a venomous caterpillar. The diamondback took the shortest route to my side of the road, perpendicular to the flow of traffic had there been a flow. By the time the first car appeared, the snake had nearly reached the far shoulder. Unfortunately, the driver, an ophidiophobe, veered to my side, struck the snake twenty feet in front of me, and then looped back, having ignobly transformed a five-foot rattlesnake with a head the size of a softball into "dead snake crawling," its innards spilling out.

Several years later, now employed at the Cumberland Gap National Historic Site on the border of Virginia, Kentucky, and Tennessee, I was a seasonal naturalist among a staff of history buffs. Here, in rural Appalachia, I met two types of snake people: those, like the Georgia driver, who killed snakes quickly, and those who killed them slowly, collecting venomous snakes for a local (and illegal) Pentecostal snake-handling sect, the very same sect that eventually sponsored the Wolfords, whose deaths by snakebite would one day be covered by the national media: Mack Ray in 1983 and his son Mack Randall in 2012, right here in Middlesboro, Kentucky, gateway to Cumberland Gap. Both types of snake people could be found at the national historic site in Cumberland Gap. One evening, as I followed a park ranger over Pinnacle Road, a male timber rattlesnake on a mating mission slithered out of the woods onto the edge of the road. The ranger stopped, opened his door, shot the snake, and then escorted me to the campground, where he introduced my talk on local wildlife.

The *Wall Street Journal* called biologist David Shepherd's research "the old rubber snake trick": place a fake snake where people are sure to see it and then record what happens when they see it. People in this case were Louisiana motorists, and Shepherd, a professor at Southeastern Louisiana State University with an academic interest in herpetology and Cajun driving habits, wanted to quantify the "intentional kill behavior" of his neighbors. Shepherd also deployed a rubber turtle to see if drivers had a preference for what they hit.

Shepherd and his students placed their toys on the highway, hid in nearby bushes, and then watched cars go by. You can replicate the study yourself; it's not quantum physics. But why bother? Very likely your results won't deviate from Shepherd's, which illuminated two universal truths: first, that snakes evoke a dark pathology in otherwise normal people, and second, that for snakes, crossing a road is risky business. Over the course of three years, twenty-two thousand motorists approached the faux reptiles, which were alternately placed midlane, on the double yellow line, or on the shoulder. Although eighty-seven percent of the drivers tried to avoid contact, of the twelve thousand four hundred vehicles that passed the snake, twelve hundred eighty-one (10.3 percent) hit it; of those, four hundred ninety-eight (nearly forty percent) deviated from their path to make contact, nearly triple the number that deliberately struck the turtle. Some drivers ratcheted "the intentional kill behavior" to another level. Eleven motorists struck the model more than once, including a housewife who drove over it six times. One man hit it, backed up until a rear tire was on the snake's head, and then peeled out. A sheriff struck the snake in the breakdown lane, backed over it, and then got out of his squad car, gun drawn, before Shepherd stepped from the bushes like the host of *Candid Camera* and told the officer about his research. Several drivers, including the operator of a tractor-trailer who crossed to the opposite shoulder, risked head-on collisions to make contact with the model. "Some people," Shepherd concluded, "just have a mean streak."

Building on Shepherd's research, a team from Butler Community College (BCC) in El Dorado, Kansas, set out to determine what was more provocative to a motorist: snake shape or snake color. The students alternately placed a black rubber snake, a neon-blue rubber snake, and a straight section of black hose at various spots on a well-traveled road, parallel to traffic to minimize accidental hits. Like Shepherd, they hid behind roadside bushes. Each trial ran separately and lasted ten minutes or until the target was hit. On the double yellow line, the black model was hit in twenty out of twenty-one trials and the rubber hose in three of twenty trials, and the lapse of time between placement and strike was nearly three times longer for the hose than for the snake—more than twice as many cars had passed the hose before first contact. In each trial, the dayglow-blue snake proved a more attractive target than the hose, nearly as attractive as the more authentic looking black model. When placed midlane, the black snake (the blue one wasn't used for this trial) was by far more popular than the hose. The upshot of the study: no matter what the color, if it's shaped like a snake, it's more likely to be struck by a motorist.

The Jayhawkers then sorted contact vehicles into two categories: truck drivers and non-truck drivers, with pickup trucks included in the latter category. Although eight times as many non-truck drivers used the road, truck drivers made sixty percent of the strikes. The conclusion: with a better sightline than non-truck drivers by virtue of their greater height above the roadway, truck drivers were more likely to run over a snake in the road, whether because they were bored, congenitally cruel, or woefully ignorant, or some combination of the three.

To conclude their project, the team administered a questionnaire to BCC students, which revealed that thirteen percent of male respondents (twenty-five out of one hundred ninety-three) and four percent of female respondents (seven out of one hundred seventy-one) had deliberately hit a snake crossing the road, more than double the number that had reported hitting a snake accidentally.

Not all species of snake cross a road at the same pace. A black racer, for instance, boogies nonstop, whereas the speed of a timber rattlesnake crossing a road is akin to that of leaves unfurling on a twig ... barely noticeable.

Which brings us to the premise of this chapter: Why would a thickset, methodical timber rattlesnake cross a road?

And you already know the answer. Don't you?

A number of years ago, on the way home from a Little League game, I came across a male mink courting, although "stalk" and "rape" are more apropos. He chased the female back and forth in front of my car, up and down a wooded hillside, in and out of a small wetland, around the trunks of stately white pine. Eventually, the female either gave in or gave up, at which point connubial bliss degenerated into the small-mammal version of a World Wrestling Federation Texas Death Match—the male biting, pulling, and scratching; the female screeching and squirming.

When it comes to romance, rattlesnakes set the gold standard. In the Northeast, male timber rattlesnakes reach sexual maturity in four to seven years (farther south this may happen slightly earlier); a few years later, after they've gained more size and some stature and have stored enough fat to fuel long-distance mate searches, a few of these younger, smaller males may enter the breeding pool. Most wait longer. Courtship for timber rattlesnakes is slow and involved—touching, stroking, entwining, tongue-flicking, cloacal brushing—and entails ritualistic jousting if two males are interested in the same female, which apparently happens often. Coitus is protracted—two snakes locked in serpentine bliss—which is why many biologists sedated by boredom, go home for dinner before the snakes have consummated their union. In addition, mating apparently provides so much joy for the snake that males seek, and females accept, multiple partners, but more on this later.

Like all sexually active males, the testes of a male timber rattlesnake produce sperm. Biologists call the process spermatogenesis, the "origin of sperm," which occurs in the spring. Sperm is stored in the vas deferens, the nexus between the testes and the hemipenes, until the summer mating season. Although spermatogenesis may begin as early as four or five, sperm production and breeding are not synonymous. Smaller males may not have the energy reserves to embark on a long-distance search for a mate, and besides what chance would a four-year-old male have against a robust thirty- or forty-year-old male in seducing a female?

According to William S. Brown, "None."

Brown, whose research in the Adirondack foothills entered its thirty-seventh year in 2015, has discovered that most males don't successfully mate until age ten, when they weigh approximately a thousand grams. What happens to younger males? The bigger, older males would bring them to their knees (if they had any).

Search YouTube for "timber rattlesnake combat dance" and you bring up several short clips of male-to-male aggression, a rarely witnessed ritual, the outcome of which governs breeding hierarchy. Jimmy the Greek missed a prescient opportunity here; unlike in sports, the outcome is quite predictable. Big boy wins. Males rise up, wrap once or twice around each other, inflate their bodies, mouths closed, fangs folded safely against their palates. There's *rarely* biting. Now, looped like disembodied arm wrestlers, they push and shove and spray musk. It's a test of strength and endurance and adamantine will; the winner snaps the loser down and down again, until the smaller snake backs off.

A recent study in New York's Hudson Highlands noted a high incidence of chipped and punctured belly scales on the lower two-thirds of large males, the result of shoving each other back and forth across the rocky hillsides. Females, on the other hand, which are coddled and caressed by their suitors, had undamaged ventral scales. In a 1955 issue of *Herpetologica*, a scientist named Shaw offered an alternative interpretation of the combat dance. It's "nothing more than an exhibition of, and a defense against, homosexuality," he claimed. A *gay* rattlesnake ... that's something I hadn't considered.

I have never seen male-to-male combat. Few have. Luck would be required. In more than fifty years of fieldwork, Marty Martin has seen it twice; Bill Brown, never.

On August 4, 2011, at 12:45 p.m., fellow Basher Jed Merrow hiked to a talus den in upstate New York. Merrow got lucky and witnessed the trifecta of timber rattlesnake mating behavior. A three-foot-eight-inch male Merrow called M1 courted a partially shed, nearly three-foot female whose old skin peeled back like a bonnet slipping off her head. Merrow described the event in the journal *Herpetological Review*. Although coquettish, the female aligned her cloaca with his, though he had not yet everted and engorged his hemipenes. He waited for her to complete the shed, to get "pretty" in a snaky sort of way. More than an hour later, a second, smaller male, which Merrow called M2 attracted by wafting pheromones, ascended the slope, which galvanized M1. Wrote Merrow:

> The two males began fighting immediately upon contact. The males repeatedly rose and fell, proceeding progressively further down the slope and eventually out of view (15–20 m). Meanwhile, the female crawled under nearby rocks and completed shedding. Approximately 15 minutes later, M1 returned upslope alone, searched out the female for several minutes, and upon finding her resumed courtship. Twenty minutes later, [the smaller snake] returned upslope, eventually re-encountering the courting pair. The courting male left the female and approached M2, resulting in M2's rapid retreat down the slope. M1 followed for several meters, with no contact between them.

During the dispute, the female moved closer to the den. After M2 had been vanquished, M1 spent twenty minutes searching for the female, and when he found her, he resumed courtship and, with her permission, eventually consummated the afternoon's festivities with intromission, his tumescent hemipenes everted into her cloaca. Twenty minutes later, M2 reappeared and began to make a nuisance of himself. Pertinaciously ignoring his rival, M1 attempted to moor in place; while his consort, apparently wanting no part of the intrusion, slowly dragged him upslope by his penis, spike-edged and swollen, which anchored into her cloacal wall.

When Merrow left at 4:30, the rattlesnakes had been hooked together for nearly an hour. And what became of pesky M2? He had

transformed from competitor to voyeur and hung in the background. As soon as Jed Merrow got off the mountain he called his wife. Then, he called Bill Brown, the local repository for most everything timber rattlesnake.

Recently, I spoke to Marty Martin to ask about his experiences with male timbers. Martin told me that he had once elicited a challenge at a Virginia gestating and shedding site. Martin had squatted down, leaned on an elbow, his forearm raised in the air, and made a snake head of his fist. When a large male immediately took notice, Martin quickly reassembled himself. On another occasion, after he had marked an estrous female and had become redolent with pheromones, a nearby male abandoned the snake he was courting and began to follow Martin. "Males of any species lose their heads during mating. [It wasn't] my looks." As you my recall, though their orbs are iridescently stunning, timber rattlesnakes are nearsighted.

Bob Aldridge, a tenured biologist at St. Louis University, forged an academic career of studying the sex lives of reptiles and amphibians. Over the past four decades, Aldridge has authored and coauthored myriad technical papers with titles including "Premature Sperm Ejaculation in Captive African Brown House Snake (*Lamprophis fuliginosus*)," "Reproductive Female Common Watersnake (*Nerodia sipedon*) Are Not Anorexic in the Wild," and "Controversial Snake Relationships Supported by Reproductive Anatomy." Aldridge is the world authority on a reptilian novelty known as the *sexual segment of the kidney* (SSK) and was the lead author on a paper cowritten with William S. Brown titled "Male Reproductive Cycle, Age at Maturity, and Cost of Reproduction in the Timber Rattlesnake (*Crotalus horridus*)." Reading it, I came across numerous references to the SSK. As I previously mentioned, like most everything inside a snake, the kidneys are long and thin, and from Aldridge's description, the SSK appeared to be a swelling on the distal third. I had never heard of the SSK, so I called Brown and asked him.

"Bill, I don't know what the SSK does."

Bill didn't know either, but suggested that I call Aldridge, who affably fielded my questions with rabbinical demeanor. He told me that the SSK is the end of the male's nephron (females don't have the SSK), which in other organisms acts as the kidney, but in snakes and lizards gets very thick and becomes secretory. SSK products are transferred to the female during mating. In other words, as a prelude to breeding, the distal end of the male timber rattlesnake's kidney swells with fluid that eventually contributes to the ejaculate, becomes part of the snake jizz. During the breeding season, said

Aldridge, the size of the SSK is equal to that of the testis. The fluid discharged from the SSK does not aid fertility, nor does it aid the sperm.

"What does it aid?" I asked.

"It has to do with the mating aspect, not the reproductive aspect."

In many species of snakes, said Aldridge, the products of the SSK form a *copulatory plug*, which keeps sperm from leaking out of the female's oviduct and prevents the female from easily mating again (at least for a while). From an evolutionary perspective, continued Aldridge, "If it wasn't important to male snakes, it would have been lost over time."

In the Northeast, the usual mating season for timber rattlesnakes extends from mid-July to late September. Females often have multiple paternities—litters with genetic contributions from two or more fathers—and males mate as often as they can (surprise, surprise). What then is the purpose of a copulatory plug? The plug may last only a couple of days. "Whatever its function," said Aldridge, "it is probably limited over time. When a female snake is in estrus, it's episodic, not continuous." During peak estrus, she produces more pheromones to lure roaming males; when she's off-peak she produces less, and males can't find her as easily.

I ran Aldridge's response by Brown, who replied. "Females do have their limits."

Aldridge thought Brown had described the copulatory plug of a timber rattlesnake in a technical paper. If he had, I must have missed it.

"No, I didn't describe such a plug. Although, I greatly appreciate the unintended accolade for my heretofore-unheralded accomplishment," Brown answered in an e-mail. "I wish all scientific discoveries were as easy as that." Nonetheless, no one has ever seen a copulatory plug in a timber rattlesnake. Which left me where I had started: What does the SSK do for a male timber rattlesnake? Who knows?

Like everything else in a timber rattlesnake's life, body temperature initiates the male reproductive cycle, which triggers the production of androgens, which in turn trigger the development of sperm and the buildup of fluid in the SSK. In midsummer, when the mating

urge strikes, the male stops foraging and begins orienteering, the matrix of a genetically healthy population of timber rattlesnakes. He sets a line-straight course, which biologists believe is the surest way to intersect the pheromone trail of a receptive female. The relentless search has begun.

Why don't timber rattlesnakes mate in spring when the crowd gathers around the den? Garter snakes and water snakes do. Wouldn't it be convenient? No arduous cross-country treks ... more opportunity to mate. I've thought about this. There has to be an evolutionary benefit to summer mating, when the population is scattered across the landscape. And I've come up with a number of possible explanations, all related, for why spring mating might *not* be advantageous. First, after six or seven months of hibernation, both males and females have depleted energy reserves. With so many aroused males in a small area vying for the females, mayhem would ensue, further draining energy reserves. A receptive female would know no peace; males would be coming at her from all sides, a serpentine battle royal. In addition, the long-distance mating trek favors larger, hardier males, with ample stores of fat, just the sort of partner a female would need to help provision healthy offspring. And, perhaps the most likely possibility, there could be far too many intraden trysts. Each den would become an island, limiting genetic diversity. Inbreeding would be unavoidable, and the variation in the relative frequency of different genotypes, owing to the chance disappearance of particular genes, would cause a surging phenomenon know as genetic drift. The disappearance of rare alleles (an allele is one of two or more alternative forms of a gene that have arisen by mutation and are found at the same location on a chromosome) makes an isolated population less fit, and the residents of that den slide toward genetic catastrophe. Health is all about diversity, genetic or otherwise.

Forced by circumstance, in an inbred population of rattlesnakes, patriarchs would inseminate their own ancestors, descendants, and siblings. Fathers would mate with grandmothers, mothers, aunts, daughters, and sisters, and matriarchs would bear offspring of sons, uncles, cousins, grandfathers, brothers, and so forth. Like Andy's Den, the lone den in New Hampshire, each den would become genetically compromised, a prescription for extinction. Fortunately, evolution rigged the male timber rattlesnakes to wander.

That snakes have a forked tongue (no secret here) does not mean they're duplicitous. Just the opposite. A male timber rattlesnake seeks with his tongue, which, like our ears, translates two threads of information, one from each tine. He then analyzes the chemosensory data in an organ on the roof of the mouth and orients himself based on the translation. Which is to say, a snake reads the world with its tongue and then triangulates based on the relative differences in the chemical signals picked up by each tine. Snakes can't get lost ... ever.

Aristotle thought that a snake's bifurcated tongue enhanced taste and provided twice the pleasure, "their gustatory sensation being as it were doubled." By the seventeenth century, an Italian anatomist named Hodierna had taken the nature of the forked tongue into an intellectual back alley. Hodierna claimed a snake picked its nose with its forked tongue, one tine per nostril. In 1811, Ludwig Levin Jacobson, a comparative anatomist, discovered a tiny organ on the roof of some vertebrates' mouths, the vomeronasal organ, now more widely known as "Jacobson's organ," which is present in a variety of species, but reaches a developmental apogee in animals with forked tongues: all snakes and some related lizards. (You remembered that, right?) A flicking tongue collects odiferous molecules that adhere to a veneer of salivary fluids that coat each tine. When a timber rattlesnake withdraws its tongue through a cleft in the upper lip, the chemical message from each tine is transferred to two pads on the floor of the mouth, and these pads then convey it to the Jacobson's organ, two tiny pores densely packed with sensitive olfactory cells. The tongue does not actually deliver information directly to the Jacobson's organ.

In 2011, Kurt Schwenk, a herpetologist at the University of Connecticut with an interest in biomechanics, wanted to know how each flick draws odor molecules out of the air. Schwenk suspected that

the rapid, oscillating tines somehow increased the speed with which the airborne molecules reached the tongue. Together with one of his graduate students, Bill Ryerson, Schwenk released a snake into a cloud of very fine, powdery cornstarch suspended in the air and illuminated by a sheet of laser light. They videoed the results. Schwenk and Ryerson discovered that the up-and-down motion of a flicking tongue creates two pairs of vortices, each one a little donut of fast-moving air that rotates in a circle. The donut pairs spin in opposite directions, and each tine passing along a donut edge moves against the direction of rotation. The opposing rotation creates a counter-current exchange system, which vastly increases both the quantity and rate of cornstarch that collects on the wet surface of each tine.

"My brain nearly exploded," wrote Schwenk, "when I saw the preliminary results. While I expected some kind of air moving, the extremely organized pattern of airflow came as a real shock to me."

Recently, I was clearing limbs off a dirt road in Virginia when I came across a five-foot female black rat snake. In fact, I had mistaken the snake for a charred branch as it soaked up the morning sun. Ten minutes later, I caught a second, longer snake in exactly the same place, a male. This snake was on a quest; slowly gliding across the road, tongue flicking, reading the breeze.

I put both rat snakes on the trunk of a black locust, the female first, and they flowed up the tree like antigravity candle wax, black wavy lines, parallel and progressing, wedged in furrows of bark, upward beyond my view. Rat snakes are quintessential arboreal predators. They feast on mammals and birds and bird eggs. To discourage nest-raiding rat snakes, the red-cockaded woodpecker, a colonial nester in southern pinewoods, nests only in living trees. The woodpecker drills a series of shallow holes below its nest, and the skirt of sap that oozes from the holes arrests the snakes.

The rat snakes eventually gathered themselves along a horizontal limb thirty feet above my head, braided together like an exquisite pigtail, tail tips rhythmically gyrating, and consummated their coming together.

To recapitulate: When a timber rattlesnake flicks his tongue, the two tines oscillate in opposite directions, creating two pairs of swirling vortices that draw chemical clues onto the tongue tips. The tips rotate so fast we can't see them move. All we see is the tongue, arched, and seemingly suspended in midflick like those of David Shepherd's rubber snakes. For a screen saver on my home computer, I have a headshot of a Georgia timber rattlesnake taken by my son Jordan. The image is tack sharp, with the tongue frozen midflick. The snake is reading the wind, a deciphering of the signals carried on the wind, with the snake generating its own wind, unscrambling for his own enlightenment what for us is a long-forgotten language, an invisible, but exquisite message written in lipids.

After decades of marking and releasing rattlesnakes, Bill Brown believes that less than one percent of snakes switch dens (fewer than one per generation). Natal philopatry is the social lynchpin; no matter where in relationship to a neighboring den a pregnant female chooses to gestate, she'll eventually move closer to her own natal den to give birth; then, as days shorten, the tongue-sensitive snakelets follow their mother (or another adult) home for the winter, adapting her den as their own, a binding social contract based on matriarchal den loyalty. This bears repeating (again and again): it is the philandering males that keep the snake den from genetically stagnating; reproductive articulation keeps the dens genetically vigorous. A recent study suggests that in undisturbed habitat, males from dens separated by up to five miles find and mate with females from neighboring dens and that one-third of all paternities can be traced to fathers from another den. One of Brown's scale-clipped males appeared DOR three and a half miles from its den, a record distance for a snake in his study population.

Historically, amorous male rattlesnakes routinely swam Lake Champlain between the foothills of the Adirondacks and the Taconic Mountains, as well as between islands in Lake Erie and across the widest of rivers—the Mississippi. They crossed marshes, peatlands, coastal prairies and sand dunes, primeval forests of innumerable types, and the rocky escarpments of the Appalachian Mountains. Except for a hungry red-tailed hawk or a great horned owl, both birds cued to movement, almost nothing interrupted the male's annual mission of lust, single-minded and heedless. As we've already learned, south of the White Mountains, the Merrimack River valley of New Hampshire was once the epicenter of a robust metapopulation of timber rattlesnakes, and, as you will soon find out, central Connecticut, hard by Hartford, remarkably still supports a small population. Outbreeding, an antidote against reproductive isolation and eventual extinction, is the key to genetic diversity, which itself is the key to population fitness. Wandering sperm-donor males are the hedge against life's vicissitudes, which for a timber rattlesnake include climate change and the keratin-eating fungus, *Ophidiomyces ophidiicola*.

Think of a snake den as a seasonal high-rise apartment, from which all residents leave for work in spring and return in autumn. Midway through their time away from home, some portion of adult females and nearly every adult male that isn't malnourished are cued to mate. Females convene at a couple of basking sites, analogous to ancestral singles' bars. Here, they discard their old skins and perfume the air and the ground, broadcasting their availability. Males, with their responsive, bifurcated tongues, figure out how and where to find them. A small percentage of adventurous males, typically the oldest, largest males, which Randy Stechert calls patriarchs (every robust population has them), visit distant bars at the outer edges of their territories and engage with new females, that is, not from their own apartment. Seed is spread, and eventually males return home to their natal den for the winter.

The ancestral breeding system has worked for timber rattlesnakes for several million years. Even the Pleistocene epoch, during which the Northeast was scrubbed bare four times, didn't eliminate the genetic integrity of timber rattlesnakes. Glaciers obliterated dens and forced rattlesnakes into climatically tolerable refuges in the Southeast, including on the continental shelf off the current shoreline of the Carolinas and Georgia, when sea level was several hundred feet lower than it is today. Wedges of inhospitable terrain separated the surviving metapopulations (just like today), but dens within the same genetic neighborhoods remained linked; the boys still found the girls.

In 1967, Robert MacArthur and Edward O. Wilson published a seminal book titled *The Theory of Island Biogeography*, which David Quammen referred to as "a dense little volume ... daring, fruitful, and provocative." MacArthur and Wilson mathematically presented the pattern of species distribution on offshore islands as the species-area equation $S = cA^z$, which Quammen considers "almost as familiar to an island biogeographer as $E = mc^2$ is to a physicist." S stands for the number of species. A is the area of the island, c is a constant that varies with the taxa (groups of related animals or plants, e.g., palm trees, ants, beetles, rattlesnakes, etc.), and z is the slope of the species-area relationship in log-log space, a mathematical constant beyond my understanding. The gist of the theory is this: the diversity of species on an island depends on the size and shape of the island and its distance from the mainland, which is the source of immigrant species, a counterbalance for extinction. A big island close to shore supports a greater diversity of life than a big island farther from shore and greater too than that of a small island close to shore.

Jared Diamond took the essence of MacArthur and Wilson and wrote the equally important "The Island Theory Dilemma: Lessons of Modern Biogeographic Studies for the Design of Natural Reserves," which appeared in *Biological Conservation* in 1975. Wrote Diamond:

A system of natural reserves, each surrounded by altered habitat, resembles a system of islands from the point of view of species restricted to natural habitats ...

The number of species that a reserve can hold at equilibrium is a function of its area and its isolation. Larger reserves, and reserves located close to other reserves, can hold more species ...

Different species require different minimum areas to have a reasonable chance of survival.

In other words, in order to survive on an island in a vicariant sea a population of grizzly bears and short-tailed shrews would have vastly different requirements. The amount of unbroken landscape required for a healthy population of bears is astronomically larger than that required by shrews. Big animals (elk and bison) and animals on top of the food pyramid (harpy eagles, jaguars, timber rattlesnakes) would become locally extinct before animals on lower trophic levels (white-footed mice, gray squirrels, garter snakes), because big animals require broad swaths of wildlands, untainted by development or agriculture. And, because of its much slower metabolism, a timber rattlesnake requires far less food, and thus, all else equal, likely far less area than mammalian or avian carnivore of equal mass, say a mink or a goshawk. To paraphrase Diamond, the bigger the park, the better; small, clustered parks are better than small, scattered parks; and corridors connecting undeveloped land are better than preservation of isolated parcels. Success is achieved by the amount and variety of genetically healthy species that survive on these protected lands.

Let's get back to timber rattlesnakes: In two articles, Rulon Clark, William S. Brown, Randy Stechert, and the geneticist Kelly Zamudio looked at nineteen different timber rattlesnake dens in eastern New York State through the lens of a new discipline called *landscape genetics*, which takes the work of MacArthur, Wilson, and Diamond to the next level. Landscape genetics fuses landscape ecology and population genetics to determine the future of a particular species in a particular region. "Wearing the snakes' shoes," Clark and company looked at how geographic and environmental features affect genetic variation within populations of rattlesnakes. What, besides food, water, and a safe haven below the frost line, does a timber rattlesnake require? The answer: a network of genetically articulated dens. (You knew this, didn't you?) Between adjacent dens, a corridor of habitat punctuated by basking rocks (the singles' bar) is good, whereas a road separating adjacent dens is bad, very bad. Even the smallest road may deprive males of their access to nearby dens, thus creating rattlesnake "islands."

Although they may occasionally move more quickly, Marty Martin told me that most timber rattlesnakes take half an hour to cross a hard-surface road in Shenandoah National Park. "How many park service roads don't have a car cross every fifteen minutes?" In Vermont, a north-south state road cuts through prime rattlesnake habitat a quarter mile west of four dens. More than fifty-two hundred cars pass the dens each day. Even though the metapopulation is healthy, snakes no longer disperse westward. They head east, up and over the ridgeline. Alcott Smith, who twenty years ago was still transporting snakes from yards and farmhouses on the west side of the road to the east side, reported that in the fifties and sixties, when traffic was much lighter, rattlesnakes routinely dispersed west. As traffic increased, however, the number of westward-heading snakes decreased. In the past ten years, there have been only six mortalities on the road. Without pheromone trails leading west, the snakes have all but abandoned the old dispersal route; young snakes follow in the paths of their elders.

Bob Fritsch, a retired policeman from Wethersfield, Connecticut (and an original Basher), who has studied and protected central Connecticut rattlesnakes for forty years, has been bitten twice in the line of duty, featured in articles in *Yankee*, the *New York Times*, and the *Hartford Courant*, and remains invariantly devoted to the cause of rattlesnake protection, now focuses on poaching and wanton snake killing, particularly on roads in Connecticut, even though he moved to Maine several years ago. Route 2 directs traffic in and out of Hartford, cleaving the central Connecticut population of timber rattlesnakes in two. I've driven along Route 2, and it's a tremendous deterrent for a snake, just as it is for me. During the middle of the day, when traffic is thin, the road becomes grill-top hot, much too hot for a snake. During rush hour, when the temperature of the highway is ideal for a snake belly, traffic is heavy and pounding trucks and cladding engine fluids make Route 2 just as unforgiving. Rattlesnakes know this. Males often enter the tall grass along the edge of the road, but they don't attempt to cross. The last time Fritsch picked a DOR off Route 2, an adult male, which he sent to Bill Brown, was in 1986.

"During peak times, small country lanes are a bigger death dealer than the big, black, rumbling, smelly limited-access highways."

In 2013, Fritsch picked up nine DOR Connecticut snakes on back roads.

"Some people hit them deliberately. Those back lanes handle a lot of commuter–housewife–school kid traffic. Much of it happens when [the snakes] are looking for sex."

Landscape modifications that don't destroy a large amount of habitat can still impede gene flow. A road, interrupting those straight-line, testosterone-fueled questings of the male snakes, restricts the gene flow between neighboring dens, both by deterring the snake from his single-mined mission and, in many cases, by actually killing him, which has already left many escarpments snakeless. From Clark et al.,

> The effects of roads on connectivity will result in the eventual isolation of subpopulations, with negative genetic consequences that may only become evident after a number of generations. During the short time paved roads have been in place, it is alarming that even smaller roads in our study have already had a detectable effect on the genetic connectivity of a long-lived, late-maturing vertebrate species.

A 2015 multiauthored paper in the journal *Herpetologica* divided New Jersey's Pine Barrens timber rattlesnakes into six discrete populations, subgroups each with reduced genetic diversity. The six populations were isolated from each other and without a guiding hand each would continue to slip-slide toward oblivion. None, however, were cutoff by geography alone. The principal barrier to gene flow between the subgroups? Roads, roads, roads. The principal victims? Randy males.

In 2005, the journal *Copeia* published Kimberley Andrews and J. Whitfield Gibbons's article "How Do Highways Influence Snake Movement? Behavioral Responses to Roads and Vehicles." Unlike David Shepherd, Andrews and Gibbons did *not* use rubber models. Their trials included nine different species of live snakes on a two-lane asphalt road that was closed to traffic. They checked the snakes' responses both to the road and to a moving Chevrolet Silverado 1500 pickup, without endangering the snakes. Like Route 2 in central Connecticut, timber rattlesnakes tended to avoid the road—over seventy percent made no attempt to cross. The few individuals that did moved very slowly and then froze in response to the pickup, fifty percent of them before it passed and fifty percent as it was passing. Once the Silverado had passed, seventy percent of the stationary rattlesnakes resumed crossing; the other thirty percent stayed put for a minute or more. Andrews and Gibbons estimated that the slow-moving, freeze-oriented timber rattlesnake populations would suffer eighty percent mortality attempting to cross a road with a traffic density of two thousand cars per day and one hundred percent mortality at nine thousand cars per day. The authors considered a road that hosted between five thousand to ten thousand vehicles per day medium traffic density.

When Brown and Aldridge published their article on the male reproductive cycle, which mentioned the fascinating and mysterious accessory organ, the SSK, all twenty-six necropsied specimens had been questing males: sixteen were either DOR or maliciously killed by humans in the foothills of the Adirondacks; two others were peeled off Route 2 in central Connecticut and donated by Bob Fritsch; the remaining eight were accidental processing mortalities, a regrettable price of field research, and a compulsory obstacle for an aroused male rattlesnake that happens to live near a university.

8

The Dangers of
Being Female

*After all, being predictable ... makes us easily managed by others who know
how we'll react.*

KATHLEEN REARDON

On June 10, 1995, Rudolf Komarek, who is thought of among the academic rattlesnake cabal as snake hunting's antichrist and is known simply as Rudy—a one-word sobriquet that for herpetologists is the embodiment of infamy—led a camera crew across an oak-hickory ridge west of the Susquehanna River near Shippensburg, Pennsylvania. It was opening day of rattlesnake season and the crew was there to film Komarek as he captured two gravids, for a pair of CNN documentaries. At that time, the bag limit in Pennsylvania was one snake per day. The following March, immediately after the first segment aired nationally on the weekly environmental program *Network Earth*, Komarek was arrested for exceeding the state's bag limit, his seventh bust in Pennsylvania, which resulted in a fine and the loss of his snake-hunting license, again. Said Komarek: "The arresting officer's comment indicating I have a disregard for the laws protecting timber rattlesnakes is quite factual. These laws are ridiculous.... [Pennsylvania Fish and Boat Commission] was mad because they weren't in my CNN thing. That's not my fault, but now they want a piece of the action. They told me I was just too popular."

Of Czech descent, Komarek was born in 1929, grew up in Little Ferry, New Jersey, and died seventy-nine years later in Daytona

Beach, Florida. In between, he never married, never had children, never owned a home or a telephone, never held a long-term job. An itinerant carny and dishwasher, Komarek drifted from state to state, town to town, restaurant to restaurant. His passion for snakes developed at an early age, and he began snake hunting in the late thirties, mostly around his home, where he kept a menagerie of fifty nonvenomous snakes in the garage. Unbeknownst to his parents, Rudy and his brother cached copperheads and timber rattlers in plastic bags under the eaves of the house. Although he never took field notes, Komarek claimed to know the location of three hundred rattlesnake dens and to have taken nine thousand rattlesnakes from New York, Massachusetts, Connecticut, Vermont, West Virginia, and New Jersey while attempting to actualize a lifelong goal of starring in a porn film with a group of snake-eating lesbians.

Rudy Komarek stood five foot five, muscular and rough-hewn, a sort of Napoleon with a snake hook. He had no formal education. His face, tight like a clenched fist, featured an aquiline nose, deep-set, narrow eyes, lunular eyebrows, shallow forehead, and a shock of thick Ronald Reagan hair that began to gray and recede only near the end of his life. Komarek had been bitten seventeen times by venomous snakes. In 1982, an Asiatic cobra had him down for the count in a Bronx hospital, where he was mistakenly pronounced brain dead. When doctors discussed harvesting his organs, recalled the late John Behler, curator of reptiles at the Bronx Zoo, who had been called in to advise the medical staff, "right over his naked, paralyzed body," a single life-saving tear rolled down his cheek. The near death of Rudy Komarek had been an opportunity lost, lamented Bill Brown.

Komarek billed himself as the Cobra King and spent much of his life in search of the Cobra Queen. Along the way, he collected rattlesnakes for bounties, delivering severed tails to participating town clerks and then selling hides to leather merchants, and the meat to food brokers in Chinatown. He also sold live rattlesnakes to Pentecostal snake handlers, to roadside zoos, and for educational displays at summer camps, where they were kept in inadequate pits and died en masse, which gave Komarek a rapid turnover on his investment. In 1968, Komarek brought the tails of one hundred twenty-one rattlesnakes pilfered from the Hudson Highlands to a town clerk in northern New York, two hundred miles away, where he hoped to bilk Washington County out of the five-dollars-per-snake bounty payment. The snakes had been stomped to death: neonates, post-partums, and gravids from which he cut out the embryos, docking each of their rubbery baby tails. When asked by a local snake hunter where he had caught the snakes Komarek pointed to a mountain known to be snakeless. His claim was denied.

By the mid-seventies and early eighties, when timber rattlesnakes began to be listed by a number of states as either threatened or endangered, Komarek forged a lucrative poaching business. (Colorful and boldly patterned northeastern timber rattlesnakes still sell for more than a hundred dollars on the international black market.) That singular trait that had made them susceptible to bounty hunters, also made them susceptible to poachers: predictability.

A pioneer in Internet marketing, Komarek developed a website, in 1998, Cobraking.com, from which he sold maps to snake dens (forty dollars a map), offered private rattlesnake-collecting trips (prices started at five thousand dollars), tracked UFO sightings, discussed government cover-ups, and posted pictures of ex-girlfriends, some dating back more than forty years. He also sold Cobra King T-shirts and ran a personal ad that read:

> young, slender female who enjoys travel, adventure, hiking, and mountains. Must be unattached. LONG-TERM COMMITMENT FOR RIGHT PERSON.

In addition to the CNN documentaries, Rudy Komarek appeared on *The Tonight Show* with Johnny Carson and *What's My Line*, performed at the Steel Pier in Atlantic City and various international sport and camping shows, and was featured on the front page of the *Wall Street Journal* and in the *New York Times*. He also appeared on wanted posters in New York, where he skipped bail and fled the state after being indicted for poaching in Bear Mountain–Harriman State Park. Besides having his snake-collecting license revoked seven times in Pennsylvania, Komarek's rapacious appetite for poaching

timber rattlesnakes led to his deportation from Kansas, numerous stints in jail and numerous fines, including four months in a federal prison and a twenty-five-hundred-dollar fine for trafficking timber rattlesnakes. He was also arrested for possession of illegal weapons in connection with the placement of baby rattlesnakes in the home of a woman who was at odds with a friend of his. The woman survived, the snakes didn't, and Komarek spent three months in jail.

Rudy Komarek delighted in poaching snakes from on-going research sites, where he moved basking rocks to disrupt and destroy habitat, and then would leave a note to the incumbent biologist tacked to a tree. His principal target was William S. Brown, whom he harried for decades, once sending the Skidmore College herpetologist a photograph of two boxes of captive timber rattlesnakes. A handwritten note on the reverse side of the picture read:

> Bill, We heard about your problem being bitten in the leg the SECOND time, so we ellimated [*sic*] the problem! In the future leave the collecting of poisonus [*sic*] snakes to the experts.
>
> Speaking of the future, we will be collecting many poisonus [*sic*] snakes in the future as there's much money to be made! In case you're wondering, ther [*sic*] are 62 snakes in the boxes (2 days hunting).

In retaliation, Brown distributed "wanted" posters of Komarek across the Northeast, gave talks, including one in East Africa, about Komarek's illegal activities, and with Randy Stechert cowrote an article about the poacher that appeared in the *Bulletin of the Chicago Herpetological Society* illustrated with a caricature of Rudy in prison stripes. Brown's most influential publication, the monograph "Biology, Status, and Management of the Timber Rattlesnake (*Crotalus horridus*): A Guide for Conservation," which ran to seventy-seven pages and was underwritten and published by the Society for the Study of Amphibians and Reptiles, has been derisively called "The Biology, Status, and Management of Rudy Komarek," by a well-known North Carolina snake breeder.

Near the end of his life, Komarek trolled for young women in local diners around Daytona Beach. He still preferred twenty- and thirty-year-olds, "sometimes younger," reported the *Daytona Beach News Journal*. "It depends on the individual. I have an open mind." He told the newspaper he had been "mysteriously fired" from his job as a dishwasher at a popular local restaurant for grinding waitresses, which helped to dispel the mystery.

Map sales from Cobraking.com lured so many poachers into northwestern Connecticut that private landowners were forced to post their property. One enthusiast from Massachusetts bought a map and paid for a weekend visit by Komarek. After a rewarding day in the field, the man asked him rather earnestly, "What else can I do for you?" When Rudy answered, "Give me your daughter," the relationship carbonized. Near the end of his life, Komarek threatened to write a field guide to dens and basking sites in New York and adjacent states, a collector's dream come true. Fortunately, on a muggy Florida morning in May of 2008, Rudy's heart gave out before the first word of the guide had been written. Quipped Randy Stechert, who had a *very* long history with Komarek, "He's finally productive now as carbon storage."

Although timber rattlesnakes eat and breed in captivity, keeping one in a terrarium would be like trimming the Mona Lisa to fit the frame or Willie Mays playing centerfield in a straightjacket.

Rattlesnake Economics

In 1991, after combing the Adirondack foothills for thirteen years, catching and processing timber rattlesnakes while simultaneously attempting to minimize the depredations of Rudy Komarek, William S. Brown published an article in the journal *Herpetologica* titled "Female Reproductive Ecology in a Northern Population of the Timber Rattlesnake, *Crotalus horridus*," which placed the timber rattlesnake with other slow-growing, long-lived, late-reproducing, small-brooded beasts such as the humpback whale, the great white shark, the bald eagle ... and, of course, us. Now well into his fourth decade studying the same metapopulation, Brown's gorgeous color-coded data continue to morph into knowledge. The depth of his insight into the nature of timber rattlesnakes has forever altered our own view of the nature of snakes in general.

In northeastern New York, Brown's research has shown that more than half of all female timber rattlesnakes breed for the first time at age eight or nine (many at eleven or twelve), none before the age of six, and that nearly sixty percent of adult females breed every third year, the remaining forty percent every fourth, fifth, or sixth year. To say that the species is an infrequent breeder (more so along the northern frontier of their range and on the High Allegheny Plateau of West Virginia than in southern and western portions, where a longer growing season and more abundant prey support breeding every second or third year) is a colossal understatement, akin to saying a cheetah can run. As lead author, Rulon Clark demonstrated that questing males are the lynchpin for the genetic well-being of a timber rattlesnake den colony, *but* it is the female's late age of first reproduction, low frequency of birthing, and small number of young per clutch—averaging nine snakelets—that form the holy trinity of crotaline demography, the joists supporting a den's population structure. To evaluate the population dynamics of any species anywhere, a biologist considers two parameters: recruitment and mortality, which for the timber rattlesnake means birth and death. The number of females of reproductive age is the fulcrum of a den's stability, and when someone like Rudy Komarek avariciously poaches gravid rattlesnakes, the population structure of the den begins to implode, which galvanizes biologists like Brown and Stechert to watch over breeding females like a lioness over her cubs.

During more than fifty years of fieldwork in the central Appalachian Mountains, a hotbed of timber rattlesnakes, W. H. Martin, who at seventy-four has visited more dens and more birthing sites than anybody else, states that in average years a den of one hundred snakes older than young-of-the-year supports *only* four reproducing females. Four, as in one ... two ... three ... four. During a really *big* birthing year, what Martin calls a "gravid pileup," the number of birthing females may jump to twelve, but typically fluctuates between two and seven per year. Since births are the counterbalance for mortality, poaching gravids stifles recruitment. Unlike mammals and birds, which have the ability to relocate from outlying areas to fill a demographic vacuum, timber rattlesnakes are homebodies; emigration isn't a factor in their population dynamics, at least not in the short term, and this is even truer in the twenty-first century, since roads have become genetic nooses.

"The cause of the collapse of this resource," wrote Martin, "is the high exploitation of reproductive females during the summer months," depredations that have slowed but not stopped since Komarek's death. On November 14, 2012, the Justice Department convicted Robroy MacInnes and television personality Robert "Robbie" Keszey, a cohost of Discovery Channel's *Swamp Brothers*, and their business, Glades Herp Farm, in Bushnell, Florida, for conspiracy to traffic in endangered and threatened reptiles and for trafficking timber rattlesnakes in violation of the Lacey Act, which bans interstate commerce of illegally obtained wildlife. MacInnes and Keszey had collected protected rattlesnakes in Pennsylvania and New Jersey and had purchased wild-caught New York rattlesnakes, shipping them to their Florida business, where, according to federal court papers, a single timber rattlesnake might fetch hundreds of dollars at reptile shows in Europe. Snakes not sold overseas were offered for domestic sale at Glades Herp Farm. (On December 8, 2014, Keszey was sentenced to a year in federal prison and fined two thousand dollars; MacInnes received eighteen months and a fine of four thousand dollars.)

The proportion of a den's reproductive-aged females that breed each year varies between zero and forty-five percent, depending on both weather and prey availability, which itself varies in accordance with cycles of mast production. On rare occasions, writes Martin, during a "perfect storm year" the percentage of breeding females may approach ninety percent, the so-called gravid pileup. Because a male reaches sexual maturity several years before a female, the sex ratio of adult timber rattlesnakes may skew toward males at a ratio of approximately 5:3. (Since most of the younger adult males may be excluded from breeding, the actual proportions may be roughly equal.) In other words, in most years, in a population of a hundred snakes older than young-of-the-year, where four gravids represent (for mathematical ease) twenty-five percent of the number of adult females, then the total number of breeding-age females would be sixteen. Using Martin's ratio of five adult males for every three adult females, in an average summer roughly twenty-seven questing males will attempt to satisfy the needs of four females, which boils down to approximately seven males for every receptive female, ironically rendering all but the biggest or luckiest males superfluous. (More than one male may service a single female, resulting in neonates of mixed parentage.)

According to Martin, in a *very* large, very remote den complex, which may shelter four hundred to seven hundred snakes, approximately one-third of the snakes are in the reproductive pool, of which sixty-four are adult females. Like an economist prognosticating commodity futures, Martin classifies each year's reproductive potential by the number of gravids per den. One of the dens Alcott and I monitor, he says, in terms of annual birthing potential, twenty-eight to thirty-six gravids forecasts a high, twelve to sixteen a moderate, and four a low year. Based on the "Martin system," 2013 had been a moderate birth year for timber rattlesnakes at our site. Alcott and I logged our highest gravid count, thirteen, on a rainy morning in early July, both the date and weather conditions considered by Brown and Martin ideal for a gravid census, because snakes are prone to lounge in the open. Once the sky dried out, all afternoon rattle-snakes tracked the sun, loop by loop, snake by snake, either knotted together or stretched alone. If we missed any snakes that day, it was *only* because we missed a birthing site. On all of our other visits that summer, we routinely missed a few snakes that had tucked themselves deep in the rocks to avoid the scorching heat.

Like males, nongravid females forage away from the den (although not as far away), where they wander through miles of leaf litter and outcrop sedge, stopping here and there to ambush rodents, on their active-season circuit away from the den and then back again. In the Northeast, the gap of three to four years (sometimes as many as six) between breeding events for an adult female is the time required to replenish her principal investment, measured in the deposition of expendable bodily resources (stored fat and protein) with which to provision developing embryos.

A female timber rattlesnake maintains a strict energy budget, and the cost associated with reproduction is high. As a rule, smaller mothers give birth to fewer young than bigger mothers. Since a gravid snake rarely has the opportunity to feed, replenishing depleted reserves during a two-and-a-half- to four-month gestation isn't an option. To breed, an adult female must have an adequate reserve of fat and protein, that is, she must have invested wisely by catching enough rodents during the preceding three to six summers. A geographically synchronized recession in the production of acorns, which happens periodically, is usually followed by a marked reduction in the number of mast-dependent rodents, which in turn is followed by a marked reduction in the number of breeding female rattlesnakes, an example of natural history dominos where one of the last tiles to fall is the umbrella predator, the timber rattlesnake, which results in a *very* low recruitment year. In fact, during the summer after a very poor acorn crop, when white-footed mice and chipmunks are in short supply, not only do females skip the breeding cycle, capital-depleted males may forgo searching for them.

To conserve energy, a gravid rattlesnake doesn't move around too much. She basks, basks, basks, shortening gestation and enhancing embryonic development. An incubating female needs to maintain her body temperature between eighty-five and ninety degrees Fahrenheit for the equivalent of seven hours a day for ninety days. If the day is hot, the sky cloudless, she basks early in morning and again late in the afternoon; the rest of the day she remains in the shade, tucked into the shelter rocks. If the day is cool and overcast, as it was that early July day in 2013, she may bask in the open all day, with raindrops beading on her scales to form a thousand aqueous prisms. When the angle of midsummer sun becomes shallower with the approach of autumn, neonates are born in ancestral rock nurseries. If July and August have been wet, births may occur as late as early October. Having fasted for perhaps an entire year, a postpartum female may have lost half her body weight constructing snakelets, which leaves her emaciated, simply an empty shell, sheets of skin hanging on ribs like underwear on the clothesline.

If a postpartum female doesn't feed before she hibernates, she risks dying in her sleep or emerging without any vitality and wasting away before she can strike a rodent. One fine June morning, several years ago, I was in the field with Bill Brown when he caught a thin female that had given birth the previous fall. "See the redundant skin," he said, pointing to loose skin on the side of her body, just anterior to her anal scute. The more she feeds, Brown said, the sooner the folds disappear; by late summer, after gorging on a chipmunk or two, they'll be gone. "A timber rattlesnake has an incredible tenacity to suffer and live in a dark, dank cavernous space," Brown announced admiringly. Then, he spat on his fingers and wiped off dried birthing fluid from the snake's back, which may attract blue or green iridescent blowflies, themselves gravid, that lay eggs on the snake's back; at no expense to the snake, maggots clean up whatever putrefying protein the biologist missed. Brown then scale-clipped the snake leaving an indelible ventral scar that read number 5782—if you understand the numbering system—which was entered into an ever-expanding data pool. "OK, girl you're off."

Several years ago, Brown gave me a carton of correspondences collected over thirty years that he no longer wanted. Four three-inch-thick files labeled "Steve Harwig" were letters, field notes, and newspaper articles pertaining to this self-taught rattlesnake biologist from western Pennsylvania who had discovered and monitored many of the state's timber rattlesnake dens beginning in the late thirties. One of his hundreds of letters to Brown contained a note about an absentminded outdoor writer named Elmer Cheuvront, Harwig's snake-hunting companion. Just before Christmas of 1948, Cheuvront had dropped two rattlesnakes into a bag and placed the bag in a box, which he then consigned to his cellar ... and forgot. "On September 14, 1949, [Cheuvront] thought he'd dump the dead snakes out and wash the bag." He found one dead, mummified rattler (it had been weak back in December 1948) along with "one 36-inch angry, dark female, thin but alive, and eleven perfect live button rattlers, between two and seven weeks old." The gravid had survived without food and water and light for ten months in a bag, in a box, in a cool, dank basement, and then given birth to healthy babies. There were no stillbirths. Cheuvront shipped the snakes to Ross Allen in Silver Springs, Florida.

As Brown noted in his 1991 article, because of a short foraging season, a breeding system in which an adult female expends as little energy as possible has a strong selective advantage. An estrus female hedges her reproductive bet by conserving critical, irreplaceable energy. Hence, aroused males must find *her*, and their long movements, both purposeful and random, ensure that by chance alone at least one male will cross her pheromone-laden path. Brown concluded his 1991 article with this quote from a vertebrate biologist named M. F. Willson: "Within the limits set by minimum requirements for growth, maintenance, and reproduction, what might be evolutionarily advantageous is constrained by what is ecologically possible."

What then is "ecologically possible" for rattlesnake survival in the Northeast? In ecological terms, a timber rattlesnake is a *K*-selected species: slow to reach reproductive age, with a long generational time, few young per pregnancy, low annual recruitment, long intervals between births, and, most importantly, a *very* small yearly population turnover—in other words, a high survival rate. In economic terms, a timber rattlesnake is a *capital* breeder; a gravid female invests accumulated resources into pregnancy. (By contrast, a wood frog is *r*-selected, with an early reproductive age, an annual deposition of thousands of tiny eggs, and a rapid population turnover. A wood frog is an income breeder, allotting a small amount of recently ingested resources to reproduction.) At a critical point in the summer, if a female timber rattlesnake doesn't have enough capital, she doesn't breed. All of which is to say that a timber rattlesnake may require years to recover from pregnancy. From the standpoint of a lifetime energy budget, the cost of reproduction and the low frequency of reproduction are offset by a prolonged reproductive life. If a snake gives birth three or four times during her lifetime, she has done her part to maintain a stable population; most snakes give birth less often, discovered Brown. The forty-five-year-old gravid rattlesnake that he recently identified, a snake first marked in the late seventies that had been born during the Summer of Woodstock, represents the longevity benchmark for the species, a candidate for the reptilian *Guinness Book of World Records* (if there were such a volume), but very likely this snake bore young fewer than ten times.

In the Northeast, if female timber rattlesnakes grew any slower, took any longer to reach sexual maturity, took any longer to recover from birth, the species probably couldn't survive. Ecological pioneers capitalizing on what's possible, timber rattlesnakes are calibrated to the elements, to the boom-and-bust economy of unpredictable food resources, to the long, harsh winter, wrapped together by the dozens in subterranean darkness, torpid for seven months of every year.

Once a female has eaten and converted three to six years worth of rodents into enough stored fat to sustain pregnancy, in July or early August, she will begin yolk formation, also known as yolking, and, within scientific circles, vitellogenesis. Increasing day length and elevated body temperature trigger the hypothalamus to release a hormone, which in turn triggers the liver to synthesize yolk from metabolized fat and protein. Vitellogenesis turns the liver yellow converting proteins, phospholipids, neutral fats, and glycogen into a soluble lipoprotein, which is then whisked through the circulatory system to the ovaries, where newly created yolk platelets adhere to the cells of each maturing ovum. Then, the ovaries begin to tumefy.

To determine whether they're yolking, Bill Brown palpates females, a procedure gentler than amniocentesis. The ovum, he says, is about the size of the last joint on your finger. Palpating is "a little rude, a little intrusive, but what the heck. She can deal with it." I've run my fingers down the sides of vitellogenic rattlesnakes, felt a hint of their future. Subtle and symmetrical, the contours of prefertilized eggs are the living beads of an ophidian rosary, one after the other, round and promising with yolk.

At this stage in the reproductive cycle, the yolking snake is still not ready to mate. First, she must shed. For preeminent Basher and lifelong bachelor Randy Stechert there's a clear analogy, "Like a woman, she wants to get pretty before she mates." The vitellogenic female's trip to the shedding station cues the male's conjugal journey. Although her vivid colors and bold markings are worn and dulled by time and growth, he is motivated by the lipid-based scent trail laid down by her skin glands, a durable message alerting him to the possibility of sexual congress. She heads directly to a traditional shedding site—a rock-studded opening on the floor of the forest; a sunny talus slope; a ledge; the leg of a cement picnic table at the far end of an interstate rest area; the grassy lip of dirt road in the New Jersey Pine Barrens; a jumble of rocks above a warm, subterranean pipeline; the spoils at a Kansas excavation site. Once a male crosses her trail, he sticks to it with dogged urgency. A smaller, younger adult male might get lucky and intercept a female as she returns to the forest to forage after mating. There, away from the patriarchs, the less competitive male will seduce her, his sperm eventually mingling with that of his unseen rival, a sequential ménage à trios that results in neonates of mixed parentage. Once, a friend of Marty Martin watched three males contest one female. The smallest snake dropped out and mated while bigger males carried on. Otherwise, to the big boys go the spoils of reproduction.

To an adult male rattlesnake, a pheromone trail of a yolking female must be unimaginably enticing. A number of years ago, Bill Brown relocated a post-shed, vitellogenic snake from the front yard of a summer camp. The next day, two males, one after the other appeared in the yard at the exact spot where he had captured the female; fours days later, a subadult male arrived, having been lured by a species-specific aphrodisiac so powerful that even a snake incapable of breeding found it irresistible, much as young boys find *Sports Illustrated's* yearly swimsuit issue irresistible. Don't think for a moment, though, that the pitviper pheromone-alert system is without kinks: a male can be so preoccupied by lust that he may not notice that his darling's heart has stopped beating. Marty Martin observed such a case of ophidian necrophilia when a male copperhead, unrequited and confused, courted a female carcass.

In early September 2013, I stood stone still as a big male passed within inches of my foot, tongue to the ground. A freshly shed skin, greasy and redolent with pheromones, trimmed a nearby rock. Preoccupied, the snake moved along a corridor invisible to me, never pausing to check the source of the vibrations radiating from each of my footsteps. The snake eventually reached the base of a small ledge, which it read with its tongue, and then turned around, backtracking several feet. Head to the ground, tongue extended, the snake returned, scaled a slight crease in the ledge, and then vanished into a weft of fallen limbs, where I imagine an estrous female waited like Sleeping Beauty to be awakened ... and fulfilled.

A female timber rattlesnake has two hormonally driven shed cycles (all others are triggered by growth, or by an injury or an infection like *Ophidiomyces*): one shed occurs at the onset of mating, and the other at the onset of ovulation. As Jed Merrow observed, a male timber rattlesnake will guard a pre-shed, vitellogenic female, either biding his time, rubbing and titillating, or jousting with competitors enticed by the same trail. Once ecdysis has been completed, courtship begins, followed by hours of leisurely mating. "They're lucky that way," sighed Randy Stechert.

Rattlesnake foreplay begins when the male rubs the female's neck with his chin, and positions his body alongside her, the two resembling a pair of muscular shoelaces. Rapidly, he jerks his head and body, and then curls his tail beneath her until vents touch, the "cloacal kiss." Inflated with blood, his hemipenis extrudes from his cloaca and everts inside hers, swelling and locking, held in place by an array of external paraphernalia. Sometimes he pumps his tail, but most often they are bound together, vent to vent, in a bizarre asymmetric outline of a spade or heart or circle. Based on metrics beyond my understanding, she determines when enough is enough.

Biologists Tom Tyning and Anne Stengle once tubed a pair of mating snakes.

"On three. One ... two ... three lift."

Because the snakes were unmarked, Tyning and Stengle PIT-tagged and scale-clipped them, a procedure that took twenty minutes, during which the rattlesnakes remained devoted to each other. On another occasion, Stengle found a pair mating along the shoulder of a class 4 road in the Berkshires. "They stayed locked for two days," she told me.

After processing two other rattlesnakes in his lab, Tyning returned them to a snake bag, where, unbeknownst to him, they began to mate. He drove an hour to Mount Abrams, climbed a steep, tortuous trail, and then poured the snakes out of the bag. "By God, they were stuck together," recalled Tyning, somewhat nonplussed.

How long did they stay hooked?

"I didn't watch. I was too embarrassed."

Each hemipenis (there are two), forked and invaginated, hides inside the tail, a paired, bilobed deflated organ adorned with barbs, spines, and fringes. The left testis supplies the left hemipenis with sperm; and the right testis, the right hemipenis. There is no crossover. If you gently insert a probe into a male's cloaca and back into the tail, the probe meets little resistance inside the hollow, balloonish hemipenis. (A female's tail, on the other hand, is solid muscle.) The ratio of the length to the diameter of a hemipenis lobe varies with each rattlesnake species. Oddly referred to as a "taxonomic tool," used to differentiate species of crotalines, these hemipenes can have long and thin or short and thick lobes depending on the species. In 1956, Lawrence Klauber published the first edition of his fourteen-hundred-seventy-six-page magnum opus *Rattlesnakes: Their Habits, Life Histories, and Influence on Mankind*, of which Harry Greene, who introduced a later edition, wrote, "I doubt anything comparable will ever again be attempted by one person." An electrical engineer and business executive, and in his spare time "father of rattlesnake biology," Klauber, who dealt himself thirty thousand hands of poker to understand the nature of probability and statistical analysis, which he numbingly applied to rattlesnake taxonomy, noted that the number of spines and fringes per hemipenial lobe depended on the size and shape of the lobe, which itself varied from species to species. A timber rattlesnake has a conservatively shaped hemipenis, not too long and not too thick, and consequently not lavishly spinulated and flounced with accoutrements relative to other rattlesnake species. According to Klauber, who had the patience and fortitude to count them, a timber rattlesnake has on average seventy-two long, thin spines and thirty-four fringes per lobe, and the crotch between each lobe lacks spines (by way of contrast, an eastern diamondback, which has a larger hemipenis than a timber rattlesnake, has many spines in the lobe crotches). As Klauber found, though, the most conspicuous ornamental feature, spines were "unexpectedly difficult to count. Even when each spine tip is marked with a color to facilitate accuracy ... the presence of numerous transitional points introduces great uncertainty." Fringes were much easier to count, which must have been a relief to Klauber's wife, who often waited upstairs with their cooling dinner.

To aid orienteering males, a yolking female may expedite her discovery by wiping her cloaca on the ground, smearing a dollop of pheromones. In Vermont, a vitellogenic female routinely observed via radio transmitter had a male train of four snakes in hot pursuit. When courtship and mating are completed, sperm overwinter in midsegment of the female's cloaca, called the urodeum. The last male to mate with her may inseminate most of the eggs. In a half century of fieldwork, Marty Martin has witnessed just nine matings, and recalls each one, jotting down both the date and location in my notebook. Most occurred in August and September, though there was one in early October. Sometimes mating snakes face away from each other resembling a mythical two-headed beast, adjoined at cloacae, and appear to pull in opposite directions in a serpentine tug-of-war.

Alcott Smith's twin brother Avery, who is also a veterinarian, sent me a series of digital images of consorting Catskill rattlesnakes, a big yellow male and a smaller black female. In each image you can clearly

see that the first four inches of the female's lower body proximal to her cloaca was distended with the male's tumescent hemipenis. Before they're hooked, a pair of courting snakes can be disturbed. Usually the female decides, confirming Robert Byrne's observation that "anybody who believes that the way to a man's heart is through his stomach has flunked geography." In mid-August 2011, biologist Kiley Briggs radio-tracked a male Vermont rattlesnake to a beaver pond, where he discovered the snake entwined, but not locked, with a female on a mat of waterlogged grass, all but their heads submerged. As Briggs stepped into the shallows, the female spooked, and the male, whom Brigg's had named "Legs" for his peripatetic nature, unspooled and followed, a testosterone-addled chaperone, his head near the midpoint of her body, never allowing her more than an inch or two of freedom. They crossed the pond in tandem, plowing a trail through the duckweed. Green dots peppered their bodies. On the far side of the pond, in the warm shade of a wooded hillside, they conjoined, which is where Briggs found them the following morning ... fastened as if with Velcro.

In northwestern Illinois, on a humid afternoon in August 2001, Brian Bielema, who entered the Bashers' illustrious orbit in 2013, came upon a pair of mating timber rattlesnakes in the limestone bluffs five miles from the Mississippi floodplain. Bielema, who had been monitoring the population for many years, was conducting an annual gravid survey. The snakes faced in the opposite direction, he told me, and when the female began to crawl north, the male had no say in the matter. Elevating her tail, she picked up speed, dragging him nearly over Bielema's boot. "I could see her cloacal tissue bulging out as she bounced [past]." After fifty feet, the male lost turgidity, the snakes separated, and the female headed into the forest. "Then," said Bielema, "the male brushed himself off, started tongue-flicking, and followed her path into the forest." When he returned the next day, Bielema found the male lounging next to a postpartum and her neonates. A year later, he identified the original female, now gravid, by her signature banding pattern, basking on a shelf of limestone. (Like James Condon in the Blue Hills, Bielema doesn't mark snakes.) A week later, she gave birth to eight grayish-pink rubbery snakelets.

Currency of Acorns: Forest Bit Coins

Red oak acorns are a keystone resource in the forests of eastern North America, particularly since the ecological extinction of the American chestnut eliminated an annual and abundant source of nuts. Red oak acorns germinate in spring, providing nutritious and sometimes abundant food throughout the winter. White oak acorns, on the other hand, germinate in autumn, soon after falling, and spoil nearly as quickly as strawberries. White oaks produce an annual acorn crop. Red oaks are biennial, with each acorn requiring two years to mature. Every decade or so, when the stars align, according to a constellation of variables including weather, soil moisture, geography, genetics, tree health, and gypsy moths (whose larvae defoliate oak trees, thereby devastating mast production), large swaths of Appalachian woodlands produce a bumper crop of acorns. Conversely, every decade or so, the crop fails.

Several years ago, red oak acorns attacked our standing-seam roof, tumbling down into the backyard, collecting below the clothesline. Day and night oaks dropped nutritious packages of protein and fat, in a sleep-depriving metallic downpour like prairie hail. One afternoon, I raked up more than a bushel and a half, yet still, the following morning, the ground was cobbled with acorns. Hanging laundry required footwear. Before I raked, there were more than a dozen per square foot below the clothesline. That fall, a couple of number-crunching biologists tallied nearly one hundred thousand per acre in western Pennsylvania.

After a big red oak acorn drop, forest animals prosper; a phenomenon ecologists refer to as a *trophic cascade*. White-footed mice, the most abundant small mammal in the oak forest, breed straight through the winter, and chipmunks breed earlier in the spring; by the following summer, rodent numbers have expanded from one or two per acre to more than fifty, and then timber rattlesnakes prosper, catching more rodents and storing more fat. Males grow bigger. Females put on weight (some more than three hundred grams), and then a very high percentage become vitellogenic. Two years after an acorn bonanza (a year after the mouse and chipmunk explosion), rattlesnakes enjoy a "gravid pileup," a two-year lag from acorn to neonate. Geographically synchronized acorn production transmogrifies into geographically synchronized rattlesnake yolking and breeding the following summer. Ed McGowen, a Basher who followed the fortunes of acorns and rattlesnakes in southeastern New York, attributes the increased production of snakelets to the number of litters, not the size of the litters.

McGowen and W. H. Martin verified the two-year lag between acorn drop and baby snakes in the central Appalachians, comparing thirty years of game department records for the Virginia acorn crop to Martin's long-term birthing records in Shenandoah National Park. The biologists concluded that the availability of mast-consuming rodents—white-footed and deer mice, chipmunks, gray squirrels—disproportionately influenced rattlesnake reproduction and governed the number of gravid snakes that appeared at the rookery outcrops. Two years after an astronomical acorn crop, baby snake production reached its zenith, and two years after a scarcity, its nadir. Big cycles, small cycles ... and cycles yet to be identified.

September 2008, I watched a four-foot-long female rattlesnake on a Vermont ridge not far from the Walmart parking lot. The snake had unspooled herself at the base of a bitternut hickory, head and neck pointed upward against the trunk as though contemplating heaven. That autumn had been a mast windfall. Half-eaten hickory nuts and acorns peppered the ground, and the snake, keyed to the subtlest vibration, the subtlest odor, the subtlest increase of infrared heat, waited for a hungry chipmunk or mouse or flying squirrel to descend the trunk.

The following summer, throughout west-central Vermont and the Adirondack foothills, the scene repeated itself as rattlesnakes culled a surplus of seed-eating rodents, whose own bonanza was a product of the previous fall's massive acorn crop. In addition to having a higher fat content than white oak acorns, red oak acorns are loaded with tannin, a bitter compound used to cure leather that happens to bind protein and prevent assimilation across the gut wall in seed predators. Rain and snowmelt leach tannin, however, making red oak acorns progressively tastier the longer they remain on the forest floor. Anyway, by early spring, there is often little else left to eat.

In August 2009, I had visited two rattlesnake birthing sites in

Vermont, rock- studded and nearly treeless, where pregnant females lay in the sun, too heavy to hunt, surviving thanks to their capital investments, metabolic proceeds of bygone summers. Half a dozen snakes stretched on warm slabs or coiled beneath rock awnings, with six to twelve oval embryos ripening inside each of them. Anything that might inhibit thermoregulation might delay parturition—cold weather, overcast skies, rain—but over the next two weeks, the weather had been ideal: warm nights and hot days, with thin, gauzy, sun-screening clouds.

When I returned on September 14, most of the snakes had given birth. They all looked deflated, loose skin slung over ribs. A total of forty-five snakelets coiled at their mothers' sides, pink and gray and... *almost* touchable. In Vermont, rattlesnake survival is a precise calibration between climate and topography and acorn production tempered by the snake's own behavior. A horned owl also took note of the production. I found baby snake scales packed in a coughed up owl pellet.

I chewed a red oak acorn once. I don't recommend it. Acrid and acidic like an oily aspirin.

Tropical Legacy

For temperate-zone pitvipers, the orderly, predictable mating cycle, honed by millions of years of natural selection, is an interpretation of the original tropical pitviper breeding cycle interrupted by winter. In the tropics, yolking, mating, and birthing occur in the same year. In the north, however, winter drives a wedge into yolk formation, and the process remains dormant for six or so months before resuming again the following May or June. Yolking ends with ovulation. Cached for the winter in the urodeum of the female's cloaca, sperm are quiescent until fertilization in mid- to late June or early July. "Long-term sperm storage," or LTSS, as its called around the herpetological coffeepot, is the hallmark of the temperate pitviper breeding cycle. Timber rattlesnakes are no exception.

Ovulation occurs in late May or June, and when the ripened eggs pass from the ovaries into the oviducts, and eventually into the uteruses, fertilization occurs, almost a year after mating. Bob Aldridge, the University of St. Louis biologist who collaborated with William S. Brown on the article on the reproductive cycle in male timbers (remember the mysterious SSK, the sexual segment of the kidney), coauthored an articled titled "The Evolution of the Mating Season in the Pitvipers of North America," which appeared in *Herpetological Monograms* in 2002. Aldridge and his coauthor, David Duvall, emphasized that natural selection favors any mating game plan that enhances a female's fitness. "For males to be successful at reproduction they must be prepared to mate when females enter estrus," said Aldridge and Duvall, and they defined this as the *mating season*, which begins, as we've seen, with the vitellogenic shed. For male timber rattlesnakes, preparation starts in late spring with synchronized spermatogenesis, followed in early summer by the hypertrophy of the SSK. Wrote Aldridge,

> As pitvipers adapted to temperate climates, due to the species expanding their range into the temperate regions or the change in climate in their existing ranges, the seasonal vitellogenic cycle was interrupted by winter. Two hypotheses may explain why.... The first is that the period of birth (late summer) was temporarily fixed, so that survival of the young was maximized. Consequently, if the duration of gestation and vitellogenesis was also fixed, vitellogenesis would have to begin the previous summer for birth at the most opportune time. The second hypothesis is that the proximate clues that stimulated vitellogenesis in the tropical populations were retained in the temperate species.

Regardless of the scenario, cold weather interrupts vitellogenesis and the sperm pass the winter in dormancy, waiting for life's first great competition, which holds dire consequences for the losers. Spermatogenesis is synchronous. Vitellogenesis is synchronous. The two, however, are asynchronous. Aldridge believes that the shedding cycle evolved to ensure the best time to mate, for both females and males. Like a female, a male maintains an energy budget; he needs invested capital—rodents—to sustain the annual, long-distance, six-week, all-day mate search. He needs consistent midsummer heat to most efficiently run his physiological machinery. "The timing of the mating season within species tends to be similar," wrote Aldridge and Duvall, "but among species it is variable, suggesting that the mating season is an ecologically plastic trait. Since estrus, in females, and being prepared to mate, incur fitness costs to those individuals, the mating season should be long enough to ensure fertilization and should occur at a time of the year that has the least fitness costs to the female." Some biologists believe summer mating may be a pre–Ice Age vestige from an era when Earth was uniformly warmer and knitting together with your kin for the winter wasn't part of the timber rattlesnake's survival tool kit.

Highlands of the Hudson

Planted in the shoulder of the road, an electric sign flashes,

TROOPS ON ROAD AHEAD

TROOPS ON ROADWAY AHEAD

USE CAUTION AHEAD

I'm in West Point, waiting in a shady, roadside pullout for Randy Stechert, who has been commissioned by the natural resource manager to monitor rattlesnakes at the U.S. Military Academy (USMA), better know as West Point or simply The Point. Randy's undated permit is open-ended, granting limited access into all restricted areas, which more or less is the whole place, sixteen thousand wooded, outcrop-punctuated acres beyond the west bank of the Hudson River, thirty-five miles north of the George Washington Bridge. For rattlesnakes to exist this close to New York City we have to want them to exist.

Established in 1802 by Thomas Jefferson, USMA occupies a prominent S-curve in the Hudson River and is home to, besides forty-five hundred cadets, three timber rattlesnake dens named for generals, who may or may not have liked snakes: Grant, Eisenhower, and MacArthur. All three are recovering from Rudy Komarek's immoral, unregulated, and, by June 12, 1983—the day the timber rattlesnake was officially listed as a "threatened species" by the DEC—illegal depredations.

It's seventy-five degrees Fahrenheit, cloud-strafed (decent snake weather). Several sunbeams spotlight the forest canopy. I'm an hour early and passing time, windows open, listening to the hum of mid-morning: birdsongs floating up from a pond below the pullout—wood pewees, an agitated pileated woodpecker, white-eyed vireos, a yellow throat, blue jays. From all quadrants comes the mechanical, high-pitched drone of cicadas, the sylvan equivalent of elevator music and a counterpoint to birdsong.

Randy arrives on time, prompt and loquacious as ever. I leave my car and get into his, which is spotless, an unusual trait among field biologists. Almost immediately, he begins a soliloquy about cadet training maneuvers adjacent to environmentally restricted areas, which results in abandoned military trash and eroded paths, and opines that very likely the student soldiers molest snakes. Randy's philippic continues as we turn onto a serpentine dirt road, where trees progressively pinch the edges until the road morphs into a footpath.

Randy Stechert, sixty-two and single, is five nine, sturdy, resembles an immaculate Yogi Berra, and has a propensity to sweat. He became very seriously involved with snakes after he bought a black rat snake from Rudy Komarek, who sold wild-caught reptiles to a Ridgewood, New Jersey, pet and hobby shop owned by Stechert's father. One afternoon in 1963, Komarek, who rented a room in a boarding house to accommodate his snakes, took Randy there to see a tangle of rattlesnakes recently taken from the Eisenhower Den. "A big pile of two dozen," Randy recalls, an unvarnished twinkle in his eyes that belies the intervening decades. "It was 1963. I was eleven. I didn't know any better."

Stechert can recall individual snakes, dates he saw them, who he was with, scraps of conversations—Bashers routinely defer to Randy's prodigious memory regarding historic matters—all the way back to the foggy, pot-smoking sixties. On October 16, 1965, Komarek took Randy to a den in the Delaware River valley. They saw four rattlesnakes. "I was hooked." For the next four years, Randy tagged along with Komarek, joining snake-hunting escapades whenever he could. By the time he was seventeen, after four years in the role of impressionable sidekick, Randy realized that Komarek had no conservation ethic whatsoever, no code to govern his collecting. Some dens that Komarek repeatedly visited no longer had snakes. "Rudy took everything. In my primitive mind, I thought 'Why don't you practice some form of conservation?' He didn't have a conservation molecule in his body. [At least], leave the gravid females, the neonates."

"No. No," Komarek obstinately insisted. "I have to take them all. If I don't take them somebody else will," an obscure conformation of economist Garret Hardin's theory expounded in his article "Tragedy of the Commons." Gorging his own self-interests, Komarek took part-time jobs in the vicinity of snake dens in the Berkshires, where he removed snakes from a golf course, and in a roadside zoo in the Adirondacks. "If you think I'm thick," Randy told me, "you should have seen Rudy," to whom herpetologists still refer by his first name, seven years after his death, with a sense of familiarity and notoriety usually reserved for baseball players and rock stars, except that herpetologists don't idolize Komarek; they despise him ... still.

We park. Disembark. Once we're out of the car, Randy peels off his shirt, which he neatly folds and leaves in the front seat. "I'm compulsive. I get it from my mother. If you were taking pictures, I'd make a concession and keep it on." Then, he slips into a sweat-stained backpack and we're off. The only troops I see are platoons of fly-size American toadlets, which perform their own species-specific maneuvers across what's left of the road, swarming the forest floor like a nascent plague. A quarter mile into the woods, Randy's dripping wet, rivulets of sweat flowing between creases in his skin.

Randy Stechert's an inveterate naturalist, a maverick biologist who never attended college (except as a guest lecturer) and unfortunately has spent decades (by his own admission) feeling slighted by the academic establishment, even though he has written several articles on the historic distribution of rattlesnakes in New York and coauthored technical articles with both William S. Brown and Rulon Clark, among others; has been an independent contractor for DEC and its

New Jersey analog, the Department of Environmental Protection (DEP); has consulted with quarry owners and developers to protect critical Pennsylvania rattlesnake habitat; and has amassed forty-nine years of firsthand rattlesnake knowledge. Stechert was Brown's field and laboratory assistant during the initial years of Brown's research. "When I met Bill in 1981, we had a symbiotic relationship. He was an academic, an assistant professor at the time. I was an experienced field man. He had three years experience [with timber rattlesnakes]; I had fifteen. I taught him to track snakes from dens to summer basking areas. He taught me the importance of keeping meticulous notes. You can't use data you don't have; write everything down." Stechert's oracular knowledge of timber rattlesnakes began here at West Point, in the company of Brown's archenemy, Rudy Komarek.

Randy flips everything we pass, a sheet of plywood, logs, rocks, and rotted planks, examines the loose bark of dead trees. Eventually, he uncovers a small spotted salamander and a Fowler's toadlet. He's delighted. Occupying the northern end of the Hudson Highlands, West Point is predominantly deciduous: red oak, white oak, chestnut oak, sugar maple, shagbark hickory, black birch, and tulip tree, the latter straight and tall, rising here and there out of the canopy like leafy radar towers. White pine makes a sporadic appearance; thickets of low-bush blueberry grow everywhere, rough and curly like vegetative steel wool. Although the USMA has logged the woods several times, the evidence is minimal, until later in the day, when we cross into state-owned land, where old-growth trees reveal the true potential of a deciduous forest.

We head toward a gestating and shedding site, where snakes from all three West Point dens reshuffle the genetic deck. Grant Den is three-quarters of a mile away, MacArthur about a quarter mile, and Eisenhower a mere five hundred feet, in a nearby bedrock outcrop. The knoll lures snakes, and snakes lure Stechert, who has driven over from his home in Narrowsburg, which he sarcastically calls "Narrows-minded-burg," a conservative backwater on the New York side of the Delaware, the next large river to the west. "Rattlesnakes are like my neighbors," Randy says, grinning. "When guys get tired of one bar, they look for women at another." No bigger than the average movie theater, the knoll serves multiple purposes for rattlesnakes—as a basking, shedding, mating, gestating, and orientation site. "I believe rattlesnakes recognize where they are. Memory plays a role in orientation."

Apparently, they also recognize who they're with. In an article published in *Biology Letters* titled "Cryptic Sociability in Rattlesnakes (*Crotalus horridus*) Detected by Kinship Analysis," coauthored by, among others, Randy Stechert, lead author Rulon Clark states that under certain conditions, pregnant females prefer to aggregate with kin. This determination was made by molecular analysis of scale clippings and shed skin that had been collected from various sites in New York, many of which Stechert supplied from snakes that use this very knoll. The gist of the article is that hanging with relatives may be good for survival and that a knot of timber rattlesnakes is not necessarily a random occurrence. In this hydra-like tangle are more eyes to spot danger, like a marauding red-tailed hawk; more belly scales to feel vibrations, like the footfalls of a poacher (or a biologist); more bodies coiled together to conserve heat.

We step out of the woods into a bright and bustling scene, a grassy knoll studded with outcrops nearly as old as Earth itself and a trio of happy, amorous-minded snakes. Fire-hose thick, a big yellow male heads toward a pair of black-morph females that bask, partially exposed, on the shelter rock outcrop, what the late Steve Harwig, the Pennsylvanian rattlesnake biologist, referred to as "hotel rocks." Randy, who wants a close look at the females, gives me his snake hook, and says, "Keep it out of the rocks," which is *far* more difficult than it sounds. Big Yellow is in perpetual muscular motion and immediately ties himself into a sheepshank in the blueberries.

He is not pleased with my shepherding style, a hesitant hook, a light pull, actually more tug than pull, a deliberate compromise between not losing the snake in the nearby rocks and tossing it around my neck. There is no time to admire chocolate crossbands, bordered in black, offset by an undercoat of vivid yellow. I hook the snake midway around the body and give a sissy-ass pull, which encourages Big Yellow to retreat in my direction. I wonder, "What the fuck am I doing?" Annoyed and inconvenienced, Big Yellow sprays musk,

which adds to a landscape that must be already redolent with pheromones. Fifty feet away, Randy smells the musk and then says, "I once
had a girlfriend, a sweet elementary school teacher from Brooklyn,
who made me wash my hands every time I handled a rattlesnake."

Stechert unknots Big Yellow from the blueberries, and then
balances the snake on the snake hook, his arms sagging under the
weight of fifty-four inches of rattlesnake, an eyeball measurement that
doesn't include the rattle. (Herpetologists never include the length of
the rattle in a measurement; rattles are fragile, easily broken.) Randy
wants to weigh Big Yellow, but unlike Bill Brown, he doesn't carry
high-end snake-bag technology. He left home his telescoping forked
frame that would have conveniently held open the bag. I scavenge a
few stout sticks, jerry-rig a frame, which proves easier than herding
Big Yellow through the blueberries.

No matter how inwardly content it may be, no rattlesnake ever
appears truly happy. Big Yellow's eyes are stone-cold yellow, bright
like polished gems, its stare lidless. The angle of its mouth cuts
straight back and lacks the upturned jaw line of an alligator, which
always evokes the image of amusement at a private joke. As soon
as Big Yellow's fist-sized head is in the snake bag, Randy pours the
body in. Then, the head pops out. After several more unsuccessful
attempts, the snake settles securely in the bottom of the bag, which
Randy ties shut. Big Yellow is so big and so powerful that his flicking
tongue pushes against the snake bag. The snake weighs twenty-three
hundred fifty-seven grams, less thirty grams for the bag. No wonder
Stechert's arms were sagging; Big Yellow weighs over five pounds,
more than a half gallon of water.

"One thing I've learned [working with timber rattlesnakes]: if you
take your time, you can get things done," which for Stechert is a pre-
requisite. He suffers mild Parkinson's tremors, a palsy brought on by a
combination of unabashed excitement at handling a big timber rattlesnake, too much coffee, nine bouts of Lyme disease, and a chronic
nerve disorder. "I have to calm down when I write [field] notes," and
as I peek over Randy's shoulder, I see why. His notes would leave
even a pharmacist scratching his head, the "82° Fahrenheit, humidity
66%" looking more like Chinese symbols than Arabic numerals.

Finished with his notebook, Randy turns to the black females. To keep Big Yellow from cooking, I move the snake bag into the shade. Although Randy estimates she's thirty-nine inches long, the gravid gets a free pass; Randy won't disturb her. He catches the vitellogenic snake, feels her swollen ova, and counts aloud fourteen rattle segments, marking the base of the first segment with a red sharpie. "I use a different color every year. Electric blue and black sharpies last longer." Not thrilled by the attention, the snake sprays Randy in the right eye.

When Stechert's finished, he releases the snake into the rocks, and then we liberate Big Yellow, who weaves himself through the blueberry maze, methodically progressing up knoll, with an urgency to quench. The yolking female waits in the shelter rocks, her painted rattle segment barely visible from where I stand. Does she sense Big Yellow's imminent arrival? I won't know how this resolves (though I can guess), because we still have two dens and another basking site to visit. In the west, a rain-threatening, gunmetal sky prompts us to quicken our pace.

Eisenhower Den, a horizontal crack in a granite outcrop, is just around the woodland corner. Rudy discovered Eisenhower in 1965, when the den supported sixty or seventy timber rattlesnakes. He took fourteen snakes that day, and as his raids continued for the next fifteen years, Eisenhower's population began to collapse.

Stechert tells the story of another den in an undisclosed location, where he and Rudy would walk for miles down a dirt road and up a mountain and find four snakes, where Rudy routinely found twenty-five or thirty. "What happened, Rudy?"

"And he'd respond, 'I wiped out the population. I took them all, Randy.'

"That's bullshit.

"Rudy would tell me, 'Hey, Randy, I went to the den and took ten more snakes.'" Eventually, the pupil ditched the master. "That pissed me off so much that we went from being friends to being bitter enemies."

At a basking site near MacArthur Den, a jumble of rocks in open oak woodland, three snakes, two black and one yellow, one gravid and two possibly yolking, sprawl on the rocks. Two snakes begin to rattle and, not wanting to pester them, we back away. In the heat of the day, the delirious sound of kids on summer vacation wafts across to us from a nearby military installation. Birdsong has tapered off, cicada song is at its peak, and as we retreat, the sound of agitated rattlesnakes dissolves into the early afternoon.

"A snake area right here and they fucked it up. Fucked it up right here," Randy carps, referring to cadet training. A piece of military trash reads:

Meatloaf 97.
U.S. Government property.
Commercial resale is unlawful.

An orange glow stick hangs from a branch, detritus of an overnight maneuver. Another lies on the ground. Critical rattlesnake habitat, the area has been zoned as off-limits for military operations, cadet training verboten. Apparently, no one read the natural resources guidelines.

Grant Den lies just over the border on state property. Each fall, for three years, Stechert took home a gravid female from a nearby basking site, birthed and head-started the neonates, and then released them all back at the den in October. Of the twenty-five snakelets, fourteen males and eleven females, two survived to adulthood: one male and one female. The male pulled a Benedict Arnold and switched dens, as rare among rattlesnakes as it is among generals.

Immense, columnar tulip trees anchored in thick loam rise above black birches and sugar maples and red oak, themselves big trees. The rocks are slippery, and I'm all over the place. Bears have overturned several large rocks to excavate yellow jackets. Randy finds a scrap of shed skin, one step removed from dust, and declares the snake was a yellow morph. "See these faint traces of banding?"

"No."

Grant Den is an imposing granite block that dwarfs everything around it—the trees, the rockslide, even the adjacent hillside. In 1994, Randy found a forty-eight-inch black morph coiled on a stone

shelf. "I had the feeling something was watching me." Fifty feet away, on a branch halfway up a white oak—he points out the tree to me—a red-tailed hawk stared down at the snake. Three years later, he found a female redtail and her airborne young circling above the birth site. "She taught her chicks to hunt rattlesnakes. And then, for a number of years, every August, just when snakes were birthing, I'd see her above the ridge." Every now and then, in October, about the time snakelets make their way down to the den, Randy sees a hawk below the canopy, perched on the limb ... waiting. Once, he found a headless rattlesnake. Hawk-bit.

The crevice, on the point of the granite slab, leads into subterranean darkness and opens onto a rock shelf, where emerging rattlesnakes bask in leafless April sunshine. Scattered on the shelf are rattlesnake ribs, long and sharp, a new medium by which to divine the future. A pair adorns my desk. Snakes use the little rock pile just below the den as a transient site during ingress and egress, where they gather in groups of twos, threes, and fours, having come down from the basking ledge or up from the forest to assemble on the sun-heated rocks, pods of relatives according to Rulon Clark's thesis, which was supported by the scale clippings Randy provided from these snakes. "They can't get below the frost line from these rocks," he tells me. Which is to say, Gertrude Stein's "a rose is a rose is a rose" doesn't apply to rock piles, something every rattlesnake knows.

September 10, 2011.

Today, I wait with Alcott Smith by a derelict building on the east side of a local stream for Randy and his friend Bob Pitler, a devout amateur herpetologist. At eleven o'clock, the heat, along with the vultures, rises like brume out of the woods. Beyond the riparian forest, an outcropped mountain ridge, running north from northern New Jersey, juts above the horizon and forms the prominent southern end of Bear Mountain–Harriman State Park, fifty-four thousand acres of wildness along the Hudson, mostly in southeastern New York. The park was established in 1900 as part of the Palisades Interstate Park Commission, initially through a gift of cash and land from philanthropist and railroad tycoon Edward Henry Harriman. In 1910, Harriman's widow donated an additional three thousand acres on the east side of the Hudson River, the Bear Mountain section. Bear Mountain–Harriman, the second largest state park in New York, together with the USMA and Sterling State Forest, protects more than ninety thousand acres of the Hudson Highlands (about sixty percent), and is the epicenter of the timber rattlesnake population in lower New York State. The core of the Highlands are composed of Precambrian granite and gneiss, formations that date back to the dawn of life, more than one billion years ago. Two hundred million years ago, the Highlands were higher than the modern-day Himalayas; today, they're nowhere higher than sixteen hundred fifty feet, although from a couple of choice vistas you can see the jagged silhouette of New York City punctuating the horizon. Rattlesnakes, of course, don't give a shit about antiquity or great views; it's the thickness of the loose rocks and the spaces between them that draw snakes. When it rains, runoff from the west side of the park reaches the Ramapo River, which flows southwest through a fault plain into New Jersey before unloading into New York Harbor, not far from Newark. Runoff from the east side joins the Hudson, eventually sweeping past Chelsea Piers and mixing with everything imaginable before entering the Atlantic Ocean.

The southeastern face of the Hudson Highlands has eroded into a prehistoric staircase, a series of bedrock treads, which are more or less forested plateaus separated by vertical outcrops and embedded talus. Each bedrock tread is composed of an understory of rock-studded patches of grass, blueberries, sweet fern, and scrub oak that is dwarfed by an overstory of chestnut and red oak, bitternut and shagbark hickory, sugar maple. Tulip tree, black walnut, sassafras, and white ash grow in deeper, richer soil closer to the stream. Nowhere do I see white pine.

The prehistoric stairs are a rattlesnake paradise: basking sites from top to bottom, denning sites in baseline rockslides and in crevices in midlevel rock walls. One den Stechert named in memory of Charlie Snyder, the zookeeper, who was fatally bitten there in 1929 after bagging seventeen rattlesnakes for the Bronx Zoo. For Stechert, Snyder's Den has further significance: the ashes of a close friend with whom he spent many afternoons in the Highlands are scattered around the den portal.

Randy Stechert first visited Bear Mountain–Harriman State Park when his parents introduced him to the Hudson Highlands in the late fifties. "Some of the things my parents did made no sense to me," he says with a droll smile. "I had to be all of nine years old. They dropped me off with a friend at the base of the mountain and said, 'Have at it.' They picked us up that evening." When Stechert was ten or eleven, his park trips became more involved. His parents off-loaded him with his brother and a friend, each under the weight of a cumbersome rucksack filled with a sleeping bag the size of an ottoman and three-days worth of canned food. "We roamed the Highlands for the weekend."

Less than two weeks ago, Hurricane Irene had pushed the stream across the floodplain. Remnants of the flood are everywhere: broken limbs, scoured trunks, water channels, puddles, a three-inch veneer of silt. Garlands of debris hang from low branches like Christmas trimming. To cross the swollen stream, we crawl along barkless logs, wet and greasy, tractionless. For the remainder of the day, no matter what level of the mountain I'm on, I can hear the river.

Once we're on the other side of the stream, we muck our way through the floodplain, around several large rain-filled pools. A spotted turtle slides off a log. The ground is spongy, water seeping everywhere. At the base of the mountain, the very first step, a short talus slope runs from the cliff to where it is subsumed by the riverine woods. A lattice of grapevine covers the larger, lower boulders, in some places thick enough to obscure them; strands of poison ivy and Virginia creeper overlie the smaller rocks of the upper talus like a loose-knit fabric. On a flat rock, beneath an awning of Virginia creeper, a slimy salamander, lustrous black flecked with white and yellow, idles in a puddle.

Randy has come to the mountains today to monitor the movements of three radio-fitted rattlesnakes—a gravid, an adult male, and a three-year-old male. The snakes are easy enough to locate by waving the antenna and following the beeping signals of Stechert's transmitter: the subadult male is moving through the lower rockslide of the first tread, not far from the salamander; the forty-four-inch adult male is coiled in the open woods, two treads above snake number one; and the gravid is tucked among ancient shelter rocks along the highest ridge, where several lonely pitch pines catch the wind and beyond what seems like an ocean of empty forest. The gravid, now postpartum, apparently birthed under the rock Randy had tracked her to ten days ago. Two neonate sheds, papery and dry, a birth announcement of sorts, garnish the corner of the shelter rock, beneath which the babies must be safely clustered with their mother.

Later in the day, we find two unmarked snakes, both males; the last snake basks in the open rocky woods on midlevel tread, a black morph nearly as large as Big Yellow, and equally as recalcitrant. After scavenging sticks again to prop open Randy's snake bag, I watch a snake that doesn't want to be restrained—tail goes in, the head comes out, often simultaneously. The snake's broken three-button rattle, a remnant of its original, yields an incongruous, high-pitched buzz reminiscent more of an insect than a meaty snake; the sound reminds me of the first time I heard Mike Tyson speak.

An official record of snake-catching prattle:

"OK."

"Oops."

"If it was on the ground, would he crawl in?"

"Why would he do that?"

"Get away from that end."

"I don't want to work on that part."

"Watch your foot."

"Which foot?"

"The one by his head."

Randy Stechert was hired by the DEC to determine whether the lower level of the mountain, owned by a nearby town, is critical rattlesnake habitat. The town, which wants to develop a hundred-acre housing project, pays Stechert's salary and expenses. Randy has identified fourteen potential basking sites on the property, which he labels on his topographical map BS1, BS2, and so on. He's already marked thirty-four rattlesnakes at eleven of the fourteen sites. By the end of the day, I'm convinced that the entire mountain, both on and off state land is, if nothing else, a continuous chain of basking sites from floodplain to crest.

Although he was arrested poaching rattlesnakes in this park, Rudy Komarek did not have a long history here. "The low point of the population wasn't just because of Rudy, it was from people *like* Rudy," Randy tells me. "Back in the forties, there was a doctor that caught rattlesnakes for medical research up here. He took quite a bunch." And, for many decades, Carl Kauffeld, of the Staten Island Zoo, made yearly collecting pilgrimages to the Ramapo Mountains, of which one was covered in the November 4, 1939, issue of the *New Yorker*.

In case you don't know how a poisonous snake is captured alive, we'll tell you. Grasping an old golf putter in your left hand, you rest it lightly on the snake's head; then, with your right hand, you seize the snake just where his neck should be. There are two techniques for snake-seizing, one calling for the use of thumb and first finger, the other for thumb and third finger. Mr. Kauffeld favors the latter method. "With it you can depress the snake's head with the forefinger and reduce the chance of lightning whip-around," he explained.

Randy's friend, the late Roy Pinney, whose ashes he scattered at Snyder Den, would tease Randy whenever they saw a beautiful yellow morph. "Can I take it, Randy? Can I take it?" Even though you can legally catch timber rattlesnakes in the Southeast and Arkansas, northeastern snakes have the most diverse colors and patterns within their range. Poaching is more of a problem today than it was four or five years ago, particularly for gravid females. You never know when you'll hit the jackpot and capture a female that births an albino or leucistic neonate. In the seventies, Rudy removed a snake from the Berkshires that gave birth to five cream-white, pink-eyed snakelets. Unfortunately, when given proper food, water, and warmth, timber rattlesnakes thrive in a cage. "The point these herpetoculturists don't realize," laments Randy, "[is that] no snake *ever* looks as good in captivity as it does in the wild." You're missing the wooded, rocky context that gave its color meaning.

Rudy Komarek collected more than four thousand rattlesnakes in New York, fifty percent of today's population; he poached twenty-seven dens in the Hudson Highlands, some (like those in West Point) heavily. As Randy says, Komarek was the most diabolically clever and incredibly stupid man he has ever met.

When Stechert was sixteen, he joined Komarek in a New Jersey diner, where his mother's proviso that "there's something weird about reptile people" was confirmed for him. Rudy, who thought himself a cocksman, saw two young women and announced, "Those women are looking at me."

"I tried to discourage him from going to their table," says Randy.

Heedless, Komarek walked over to the women and said, without a hint of self-effacement, "I've been watching you watch me."

"I wouldn't go out with you if you were the last man on Earth," the duo responded in unison.

Strolling back to his table, Komarek broadcast to anyone else who might have been listening, "Those women are fucked up."

9

A Commodity of Rattlesnakes

SNAKES TODAY, BIBLES TOMORROW

Cursed are you among all animals

Genesis 3:14

*They sharpen their tongues as a serpent; Poison
of a viper is under their lips. Selah.*

Psalms 140:3

You serpents, brood of vipers, how will you escape the sentence of hell?

Matthew 23:33

*And he laid hold of the dragon, the serpent of old, who is the
devil and Satan, and bound him for a thousand years*

Revelation 20:2

*Now the serpent was more subtle than any beast of
the field which the LORD God had made.*

Genesis 3:1

*They have venom like the venom of a serpent; Like a deaf cobra that stops up its
ear, So that it does not hear the voice of charmers, Or a skillful caster of spells.*

Psalms 58:4–5

Their wine is the venom of serpents, and the deadly poisons of cobras.

Deuteronomy 32:33

*The wolf and the lamb will graze together, and the lion will
eat straw like the ox; and dust will be the serpent's food.*

Isaiah 65:25

Behold, I give unto you power to tread on serpents.

Luke 10:19

My friend Digger gave me a three-foot-long eastern king snake during my junior year in college. He had found it coiled beneath a tractor tire on a west Louisiana farm close to the Sabine River. I named the snake Bowie and kept him for more than a year. He entertained my friends and dates, for he tolerated handling and never bit anything larger than laboratory mice, five-lined skinks, garter snakes, or fledgling house sparrows. Bowie crawled through hands and around wrists, and, unless restrained, continued up under shirts to coil in the dark warmth of upper arms. Whenever biceps flexed, his grip gently tightened like a blood-pressure cuff.

Bowie and I had shared our upstairs apartment with several housemates. One of these had a girlfriend with a two-year-old daughter, who was drawn to the king snake. For her, Bowie was animated and rubbery, a black toy banded in thin iterations of yellow, like the links of a chain-link fence. Gentle and unassuming, he brushed her hand with his dark bifurcated tongue, and she broadcast euphoria. One evening when I was away at class, she left the terrarium lid open, and Bowie escaped. Several days later, after a grueling session in organic chemistry lab, I pulled up our driveway in suburban Muncie, Indiana, to find four prepubescent boys, marching across the front lawn, a limp king snake slung in the crotch of their forked stick. Like pagans, they chanted and glorified their deed—a scene straight out of *Lord of the Flies*—as, watching, I felt like vomiting.

Several years later, on a sultry August morning in the 1972, I made my first pilgrimage to Walden Pond. Along the hiking trail that circumnavigates the pond and not far from the jumble of stones that marks the site of Thoreau's historic cabin there was the sound of commotion, then the sound of a gunshot. As I rounded a corner I saw a bumptious, middle-aged ranger, pistol in hand, directing hikers around a writhing milk snake. The man, clearly full of himself, as though he had rid Concord, Massachusetts, of a great evil, gloated like a wide receiver who had just caught a touchdown. "Damn copperheads," he repeated over and over. Before the crowd dispersed I grabbed the dying milk snake, pointed out the distinguishing features, and referred the ranger to Roger Conant's *Field Guide to Reptiles and Amphibians of Eastern North America.* His face tightened and reddened, not the deep red of a red-bellied snake, but rather a lighter shade of human humiliation, more like the pink flush in the eye of a northern water snake.

On a recent late April afternoon, the talus is cool and barren, an empty temple of rocks, while the sun peeks through a translucent haze of clouds. Beyond the rocks, a spacious world stirs—songbirds sing, ducks dabble, red admiral butterflies flit above sloping meadows, and spilling down the face of a cliff, a brook transitions into mist. Snakes wait below me patiently hidden underground, dozens of immobile braids. Then, two days later, while the sun heats the rocks, rills of warm air loosen the snakes, which emerge from inhospitable crooks in the rockslide as if plucked from a magician's hat, measured, solemn, unfettered by cold. One snake . . . two snakes . . . three snakes . . . ten . . . sixteen. Black morphs, yellow morphs, wrist-thick snakes, their jeweled eyes set in expressionless faces, their scales worn and dulled by dust, heat-seeking reptiles trivializing everything else in the valley not born without limbs. Time passes untethered by thought, minutes become hours as in childhood, when salt water and sand rendered me deliciously insignificant. I teeter on stone steps, one foot up, one foot down, on the verge of either falling upon what I'm looking at or dissolving into a slow-motion afternoon. The sun warms my face and the rocks, orchestrates the landscape, a maestro on the dais of the sky casting out its rays. And here am I, amid rocks and snakes . . . exactly where I prefer to be.

Unlike the Walden ranger or the boys who killed Bowie, or so many other people we both know for whom snakes are anathema, I cannot recall having ever been afraid of them. Respect, well, yes, of course. I was drawn to their mellow, predictable pace long before I realized that hillside rattlesnakes had a pace or that their behavior could be quantified. It took getting close to them (often) to see they were calibrated to the seasons, the rocks, and the grand forest cycles of feast and famine, which for me has been both a privilege and a revelation. Sadly, however, most humans are loath to accept rattlesnakes; it would take a miracle (or at least faith in an enlightened guide) for the unwashed to realize that venomous serpents aren't errors in God's benevolence, all surprise and threat, hardwired by a tiny, inflexible pea brain; they are living, evolving beings as finely tuned to their respective corners of the planet as is a sugar maple or a royal palm.

Recently, I googled "phobia" and was directed to a website that alphabetically listed more than five hundred of our most common fears, many of which had never occurred to me, among them fear of garlic, fear of chopsticks, fear of figure eights, fear of long words, fear of sex, fear of men, fear of women, fear of peanut butter sticking to the roof of your mouth, fear of your mother-in-law, and *zemmiphobia*, fear of the great mole rat. Of course, *ophidiophobia* was there, too, our most common, most debilitating, and perhaps most ancient fear—to an ophidiophobe one rattlesnake is an infestation—born in the deep past long before the written word . . . in fact, long before we walked upright, claims Lynne Isbell, a professor of anthropology and animal behavior at the University of California, Davis, and author of *The Fruit, the Tree, and the Serpent: Why We See So Well*.

That hair on the nape of the neck rises for one-third of all human beings when they see a snake, even a picture of a snake, may be more than mere gut reactions. According to Isbell, more than one hundred sixty million years ago, when the earliest primates appeared in the jungles of Gondwana, the Southern Hemisphere supercontinent that would eventually splinter into Africa, Australia, Antarctica, Madagascar, and the Indian subcontinent (India and Pakistan), their principal predators were neither leopards nor eagles nor goofy, drooling hyenas, all of which had yet to evolve, but constricting snakes—

big, fire-hose-thick snakes that hid in plain sight—that chilled the blood and frosted the nerves of primordial lemurs; death-dealing serpents—motionless, patient, reflexive—that struck out of seemingly shapeless carpets of mulch. To survive, stem primates had to be able to spot constrictors. Then, about forty million years ago, when anthropoid primates (forerunners of modern monkeys and apes) appeared in the southern jungles, venomous snakes on hand to greet them pushed the second *big* pulse in primate visual evolution. That snakes hit a nerve in people before people were people is not news. In 1872, Charles Darwin brought a stuffed snake into the monkey house at the London Zoo and noted that the hair on several species instantly became erect, and field biologists have reported that chimpanzees and baboons shout and scream whenever they see a fat viper lounging in the shade of an acacia. However, Isbell believes that our long history with snakes, particularly venomous snakes, has been a double-edged sword: our primordial fear may be antediluvian, but to snakes we owe a debt of gratitude for the evolution of our exquisite, pattern-finding vision. Venomous snakes, as predators, Isbell says, were the natural-selection agent that shaped the way we perceive the world ... our big-eyed, three-dimensional, color vision, best on the planet by far, is essential for spotting a camouflaged snake.

Lynne Isbell refers to our ability to discern patterns in the immediate landscape as "the snake detection theory," which she believes was responsible for the evolution of both our keen vision and our depth perception, setting primates apart from all other mammalian orders—our forward-facing eyes, our visual acuity, both very helpful for spotting and avoiding dangerous snakes, a life-and-death game of *Where's Waldo.*

Writing in the *New York Times,* Isbell lucidly explains that earlier theories, which suggested that fruit eating was the catalyst for the evolution of our color vision, do not hold up:

Now, it's worth pointing out that other creatures began eating fruit, too—tree shrews and neotropical fruit bats, for example— and that these animals did not develop great eyesight. It follows, then, that there had to be some further incentive for primates to develop superior vision. My contention is that the push may have come from snakes.

Isbell's "snake detection theory" hangs on two primary observations:

First, all animals have early warning networks, neurological wiring that tells them there's danger. These networks, however, are more hard-wired to the visual system in primates than they are in other vertebrates ... over time, the visual component of the primate warning system has grown more than it has in other creatures.

Second, monkeys with the sharpest eyesight tend to be those who live in greatest proximity to venomous snakes.... The Old World monkeys and apes were the ones most exposed to venomous snakes, and of the three major groups, the Old World monkeys and apes have the best vision.

In October 2013, in a research paper published in *Proceedings of the National Academy of Sciences*, Lynne Isbell, collaborating with a team of neuroscientists and psychologists from Brazil and Japan, found further evidence that snakes exerted strong evolutionary pressure on primates. Using Japanese macaques, the scientists identified neural circuits in the brain region known as the pulvinar, which is disproportionately large in monkeys and apes compared to other species of mammal and observed that these circuits *only* responded to images of snakes. Presenting a series of pictures to the macaques, including facial expressions of other monkeys, the scientists found that photographs of snakes triggered the strongest, fastest responses. Isbell points out that we have the very same genetically coded, snake-sensitive neural circuits in our own pulvinars ... a deep-seated, primal revulsion for snakes and snake-shaped objects that for many of us may be beyond our conscious control. But, says Isbell, we may owe more than our vision to the existence of venomous snakes. The development of language—what sets us apart from all other creatures—may have been kick-started by attempts to avoid stepping on puff adders and gaboon vipers and black mambas.

I offer that snakes gave bipedal hominins, who were already equipped with a non-human primate communication system, the evolutionary nudge to begin pointing to communicate social good, a critical step toward the evolution of language, and all that followed to make us who we are today.

If Lynne Isbell is right that we evolved with snakes on the brain, that snakes make us who we are, make us human, then I wonder: Were there clowns in those ancient, snake-congested jungles?

Sometime during the early years of the twentieth century, an itinerant moonshiner named George Went Hensley introduced snake handling as a show of faith to a Pentecostal congregation near Grasshopper Valley, Tennessee, which amounted to his personal proclamation of victory over the devil. Hensley, who had grown up watching faith-based snake handlers perform at coal mining camps in Wise County, Virginia, became the Johnny Appleseed of snake handling, planting his bizarre brand of Christianity in Pentecostal churches in Kentucky, Tennessee, Indiana, Ohio, North Carolina, South Carolina, Alabama, Georgia, Florida, Virginia, and West Virginia. He based his belief on a literal interpretation of the King James Version of scriptures, Mark 16:17–18, which most biblical scholars believe actually *never* appeared in the original version of the New Testament but were shoehorned into Mark much later. The verses read:

> And these signs shall follow them that believe: in my name they cast out devils; they shall speak with new tongues; they shall take up serpents; and they shall drink any deadly thing, it shall not hurt them; they shall lay hands on the sick, and they shall recover.

Understanding the launching of serpent handling is like understanding a crime scene. Everyone who witnessed the event disagrees about what they saw, where they saw it, and when they saw it, sometime between 1899 and 1914. Appealing to poor southern whites, at home deep in the hollows of southern Appalachia, holding timber rattlesnakes, copperheads, cottonmouths, and, less commonly, eastern diamondbacks as a sign of faith peaked just after World War II, when Dust Bowl immigrants brought the ritual to southern California, where it is still practiced (to a limited extent) in Los Angeles. Estimates of the number of snake-handling churches varies between two and three hundred, though Paul Williamson, a psychology professor at Henderson State University, in Arkadelphia, Arkansas, and an authority on religious fundamentalism, believes the number may be closer to one hundred twenty-five. Whatever the number, the majority of churches in which snake handling is practiced have very small memberships. "In Appalachia, religion is serious stuff," says Henderson. "It's not Sunday morning stuff; it's every day of the week stuff."

In just over a century, approximately a hundred people have been fatally bitten during these energized religious spectacles (lots of dancing and trances and speaking in tongues), most recently in Middlesboro, Kentucky, on February 15, 2014, when pastor Jamie Coots, of the Full Gospel Tabernacle in Jesus Name Church, who starred on the National Geographic Channel's reality show *Snake Salvation*, was bitten during a service by a timber rattlesnake and, as a demonstration of faith, refused medical attention. Coots's son Cody, who has taken over the congregation, said his father had been bitten eight times before but had never experienced such a severe reaction. "We're going home, he's going to lay on the couch, he's going to hurt, he's going to pray for a while, and he's going to get better. That's what happened every other time, except this time was just so quick and it was crazy, it was really crazy."

Despite multiple church members witnessing the fatal bite, the Kentucky medical examiner, for reasons only she understood, was reluctant to list snakebite as the cause of Reverend Coots's death. She contacted Dan Keyler, the authority on snakebite toxicity, from the University of Minnesota, asking him whether a test could confirm the presence of venom. There's a test, but it's expensive. Ask Coots's flock from the Full Gospel Tabernacle in Jesus Name for

donations, Keyler suggested. She did, but the hat came back empty. In an e-mail broadcast to Basher cohorts, Keyler revealed his conclusion, "God just didn't approve of the Reverend Coots dancing with timber rattlesnakes."

A hand-lettered note taped to a Jolo, West Virginia, pulpit cautions, "The pastor and congregation are not responsible for anyone that handles the serpents and gets bit. If you get bit, the church will stand by you and pray with you. And the same goes with drinking poison." On May, 28, 2012, Mack Wolford, preacher, proselytizer, and star of a widely distributed documentary film, was bitten on the thigh in a Memorial Day service in Panther State Forest, in McDowell County, West Virginia, by Sheba, a yellow timber rattlesnake Wolford had used for hundreds of services. Wolford died the following morning, an event more than a few biologists ascribed to natural selection.

Hensley himself, who occasionally inserted rattlesnakes six to eight inches into his throat, claimed to have been bitten four hundred forty-six times while preaching, including once on the earlobe by a snake he wore on his head like a yarmulke and once in the face by a big yellow morph whose fang broke off in his nose. Hensley survived the first four hundred forty-six bites. Then, in Altha, Florida, on July 24, 1955, at age seventy-five, bite number four hundred forty-seven killed him. The Calhoun County judge ruled his death suicide.

Although George Went Hensley had been the Saint Paul of snake handling, he could neither read nor write. He married four times, sired thirteen children, of whom most were raised by someone else, shuffled around like a croupier's cards, and none of whom took up serpent handling. Whenever his children visited, Hensley stored wooden snake boxes under their beds so that the buzzing would lull them to sleep. One of Hensley's grandsons married one of his stepdaughters, becoming also his son-in-law. When carloads of descendants drove from Tennessee to Georgia to attend his funeral, the Hensley clan posed for a graveside photograph, seeds and sprouts of snake handling's Johnny Appleseed.

According to one Pentecostal website, worldwide there are approximately 105 million practitioners, of which approximately two thousand handle snakes as a show of faith. (Many also fondle fire and drink poison, mostly strychnine or lye, and nine of these are known to have died over the past century.) Although snake-handling meetings are outlawed in every state but West Virginia (in Georgia, you need a permit), very few states prosecute handlers, even after a snakebite death. Rattlesnakes are sometimes brought to funerals and placed in caskets—once, a mourner at the funeral of his brother-in-law, a snakebite victim, was bitten by a snake and died the following morning. One rogue timber rattlesnake is credited with dispatching two parishioners in less than a week. Mothers have died by snake-bite, leaving behind as many as half a dozen dependent children. And in Virginia, a full-term, pregnant woman bitten three times on the wrist delivered a baby three days later. Both baby and mother died, the baby within in an hour of birth and the mother several days later while singing hymns in bed. Because he had handed his wife the snake that bit her, the woman's husband was convicted of voluntary manslaughter. Once bitten, almost none of these religious snake handlers seek medical help.

Snake handlers believe in kismet. "God administers to our heal-ing." After George Went Hensley was bitten the final time, he told his congregation, "I know I'm going. It is God's will." For the true believer, it's a win-win situation: either God spares you or he calls you home. For the snake, it's a death sentence. So odd is the practice of testing one's faith by dancing to delirium with venomous snakes that every few years, generally coinciding with a practitioner's death by snakebite, the broadcast and print media have a field day revisit-ing the Pentecostal snake handlers. Media as disparate as *Fox News*, *CNN*, the *Wall Street Journal*, the *New York Times*, *Life* magazine, and *Hustler* have covered Pentecostal snake handlers.

That snake handling has persisted into the twenty-first century is a testament to our ambivalent relationship with snakes in general and timber rattlesnakes in particular, by far the most commonly handled snake for religious purposes. It also says something about the social disenfranchisement of southern Appalachian whites, many of whom live in abject poverty, often under the thumb of coal mining goons. Perhaps the adrenaline rush triggered by dancing with rattlesnakes counteracts the monotony of working in the Stygian depths inhaling coal dust, underground like the sleeping serpents. As Lynne Isbell points out, a fear of snakes may be our evolutionary baggage, and thus innate, and when coupled with the First Amendment to the U.S. Constitution, which guarantees religious freedom, Pentecostals do as they choose with timber rattlesnakes. Who cares about snakes? We don't tolerate the use or abuse of any other creature in the pursuit of worship. Start a religious practice that handles bluebirds as a show of faith and see how far you get. In our worldview, venomous snakes are beyond expendable; they're perceived by the ecologically unwashed as thoughtless, mindless miscreants condemned to crawl on their bellies for tempting Eve in the Garden, the devil incarnate. For snake handlers, "The serpent does not have the keys to heaven and hell. We are not looking to die. We are looking for eternal life." For the poor snakes that serve the anointed purposes of these Pentecostals, their life becomes hell; overhandled, often kept without food and water, scared limp and listless, they die, sagging tubes of skin and bones, the ultimate echo of Randy Stechert's "spook factor" and Bill Brown's "listless syndrome."

In 1945, William Tuck, then governor of Virginia, insisted the public would not tolerate snake handling, and told police busting up a service to "seize and destroy" all confiscated serpents. In 1940, in Hazard, Kentucky, a judge named Baker addressed snake handlers: "If you are going to stay in the snake business, you had better get some snake hunters out, for I'm going to kill every snake in this county." Seven years later, in front of a crowd of four or five hundred protesters, outside the Harlan County Courthouse, in eastern Kentucky, an indicted snake handler stuck a copperhead in his pants pocket and challenged a police officer to search him. After a lengthy imbroglio, wrote David Kimbrough in *Taking Up Serpents: Snake Handlers of Eastern Kentucky*, the presiding officer braced himself with a heavy canvas glove and removed a three-foot copperhead. The officer then killed the snake in front of the crowd. Someone in the crowd shouted, "Let 'em kill it. We've got plenty of them. They ain't rationed." For the practitioners, life is hard. And so is getting saved.

For the snake, however, there is no redemption. What recourse does the snake have?

I've come to Sweetwater, Texas, to see the public slaughter of west-ern diamondbacks, to learn why our odd relationship with nature, though healing in many corners of America (with many different species), still festers where venomous snakes are concerned. That rattlesnakes evoke fear and revulsion in so many humans makes them all the more fascinating to me. As Lynne Isbell believes, there may very well be evolutionary reasons why most people see rattlesnakes as blight on the animated landscape. Removing one from under the front porch is prudent, but a species-specific pogrom couched as a community fund-raiser is still a throwback to the dimmest days of "wildlife management," when the U.S. Biological Survey, the fore-runner of the U.S. Fish and Wildlife Service, poisoned and trapped wolves and pumas in western national parks to protect deer and elk. I want to see the community that identifies with (and slaughters) rattlesnakes, and the snakes themselves, alert and at large in the Texas outback, their chutzpah a link to primal America.

Were it not for the Sweetwater Jaycee's World's Largest Rattle-snake Roundup, which attracts thirty-five thousand tourists each year, most to witness the public killing of rattlesnakes, few people other than Sweetwater's eleven thousand residents would know of this Texas town, hard on the southeastern rim of the panhandle and surrounded by vast, uncelebrated plains and juniper-lined draws. Gone are the buffalo and the Kiowa who hunted them. Gone too are the black-footed ferret and the swift fox. Cougar and black bear are little more than rumors, and the prairie dog town that once sprawled across one hundred thousand square miles of West Texas has been nearly eradicated, existing only in city parks and the forgotten out-back. Most of what lives out here has thorns—mesquite, prickly pear, yucca, acacia, devil's claw—or fangs. "Pasadena has its roses and its rose parade. Sweetwater doesn't have roses. We have rattle-snakes," proclaims Mayor Greg Wortham. "The roundup," he says, "is part of our identity, our community fabric."

David Quammen sees it differently. "Snake-haters kill diamondbacks for fun. They kill diamondbacks with a righteous zeal that they'd like to believe is somehow religious or patriotic, or at least neighborly. They kill diamondbacks from habit. These people gather together annually in great civic festivals of cartoonish abuse, and slaughter, and ecstatic adolescent loathing, that go by the name of rattlesnake roundups."

Mind you, Texas is not the only state that holds roundups, and western diamondbacks are not the only target. Pennsylvania sponsors six (fourteen historically). Since the eighties, however, Pennsylvania roundups have been monitored and tightly regulated by the state Fish and Boat Commission, which made benign celebrations out of ghoulish festivals that once supplied Korean markets in Queens, New York, with live baby snakes to float in bottles of vodka that sold for one hundred dollars a pint. You need a permit to take a snake in Pennsylvania. (Recall Rudy Komarek's problems?) Licensed snake hunters are limited to *one* timber rattlesnake per year, male only; after the roundup, each snake must be released where caught. Females and young were given a free pass, which wasn't always the case.

So common were rattlesnakes in Pennsylvania in the late fifties that a camp owner once complained that he mutilated two to five snakes every time he mowed his property. Between the fifties and the eighties, when Pennsylvania roundups popularized and glamorized snake hunts, females accounted for seventy percent of the catch, of which sixty percent were gravid. By the eighties, the late Steve Harwig, a volunteer assistant in the Department of Reptiles and Amphibians at the Carnegie Museum, who for more than sixty years monitored three hundred of the state's thousands of dens, believed that seventy-five percent of Pennsylvania dens were below sustainable levels. The reason: roundups. Pregnant females were easy targets. The predictability of their gathering at ancestral basking sites became synonymous with vulnerability. In response to the warnings of Harwig and others, in 2007, the Pennsylvania Fish and Boat Commission set a statewide minimum-size limit of forty-two inches long and a minimum subcaudal scale count of twenty-one for a timber rattlesnake. Since females average thirty-nine inches and have twenty-one or fewer subcaudal scales and males have twenty-one or more, the new regulation effectively eliminated gravids, as well as subadult males and snakelets, from the legal harvest. In 2013, eight hundred eighty snake hunters removed approximately one hundred seventy-five rattlesnakes from the hills, mostly for roundups, some for meat and leather. A far cry from the 1975 Morris Hunt, when three hundred snakes not sold as meat or leather or pets died in their crates.

West Texas is diamondback country, raw, wild land punctuated by uncountable gypsum outcrops, where distance is measured in hours, not miles. Over deep time, rain has worn a network of tunnels into the soft rock, portals into the netherworld of the western diamondback, chalky recesses where snakes retreat by the hundreds into tight cavities, woven together in great serpentine knots for the winter, emerging only to sunbathe at the threshold of the den, fixed and motionless, like the rock itself.

It's the second Sunday in March, 11:00 a.m. The morning is hot. I'm on the Blue Goose Ranch, south of Sweetwater, the guest of Tom Henderson, a tall, fit rancher in his late fifties, an Oklahoma transplant and thirty-year member of the junior chamber of commerce, the Jaycees, official sponsor of the World's Largest Rattlesnake Roundup. Henderson's wife owns the Blue Goose, one of two ranches that have been in her family for six generations, more than one hundred twenty-five years. Henderson wears dungarees, a snap-button shirt, and leather boots. His cream-colored cowboy hat shades an expressive face. Like the Southern Plains, Henderson's humor is dry and uncensored. Twice each March, he hunts rattlesnakes (the word "collect" is not used when gathering rattlesnakes) and, like a catch-and-release fisherman, Henderson turns most of the snakes loose, often redistributing them to various nooks on the Blue Goose. Once in a while, he barbecues rattlesnakes for curious friends but cautions, "If you invite people over for snake and French fries, you'd better have a lot of fries."

Rattlesnake tastes like chicken; all ribs and white meat, best cooked on a grill or deep-fried, as they do at the Sweetwater Roundup, which sponsors a highly competitive rattlesnake-eating contest. Unlike the preparation of chicken (or any other creature that I can think of), there are a few obstacles to consider when preparing a rattlesnake for the table. C. E. Brennan mentioned one of them in *Rattlesnake Tales from Northcentral Pennsylvania*:

> Gary Dillman skinned and dressed a freshly killed rattler during a cookout. He washed the carcass in the nearby brook, put it in a plastic bag, and placed the bag on the ground. When the pot was ready, Gary reached for the bag, only to discover it was empty. He located the elongated piece of meat, which was casually crawling through the woods, some twenty feet from camp.

Our entourage at the Blue Goose includes three other Jaycees, all middle-aged men; six boys, ages fourteen and fifteen (mostly from Missouri); an Abilene librarian and her ten-year-old daughter; an elementary school teacher from South Texas; and a portly detective, retired, and his portlier sister from Asbury Park, New Jersey. "We know we have pretty country around here, but snakes don't fly. Be looking at the ground," Henderson announces as we fixate on a flotilla of cotton-ball clouds scudding across the deep-blue sky.

Trails cut by snake bellies radiate from a den, a wide crack a third of the way up a gypsum outcropping. Pack rat scat is everywhere outside of the den (and for all I know may be deep inside the den as well), compact, oblong pellets of undigested roughage about half an inch long, black against soft-beige rock. As we approach, two diamondbacks idling in the open withdraw into the rocky foyer, just out of reach, displacing a few rat droppings, which roll away. We gather by the den, bent at the waist, and bounce sunbeams off mirrors illuminating the serpents; their lidless eyes sparkle. Henderson steps back, watches from a distance, pleased. Our collective enthusiasm ratchets upward as the snakes loiter just out of reach.

In 1965, when Texas folklorist J. Frank Dobie wrote, "I grew up understanding that a man even halfway decent would always shut any gate he had opened to go through and would always kill any rattlesnake he got a chance at," Henderson's snake-hunting paraphernalia consisted of a twenty-two, which he employed to shoot off the snake's head, an activity known as plinking. The novelist James Dickey preferred darts. In 1974, his essay in *Esquire*, "Blow Job on a Rattlesnake," describes how to make and use a pipe blowgun. After more than three decades on the Blue Goose, however, Henderson has come to respect rattlesnakes. "They're part of our heritage, as integral as cactus and mesquite," he says. "It would be a sad day if you don't have to watch where you step." Carefully, I gauge my own.

Although I had admired them for many years behind glass at the Staten Island Zoo, where Carl Kauffeld exhibited nearly every sub-species and color morph of rattlesnake, I saw my first free-ranging western diamondback in the spring of 1973. I was a graduate student at Texas Tech, doing fieldwork in the remote southwestern corner of the state, near Kermit, a hamlet so isolated that drinking water had to be trucked in twice a week. Every morning before it got too hot, I collected bobcat and coyote scat in the desert beyond our field station, part of a long-term predator-prey study. One morning, stepping over a piece of tumbleweed, I looked down in mid-stride through a screen of skeletal branches and noticed a three-foot diamondback squarely in the middle of my stride, tightly wound and numbed by cold. After I cleared the snake, I turned around and studied its texture (visibly rough) and its weather-induced quiescence from a safe distance. Too chilled to rattle, the diamondback looked through me as though I were transparent. Not so much as a tongue flick. An hour later, prodded by an ascending sun, the diamondback had disappeared down a nearby rat hole.

"Whad'ya do when ya see a snake?" someone asks Henderson, drawing my attention back to the present. Hope that it's stunned by cold weather, I think to myself. Henderson, of course, provides an alternative response. "Jump up and holler shit! Then, stand real still." Ninety-eight percent of a rattlesnake's waking life is spent sunning in front of the den or sitting by a bush, waiting. The other two percent is spent either eating or making little snakes. Like timber rattlesnakes in the wooded East, diamondbacks blend in with the vast array of western browns and are not easy to spot. I've walked right by them on numerous occasions and have always been startled when they rattled; once, I pulled a tour van over to the side of an Arizona road and opened the passenger door from the inside. As the first guest started to step out, I noticed a large diamondback luxuriating on the warm pavement below the door. Henderson has his own equally chilling story. "I took a mother and her six-year-old son on a snake hunt," he tells us. "The boy hollers, 'Mom! There's a snake.' Seven experienced snake hunters had walked right past it and the snake's inches from the boy's foot."

Everyone stops. Looks around.

Like timber rattlesnake venom, diamondback venom is not intended for us, of course; it's an elaborate adaptation to capture, subdue, and process warm-blooded animals—the result of a life without limbs in the arid fast lane. In fact, an adult diamondback is so particular about how and when it delivers venom that it will pass up a small meal (a mouse, say) and wait for a larger meal (a rabbit). Once injected, western diamondback venom digests the prey's skin, exposing the body cavity, and within twenty-four hours begins to break down the viscera.

Several years ago, someone proposed to the Texas Department of Parks and Wildlife that the western diamondback be managed as a game animal with a season and a bag limit like the timber rattlesnake in Pennsylvania. "Parks and Wildlife doesn't give a rat's ass about preserving the rattlesnake," complains Henderson. "The idea went over like a screen door in a submarine." If Texas declared the diamondback a game animal, state wildlife managers would have to set season and bag limits based on county-by-county inventories in order to maintain a sustainable population, much like they do for quail and deer. "For this reason alone, and there are many others," wrote Texas A&M professors Clark E. Adams and John K. Thomas in *Texas Rattlesnake Roundups*, "it appears highly unlikely that [the western diamondback] will ever be declared a game species in Texas." Beginning in 2003, however, the Texas Parks and Wildlife Commission required a commercial nongame collection permit for anyone who collected, sold, or purchased wildlife. The permit holder is required to submit detailed annual records of harvest and sales. Adams and Thomas call the permit "a paper tiger: toothless and ineffective." Parks and Wildlife officials don't attend most roundups, and when they do nothing meaningful happens with the data.

Henderson, who's ambivalent about hunting snakes, stopped bringing diamondbacks to the Sweetwater Roundup years ago because the Jaycees met their economic needs without his contributions. "I'm out here because of this," he says, sweeping a hand around a wide and brittle landscape, like a realtor showing a living room. "I want to preserve the ability to keep hunting my favorite areas and to increase the odds that I can show a group like this a rattlesnake." Whether Texas declares the western diamondback a game animal or not, Henderson, one of the few active Jaycees who ranches, applies the ancient concept of resource management to the Blue Goose—take what you need and leave the rest. For many years, Sweetwater held a cash contest for the smallest snake brought to the roundup, which encouraged hunters to catch every snake they found. "Why was I taking these little bitty snakes?" Henderson asks me, rhetorically. "It was like taking a fingerling home and not eating it." He stopped taking neonates and young-of-the-years and, following his lead, fifteen years ago, Sweetwater discontinued the smallest-snake contest.

Still, the prevailing attitude in West Texas is "kill the rattlesnake on sight." "Now, when a cowboy randomly kills a snake on the Blue Goose," he tells me, "[he] can't come back here. Thirty years ago, I would have said 'get rid of every rattlesnake.'" Henderson believes in personal freedom and fiscal responsibility, but practices moderation and conservation.

He hunts fifteen dens on the Blue Goose, a fraction of the number he believes are out there. If, as he claims, an acre of West Texas outback supports an average of five western diamondbacks, then the twelve-thousand-acre Blue Goose and its sister ranch may support more than sixty thousand rattlesnakes. Taking his math one step farther, Texas as a whole, at 267,339 square miles or 171,096,960 acres, of which 111,000,000 acres (about two-thirds of the state) is undeveloped, may host more than half a billion western diamondbacks. That would be one hell of a lot of rattlesnakes. But alas, Henderson's estimates are much too high, contradicted by the decline in both numbers and size of rattlesnakes brought to Sweetwater; not even primal Texas would have supported half a billion western diamondbacks.

Henderson announces, "Diamondbacks have a good sense of the weather. They know it's going to get cold and rainy tomorrow, they're going to stay back in the den." Although it seldom rains in West Texas, gypsum is soft and porous like limestone. Through the years, rain has worn grooves in the undulating landscape, tight little canyons that Texans calls "draws," one after another to the very edge of the earth. Only Interstate 20, which runs to our south between Dallas and Los Angeles, interrupts the limey corrugations. Each draw has a potential den (or two or three or four), and each den may support dozens of snakes, entwined and patient, waiting for the sun to lure them out.

Twenty-five years ago, after heavy and repeated winter rainstorms, diamondbacks on the Blue Goose were hard to find. Percolating water had caved in one den that Henderson regularly hunted and had flooded several others. "We weren't catching shit." Eventually, Henderson found several diamondbacks well off the saturated ground, wrapped around the pancake arms of a prickly pear. "We hunted cactus for the rest of that winter." Another year he took one hundred fifty pounds of rat-fat snakes out of a neighbor's hay barn.

While Henderson speaks, one of the Jaycees inserts a length of copper tube soldered to a pump-action, garden spray can into the den, and then squirts a fine mist of gasoline inside to coax the diamondbacks out, a common practice in north Texas that is outlawed in most other states to protect groundwater, if not rattlesnakes. Within moments our remote outpost smells like an Exxon station.

Henderson's annoyed. Although he stopped gassing regularly because he knows outsiders (like me) consider it barbaric and he doesn't want to draw more critical attention either to snake hunting or to the Sweetwater Roundup, he still defends the practice. "I've gassed this den twenty times over the years and it has probably been gassed fifty times. Dens don't become inert," he says. Of course, not everyone agrees.

In 1989, University of Texas herpetologist Jonathon Campbell determined from laboratory tests that gasoline fumes damage and disorient rattlesnakes, and may kill small mammals, nonvenomous snakes, box turtles, lizards, toads, scorpions, and whatever else may be shacked up in a rattlesnake den; if the morning is cool and the snakes inactive, gassing can be lethal; and if the copper tube doesn't reach the back of the den, asphyxiated snakes retreat rather than emerge, gagging on gas fumes. Patient snake hunting is usually rewarded— like timber rattlesnakes in the Northeast, diamondbacks are charmed out of hiding, slowly but surely, by the heat of the day. Fumigation, or "giving the snakes a little incentive" (Henderson's words), contaminates both the hunted and the hunter, who places capture above quest, subverting the ambience of the day.

Because it poisons aquifers, Texas outlaws the pouring of gasoline on the ground. Although no agency monitors the gassing of snake dens, in 2014, Texas Parks and Wildlife held hearings across the state to discuss banning the use of gas fumes to capture rattlesnakes, a practice already partially or completely outlawed by more than two dozen states. In San Antonio and Fort Worth, just twenty-one people showed up at the hearings. No one showed up in Houston. Sweetwater drew two hundred fifty people. If Texas banned gassing, the catch for the Sweetwater Roundup would decrease by an order of magnitude. Biologist Campbell told me that gassing taints snake meat, which the roundup crowd eats with gusto. "Could you smell a hint of gasoline at the barbeque pit?" he asked me in a recent phone conversation. I couldn't.

If Vermont permitted gassing at each of its six dens, virtually every timber rattlesnake in the state could be collected in short order. Still, Henderson makes exceptions and gasses a den when people travel a long way to see a snake. Today, he wants to guarantee the boys' success, which he's sure will be a highlight of their childhood, as hunting continues to be for him, and, short of waiting all day, gassing is the only way to do that. "Practice moderation in all things," he says. "A little bit don't hurt; a lot is bad."

When gasoline hits warm rock it vaporizes filling the den with flammable fumes. Several years ago, a careless hunter flipped a lit cigarette butt in front of a recently gassed den. The explosion that resulted tore through the hillside and wrecked his car.

Diamondbacks behave differently in South Texas than here in the northern part of the state (much like comparing the behavior of Vermont rattlesnakes to those in Mississippi). Because a milder winter makes communal denning less critical, snakes hibernate in rodent burrows, often alone, and are inactive for only a couple of months of the year at most. Hunters behave differently as well. Gassing isn't as efficient. Before Texas outlawed roadside hunting, hunters cruised public roads looking for basking snakes. Now, they search the outback one bush at a time. In South Texas, where diamondbacks spend less time sleeping and more time feeding, zaftig six-foot snakes are much more common than in North Texas.

From the abyss, I hear diamondbacks stir, heavy bodies rasping across the stone floor. To escape the gas fumes, the snakes emerge one at a time, forked tongues extended to read the air, which sizzles with youthful testosterone. Supervised by Henderson and his friends, tongs in hand, the boys safely catch fourteen snakes, trumpeting an exuberant play-by-play commentary. Each diamondback is placed in a wooden crate to be released on Monday after the weekend's festivities have ended. One Jaycee hastily closes a lid across a snake's back, breaking it, and then, just as hastily, deposits the crippled serpent on the ridge away from our activities, to cook under the Texas sun into a reptilian pretzel. Says Henderson, no longer ambivalent about this morning's activities, "They're squealing like it's their first orgasm."

Texas has nine species of rattlesnakes (including timber rattlers in the Big Thicket National Preserve east of Houston), of which the western diamondback is by far the largest and the most common, as well as the most belligerent, an animal whose imagery and folklore are inseparable from the culture of West Texas. The western diamondback, *C. atrox* (*atrox* is Latin for cruel or frightful), is easy to recognize: a series of black-and-white rhomboids—the snake's signature diamonds—run the length of its back; each eye is framed in white; and the tail is banded black and white, which accounts for one of its more common colloquial names "coontail rattler." Its subtle body color echoes the ruddy tones of the arid landscape, the better to go unnoticed by predators—hawks, eagles, owls, roadrunners, coyotes, foxes, badgers—that find the nine-inch snakelets tasty. For small diamondbacks, there's no relief from king snakes, which read the Texas breeze with their own sensitive tongues and are immune to crotaline venom. A monster diamondback, which may live more than twenty-five years, has a body as broad as my calf, a head the size of a softball, reaches over seven feet long, and weighs more than fifteen pounds, has little to fear save the snake hunter. In 2005, Snopes.com—the website dedicated to investigating and quashing rumors—reported that a nine-foot one-inch diamondback weighing ninety-seven pounds had been killed at the old Turkey Creek gas plant north of Amarillo, Texas, a tale nearly as outlandish as a polar bear sighting in New York Harbor. The piece included a digitally distorted photograph of a large snake draped over a snake hook in the hands of a smiling, balding redheaded man. Accompanying the announcement was a recipe for deep-fried rattlesnake and a note that read: "No matter what anybody else tells you, kill the snake before you try do anything else to it! It's the safest way for you and the snake doesn't care anymore."

Not all diamondbacks warn when frightened, which makes watching your step essential, and makes wandering West Texas a breathless experience in a landscape elevated above its visual shortcomings by the thought of what you cannot see, like hiking the tiger-plentiful Sundarbans of India. Though West Texans still rationalize the killing of rattlesnakes as a matter of safety, the facts suggest otherwise. According to the Centers for Disease Control and Prevention, only about five people die each year from snakebite—one or two in Texas—compared to thirty-nine thousand from poisoning, three from rabies, twenty-nine thousand from alcohol, and four in Maine from paddle-sport mishaps. Although the western diamondback is responsible for most fatal snakebites in the United States, the potential danger it poses is greatly exaggerated: neither the Rolling Plains

Memorial Hospital in Sweetwater nor the Hendrick Medical Center in Abilene, the regional trauma center, has ever seen a fatal snakebite. Hendrick treats approximately fifteen bite victims each year. In the past twenty years, only two people, both children, sustained snake-bite injuries that were serious enough to warrant transfer to a larger facility in Dallas. Both made full recoveries.

I bet you remember that between one-half and one-quarter of all rattlesnake bites are dry (no venom released), and most people bitten are either handling or attacking the snake (although a forty-four-year-old man was bitten a couple of years ago in a Texas Walmart as he reached for a bag of lava rock). If you are bitten by a rattlesnake, you are much more likely to die in an automobile accident on the way to the hospital than from the bite itself. Ironically, roundup activities are the leading cause of snakebite in Texas (and Pennsylvania). One hunter, who brought a thousand pounds of diamondbacks to a 1991 roundup, arrived in leather sandals, cut-off shorts, and a tank top shirt. He had one arm; the other had been lost to a mistreated snakebite. Bill Ransberger, the Sweetwater Jaycee who for many years ran the snake-handling demonstration (now called the "education pit"), theoretically stressing the importance of safe behavior around venomous snakes, had been bitten forty-two times, many in front of an audience.

Amid the bitter, mineral-rich creeks of what is now Nolan County, Texas, the Kiowa found what they called *mobeetee*, "sweet water," an oasis favored by buffalo and buffalo hunters, pioneers, ranchers, cotton farmers, gypsum miners, and petroleum drillers and refiners. After the railroad arrived in 1881, Sweetwater was named the seat of Nolan County. The county, which is about the size of the state of Rhode Island, has a population of fifteen thousand, and most people live right in Sweetwater. Beyond the last city street in town, short grass hugs the ground and waits for rain, which sometimes arrives in spring as a fusillade of hail. At best, twenty-five inches of precipitation fall on West Texas each year, sometimes with a hard cloudburst that Henderson compares to "a cow pissing on a flat rock." When it's not raining, a gritty haze borne on the wind covers the Southern Plains like epidermis, clogging screens even as it makes spectacular sunsets.

Sweetwater's ethnic makeup is pretty close to that of Texas as a whole: 61.5 percent white, 31.7 percent Hispanic, 5.8 percent black, and 1.0 percent "other," according to the chamber of commerce. Only about a third of all adults in town have completed high school, less than half the national average. You can rent a home in Sweetwater for less than two hundred dollars a month, buy a chrome-plated blue coffin at a yard sale (price negotiable), or swim in the community pool tiled with crucifixes on the walls and across the bottom. In large red-and-white letters above the marquee of the old limestone movie theater, the word *Texas* faces the courthouse, and more title companies than one might think a town of this size needs. Sidewalks in the center of Sweetwater are raised, the streets clean.

Sweetwater is home to the United States Gypsum Company, the country's largest producer of wallboard, and since 2005, to the world's largest wind farm, part of a multibillion-dollar regional alternative energy business. The rattlesnake roundup was once an essential source of income for the town: businesses catering to the influx of out-of-town guests—motels, restaurants, bars, liquor and convenience stores, filling stations—used to pull in as much as twenty-five percent of their gross annual income, about three million dollars, according to one study during the three-day weekend roundup, says Jaycee president Riley Sawyers. Today, most local businesses are flush with patrons drawn to Sweetwater by the restless wind, an ironic twist for the snake-killing capital of America to emerge as the country's renewable energy hub.

Still, says Sawyers, a Marlboro chain-smoker whose tattooed arm features rattlesnakes and the Davy Crockett quote—"You may all go to hell, and I will go to Texas"—the Jaycees donate more than forty thousand dollars each year to local charities—the Muscular Dystrophy Association, American Cancer Society, Special Olympics, hospice, youth baseball, Boy Scouts, Girl Scouts, and fire and police departments. The Red Cross operates the Nolan County Coliseum parking lot during roundup weekend, which pulls in an additional three thousand dollars, or about enough to provide disaster relief for fifteen people who lost their homes in tornadoes, which are common in the area. Ronnie Broadus, emergency services coordinator, who looks forward to the roundup every year, says it's a way for neighbors to help neighbors, "Without it we'd have to find another fund-raiser."

Pennsylvania roundups, which began in 1956 with the Morris Hunt, initially served as a vehicle to eliminate snakes, as well as a fund-raiser for the local fire departments. So popular was snake hunting in the rolling Pennsylvania hills that the May 1978 issue of *Strength and Health* published a training routine for snake hunters that featured bench presses, half squats, pullovers, curls, and power cleans. Today, Pennsylvania roundups are prosnake educational events that include lectures and exhibits and, when they're over, the mandatory release of snakes back into their rocky stronghold.

This weekend marks the forty-ninth annual Sweetwater Jaycee's World's Largest Rattlesnake Roundup, still the umbilicus of community identity. The festivities begin with a Thursday afternoon parade—marching band, floats, and antique cars—followed by the crowning in the Municipal Auditorium of Miss Snake Charmer, the winner of the annual beauty, talent, and scholarship pageant. Jacque McCoy, the executive director of the Sweetwater Chamber of Commerce, competed in the pageant in 1964, and two of her daughters have, too, including her youngest, Lori Yarbro, who was crowned Miss Snake Charmer in 1989. "It's a chance for our young women to come to the forefront," McCoy says. "It's their chance to shine." McCoy hopes her granddaughter, China, will enter someday. "I would love to see her in the same pageant as her mother and grandmother."

In addition to competing in the talent and congeniality rounds, each contestant has the opportunity—strictly voluntary, but many girls choose to do it—to decapitate and skin a rattlesnake, an event covered by the *Sweetwater Reporter*. As I pay for my ticket at the coliseum, I ask the Jaycee who mans the booth what motivates a contestant to decapitate and skin a rattlesnake: "If they want the scholarship money, they must earn it," he jokes, handing me my ticket stub and pointing to a newspaper photograph of a beauty queen peeling the skin off a diamondback the way you might peel a banana.

Separating a snake from its head and skin seems a strange rite of passage, yet another reminder that our relationship with venomous snakes is far from peaceable. Our struggle to make amends with creatures long feared and reviled persists in many parts of the country: in the northern Rockies, dozens of gray wolves were shot and killed within just a few weeks of the government's decision to remove them from the endangered species list. Grizzly bears following the scent of food into backyards often meet the same fate. Rattlesnakes may be as finely tuned to the arid plains as any creature, the spectacular by-product of eons of hemispheric evolution, yet here in Sweetwater, as is so often the case, fascination takes the form of fear rather than reverence. And so the snakes are killed for fun, for profit, and—though no facts support the claim that rattlesnakes are a threat to us—in the name of human safety.

I haven't yet set foot in the coliseum, but the action has already begun. There's a barbecued brisket and chicken cook-off near the parking lot; carnival rides; and stands selling barbequed corn, deep-fried Twinkies, and more. Inside, there's a gun, coin, and knife show, a flea market, and a snake-meat-eating contest and scores of vendors sell food and curios made from rattlesnake skin: cell-phone and snuff-can holders, iPod cases, hatbands, barrettes, belts, boots (which Henderson assures me are useless for ranch work), golf putters, and bikinis, as well as fang and rattle earrings. For ten dollars you can buy a snake-head key chain or a plastic paperweight with an embedded snake head; fifty dollars buys a guitar strap. There are also entire snakes—freeze-dried coiled (and still threatening somehow) or inlaid in clear plastic toilet seats, which pretty much sums up their perceived ranking in the animal kingdom. Volunteers from the Children's Christian Ministry paint kids' faces, and at a military recruiting booth, soldiers wearing T-shirts that read "U.S. Marines strike if provoked" encourage children to do chin-ups.

The live snakes are also inside. Roundup participants collect them for months leading up to the event, and when the snakes arrive at the coliseum, they're poured from their crates into the holding pit, where a red-vested Jaycee stirs the lot so they don't suffocate. Dead ones are tweezed from a pile of thousands; for them the ride ends here. From the holding pit, snakes move on to the "research pit," where they're crudely sexed and measured. From there, it's off to the milking pit, where the venom is gathered for medical research, including studies of tumor growth and heart disease. Next stop is the butcher pit, where snakes are decapitated, their heads still seething in a bucket on the floor. The last stop on the assembly line is the skinning station, where men in bloodied white jumpsuits suspend still-twitching snakes by their tails and flay them. The bodies are then deep-fried and eaten. Skins are tanned, perhaps to return to future such events as golf putters or cellphone cases.

I see why Henderson prefers the Blue Goose to the coliseum. People crowd around the various booths and the snake pits. As I approach what's now called the education pit (formerly known as the snake-handling pit) I hear the voice of Susan King, the first woman elected to the state house from Nolan and Taylor counties. "There's no place I'd rather be than a rattlesnake roundup," she yells from the center of the pit. The crowd, which also includes a remote video audience in area high schools, applauds enthusiastically. Around her, fifty or sixty sluggish diamondbacks hug the periphery of the pit, their muted browns in sharp contrast to King's red blazer. As King answers questions about school prayer and high school steroid testing, a Jaycee pushes a snake back with the edge of his boot. The diamondback disappears into the depths of an ophidian knot. When King has finished speaking, another Jaycee, David Sager, to the delight of the crowd, provokes a snake to strike at a yellow balloon. Sager, who began running the education pit when Bill Ransberger retired, recalls a diamondback striking the fly on his partner's zipper. Sager's wife, who had witnessed the strike, insisted he leave the pit immediately and go home. "You need a rest," she said.

Pennsylvania's Steve Harwig (1921–2013) believed that most of the damage sustained by snakes occurs during capture—yanking them out of crevices with tongs—or in the "education pit," where so-called naturalists stretch and measure rattlesnakes as though they were elastics, separating vertebrae, disarticulating skulls from backbones, abrading skin, breaking ribs, crippling unborn young. Harwig believed that nearly thirty percent of timber rattlesnakes on display were maimed, many severely, by mishandling. From 1985 through

1987, the Pennsylvania Boat and Fish Commission contracted biologist Howard Reinert to assess roundup rattlesnakes. Reinert, whose PhD dissertation at Lehigh University unveiled the variation in habitat use of timber rattlesnakes and copperheads, interviewed snake hunters (white males, average age thirty-four) and examined snakes. His conclusions echoed Harwig's: twenty-nine percent of the snakes had sustained injuries. Reinert advocated a closed season that wouldn't open until rattlesnakes had dispersed from their dens; prohibition of the use of a noose when catching a rattlesnake; and elimination of the stretching board as a tool for measuring snakes. He also recommended a reduction in the number of Pennsylvania roundups from thirteen to four. And make the educational pit educational, he argued, and stop *all* sacking contests.

Back in Sweetwater, Henderson, who drifts through the Nolan County Coliseum like a wraith, answering my questions whenever our paths cross, says sardonically, "You don't have to be a herpetologist to run the education pit."

More than a few visitors wear press credentials; a film crew from Taiwan commandeers the space in front of the skinning booth, where a Korean merchant from Dallas collects blood and gallbladders from a string of rattlesnakes hung by their tails. Two boys about eleven years old press against the Taiwanese crew and provide their own enthusiastic commentary.

"What's that?"

"A gallbladder."

"Awesome! What's a gallbladder?"

"I don't know."

Tom Wideman, former mayor of Sweetwater and former president of the Jaycees (Henderson, with whom he caught diamondbacks at the Blue Goose for an episode of *National Geographic Explorer*, calls him "Tommy"), is seventy years old but looks and acts a good deal younger. Wideman has broad shoulders, thick hands, and is six feet tall, about the length of a large diamondback. Dark-haired and mustached, quick to smile, Wideman wears a green long-sleeved shirt, light-brown pants and vest, and a stylish fedora—the same outfit he wears on the cover of his recently published book, *Texas Rattlesnake Tales*. A pair of metal key-chain rattles hangs from his vest pockets, one on each side. The key chain was cast from a pet diamondback named Red Rider that bit Wideman (he shows me the scar on his thumb). "A rattlesnake's bite compares to being stabbed by two red-hot ice picks." If you purchase his book, he gives you a key chain. I buy the book (and read it), get my key chain, and speak with Wideman, who is both gregarious and patient.

Like Henderson, Wideman believes roundups don't impact the rattlesnake population. "Look around you. This place is wild, totally uninhabited." Wideman hunts snakes on a seventeen-square-mile ranch outside of town. "That's seventeen square miles," he says, "seventeen square miles and I know one hundred ninety-three dens. And that's not all of them. We can't possibly hurt the snake population down here." I ask him if he's ever heard of passenger pigeons.

Wideman has been part of every Sweetwater Roundup since the beginning; he's collected tickets, run the weigh-in pit, and cochaired the roundup. In 1966, he became a snake hunter. Before 1963, the Jaycees gave a prize for the heaviest diamondback. Wideman worked the scale in those early days, weighing every large snake that was brought in. He recalls a lethargic five-footer that was too heavy to crawl. At sixteen pounds, Wideman was suspicious. When he lifted the diamondback off the scale a cataract of shotgun pellets spilled out its mouth. As roundup organizers, "We didn't know what we were doing back then," he says.

"Now the roundup perpetuates itself. You could put a monkey in charge," he says, smiling. "No one would know the difference. And there are times I'd prefer a monkey. It's total chaos. Organized confusion." Wideman is irked because the Jaycee who was in charge of positioning the floor vendors in the coliseum was three hours late on Thursday morning. Without his direction, Wideman had set his book-signing booth in the spot reserved for the lemonade guy. By the time the Jaycee finally arrived, an hour before the downtown parade was to begin, it was too late for Wideman and his classic 1957 Thunderbird ("it's a stunningly beautiful car") to participate. "I had to move all that shit."

Wideman continues to wax effusive until a man with mustard smeared on his face stops by to chat, spraying bits of hotdog in his face.

As far as roundups go, Sweetwater's is by no means the most bizarre (or grisly). Several others in Texas and Oklahoma (which legitimized its roundups in a state publication) hold a snake-stomping contest (a spectator favorite), where each contestant stomps on a diamondback secure in a burlap bag. The person who kills his or her snake with the fewest stomps wins the cash prize. Live snakes are kicked, putted, juggled, provoked, burned with cigarettes, funneled full of liquor, and separated from their rattles (by knife). For five dollars, a parent can give his child the opportunity to decapitate a snake with a hatchet. And for a fee (of course), a family can hold a live diamondback for a photo opportunity; safe in the knowledge that the snake's mouth is sewn shut.

Then, there's fifty-eight-year-old Jackie Bibby, ex–drug addict and felon, lifelong adrenaline junkie, and king of the rattlesnake daredevils, who calls himself "The Texas Snake Man." A snake handler since 1969, Bibby appears in four editions of the *Guinness World Records* for sharing a sleeping bag with one hundred twelve rattlesnakes and a transparent bathtub with eighty-four, for crawling head first into a sleeping bag already cohabited by thirty rattlesnakes, and for suspending eleven from his mouth by their tails. *Guinness* voted Bibby's snakes-in-a-mouth record sixth on its list of the top ten feats of the last fifty years. Bibby's been bitten twelve times. When he's not performing with diamondbacks, Bibby runs a boarding house for recovering alcoholics and drug addicts in Fort Worth, appears in movies (actor and stuntman), and has guest spots on television shows. He maintains a website (Texsnakeman.com) that provides far more information than you'd ever want to know about him. In November 2008, Jackie Bibby failed to break his snakes-in-a-mouth record in Cologne, Germany. "I do it for the attention," he said. Although he's missing a leg and has a deformed thumb, Bibby claims handling rattlesnakes in front of a live audience is the most fun he's had with his pants on.

Bibby also holds the world record for sacking ten diamondbacks in 17.11 seconds, a popular contest at roundups. Whenever a sacker is bitten, the judge adds five seconds to his time; if he's bitten twice, he's disqualified. In Pennsylvania, two-member sacking teams also competed for fastest time, a little like *Beat the Clock* but without Bud Collier. A sacker would stuff five timber rattlesnakes barehanded into a sack held open by his partner. After the fifth snake was in and the sack twisted shut, three judges stopped their watches and averaged their times, which were often less than four seconds. If a snake was injured, the team was assessed a five-second penalty. Even though timber rattlesnakes aren't as recalcitrant as western diamondbacks, Pennsylvania sacking contests had a home-state caveat; if you were bitten once, you were disqualified. Children were encouraged to compete in nonvenomous-snake-sacking contests. Then, in 1985, after Ralph Abele, the executive director of the Pennsylvania Boat and Fish Commission, banned the use of native timber rattlesnakes for sacking contests, Pennsylvania legislator Thomas Petrone introduced a bill in the state house to overturn the decision. One representative in favor of passage wondered what Director Abele would think of next, a "ban [on] fish hooks because it hurts the fish's mouth." To which the coordinator of endangered species for the Boat and Fish Commission responded, "If they were stuffing Bambi in a sack, there would be a tremendous public outcry." Frank Anderson, an editorial writer for the *Valley News Dispatch*, questioned whether Representative Petrone thought that was why his constituents had sent him to Harrisburg—to legalize rattlesnake-sacking contest—and suggested an alternative roundup activity: "taxpayers could compete to see which team can place a legislator in a sack the quickest."

When a reporter asked Marilyn Black, the first woman appointed to the Pennsylvania Boat and Fish Commission, her thoughts on the attempt to overturn the ban on sacking contests, Black responded, "It would be hard for me to believe that rapidly pushing a snake into a bag or other container would be advantageous to its health."

No matter. The Pennsylvania house voted one hundred fifty to fifty-one to lift the ban. Fortunately, Dick Thornburgh, then Republican governor of Pennsylvania, ignored them.

Unlike timbers, western diamondbacks joust for love in spring, the males attempting to rise up above each other—straight up, leaning, pushing, each snake straining against its rival, face pointed skyward. Tallest snake wins. The female chooses the victor and mates, an entwined affair that lasts up to twenty-five hours. She may then seek the same mate over several years, passing neighboring males to reach him. Beginning in her second or third year, the female breeds every other year, incubates the developing eggs in her oviduct, and gives birth in the fall to two to twenty-five live membrane-wrapped snakelets. Their venom, drop for drop, is more potent than that of their parents.

Like timber rattlesnakes, baby diamondbacks eat mice; but the adults, which are larger and heavier than adult timber rattlesnakes, consistently eat bigger mammals—rabbits, hares, rats, ground squirrels, and occasionally birds—all warm-blooded creatures, whose presence a diamondback detects through its heat-sensitive facial pits, those richly enervated organs midway between the eyes and nostrils on either side of the head. As you recall, from within these pits, an infrared image is projected onto the snake's optic tectum, where it is overlaid on the visual image transmitted by the optic nerves. All pitvipers see a world unknown to man, a world of radiating heat, of warm food and cool hideaways.

One late April morning, a number of years ago, I spotted a four-foot western diamondback commuting through cactus-country outside Tucson. A reptilian sleuth in the company of my nephew and son, both energetic and nosy and under the age of six, I followed the snake, which appeared to have an agenda and, as far as I could tell, a destination. Deliberately and with nonchalance, it wove through parched grasses and orange and blue wildflowers, swam a hillside stream, its tail held well above the water, and then climbed a canyon wall, passing chorusing canyon tree frogs, while oozing around sun-burned rocks. Above the canyon lip, the diamondback vanished into a weft of yellow-flowered palo verde and gnarly mesquite. The flowers were astir with honeybees and wasps. Beneath the trees, seedpods littered the ground, which attracted rodents (lots of droppings and gnawed pods), which in turn may have attracted the snake. Some-where in the filigree of shade, the rattlesnake waited, coiled, ready, and invisible.

The western diamondback ranges from sea level along the Texas Gulf Coast to ten thousand feet in the arid and semiarid mountains of Mexico. From southeastern California to central Arkansas and south through central Mexico they frequent plains and desert scrub, offshore barrier islands—where laughing gulls may appear on the menu—rocky canyons, riverine bluffs, desert mountains, and sparsely vegetated foothills. Although western diamondbacks vary in size geographically, males everywhere grow ten percent (or more) larger than females, what Darwin called "sexual size dimorphism," which is probably due to males competing for mates and feeding throughout the summer while gravid females fast. Diamondbacks grow largest in the Rio Grande valley (although a snake was recently killed south of Dallas that measured seven feet two inches), are most common in West Texas (hence the roundups), and are everywhere infamous for not backing down. An annoyed diamondback rises off the ground like an expanding spring, head held high above the coil—the eternal position of most of the freeze-dried snakes for sale at the Sweetwater Roundup.

On the continent of North America, rattlesnakes are the most widely encountered of all of the dangerous animals that remain. Though snakes rarely harm humans or domesticated animals, Americans nev-ertheless have a long history of organized efforts—like the Sweet-water Roundup—to collect and eliminate rattlers. As recently as 1989, as I mentioned earlier, Clairemont, Texas, held its forty-first and final Peace Officers Rattlesnake Shoot.

Still, Sweetwater Mayor Wortham, whose paying job is executive director of the West Texas Wind Consortium, adheres to the notion that snakes are a serious threat to people and to the well-being of the community: "There's a thousand people working in rattlesnake country [on wind turbines] on a twenty-four-hour basis. Snakes need to be controlled." Jacque McCoy, of the chamber of commerce, agrees: "We have a rattlesnake problem. Something needs to be done with all those dens." Tom Wideman agrees. "I believe our area would have been overrun with rattlesnakes if not for the roundup."

Sweetwater's inaugural roundup was held in 1958, after area ranchers and farmers complained of losing livestock to snakebite and the Board of City Development decided to purge Nolan County of western diamondbacks. This took place in spite of a survey conducted two years earlier by Lawrence Klauber, who polled all two hundred fifty-four county agents in Texas for his opus and asked them to rate livestock loss due to snakebite as serious, moderate, unimportant, or negligible. Of the one hundred thirty-four that responded, none listed snake damage as serious. More than half of the agents thought it negligible, and one claimed rattlesnakes accounted for only one-half of one percent of all livestock deaths.

Apparently, nobody paid attention to Klauber's research. During the first Sweetwater Roundup, ranchers incinerated more than three thousand live diamondbacks in fifty-five-gallon drums at an oil well outside town. "I killed every damn thing in the world," says Wideman, who was a teenager studying geology at a local university at the time, "but I couldn't do this. It was too brutal. I had to leave." To ensure that the roundup would continue to thrive, Sweetwater's development board enlisted the help of the Jaycees. The following year, in 1959, the Jaycees sent representatives to Okeene, Oklahoma, home of the country's oldest roundup, to learn how to transform their event into a full-blown festival.

If Sweetwater could ever be said to be truly "on the map," the turning point came in 1980, the year John Travolta, then Hollywood's golden boy, appeared in the box-office hit *Urban Cowboy* wearing snakeskin boots. The demand for rattlesnake products soared, and roundup boosters had new reason to keep the spectacle alive. Nobody can say how many snakes were brought in over the ensuing years; snakes are not recorded as individual animals, but rather tallied by weight of the total catch. In 1982, approximately eighteen thousand pounds of diamondbacks, double the previous record, were brought to Sweetwater, and over the next six years the roundup averaged more than twelve thousand pounds per year. The entire snake has economic value: the skin, the venom, the gallbladder, the blood, the rattle, the fangs, the head, and the meat, with its innumerable toothpicky ribs.

In 1994, *The Simpsons* lampooned the Sweetwater Roundup in the twentieth episode of the fourth season. "Whacking Day" highlights fictitious Springfield's annual snake drive and clubbing in the center of town, accompanied by a chorus of "O Whacking Day," sung to the tune of the Christmas carol "O Tannenbaum,"

CHILDREN:
   Oh, Whacking Day, Oh, Whacking Day,
   Oh, Whacking Day, Oh, Whacking Day.
SOLOIST:
   We'll break their backs,
   Gouge their eyes,
   Their evil hearts
   We'll pulverize.
CHILDREN:
   Oh, Whacking Day, Oh, Whacking Day
   May God bestow His Grace on thee.

"Whacking Day," which won a prestigious Genesis Award for Best Prime Time Animated Series, is considered a classic. Guess where Homer's allegiance lay?

All told, over three hundred thousand pounds of western diamondbacks have been weighed, measured, sexed, and killed at the Sweetwater Roundup, although Tom Henderson is skeptical about the results from the research pit. "It's a little different than opening up a *Playboy* magazine and saying 'It's a girl alright.' We used to force out the male's hemipenis. Today, they just insert a probe, quickly, and make a hasty determination." The Texas Department of Parks and Wildlife reports that snakes are captured not only from Nolan and Taylor counties but also from more than twenty other counties across the state, and even from Oklahoma. Unlike Henderson, serious hunters no longer go afield on roundup morning and return in the afternoon to weigh their catch. They begin the hunt months in advance, storing snakes in wooden crates—tangles of serpents crammed together in basements, usually with no food or water. (The 2014 statistical report for the fifty-sixth annual World's Largest Rattlesnake Roundup listed the total weight of diamondbacks at thirty-eight hundred ninety pounds, and the longest snake measured, a crotaline behemoth stretched six feet four inches.) One Jaycee told me that to appease his wife he built a separate basement for his snakes so she wouldn't have to do laundry among crates of buzzing reptiles.

🐍

No one knows how many diamondbacks still live in West Texas, though the estimation by some (as I mentioned earlier) is astronomical. Hunters, including Henderson and Wideman, believe diamondbacks are inexhaustible, that they're everywhere in a commodious and vacant terrain. "Look around you," Wideman tells me, "there are miles and miles of nothing but miles." When I ask Henderson if someone should monitor the ecological impact of the rattlesnake harvest he shrugs, his cream-colored hat frozen in place like the rattlesnake curios behind him. "One big, bad cold winter will kill more snakes than twenty-five roundups," he says. I remind him how fishermen once thought cod off the Grand Banks were inexhaustible. Smiling, he replies, "I never knew anyone who died from a cod bite."

Although the American Society of Ichthyologists and Herpetologists, a group of about two thousand scientists who specialize in the biology and conservation of fish, amphibians, and reptiles, urges an end to roundups because of their public cruelty—"It is hard to imagine subjecting any other vertebrate animal to such thoughtless and inhumane treatment"—it is equivocal about whether these events permanently deplete snake populations. "The biological ramifications of decades of rattlesnake roundups are difficult to assess, but they have great potential to affect populations negatively, and it is difficult to predict when rattlesnake harvests will push local populations beyond the point of recovery."

Still, there may be another reason to protect rattlesnakes: they eat rodents. During the occasional wet year, the plains are vibrant and rodents abound. A diamondback feeds about once every two weeks, consuming fifteen to twenty mice and rats a year. If five thousand snakes are killed in a single Sweetwater Roundup, that leaves seventy-five thousand to one hundred thousand rodents free to breed without fear of snake predation in a land of little rain, challenging cattle for what meager vegetation grows naturally, which may be a greater threat than the occasional snake that bites a cow. (An adult Norway rat, which can produce as many as ten litters in a year, with an average litter size of eight can consume more than twenty-five pounds of grain per year.) Herpetologists have proposed that the recent spread of rodent-borne diseases like hantavirus and bubonic plague could be connected to efforts to control rattlesnake populations. Still, eight states continue to hold roundups. If Sweetwater's education pit were to live up to its name, we might learn to marvel at these wondrous creatures: their legless gait, their lidless eyes, their hypodermic fangs, and their forked tongues and facial pits that read messages we cannot possibly comprehend. Instead, for twenty dollars you can undo a snake yourself at the skinning pit, have your picture taken, and then contribute a bloody handprint, signed and dated, to the wallboard behind the skinners.

Part Two

INGRESS

THE DANGERS
OF COMING HOME

10
Chaos of Rocks

To find the snake is the Challenge. To get out into the wilderness, enjoy all the beauties of nature and finally, with good fortune, to discover the creature in its natural haunts: this is the Reward.

CARL KAUFFELD

One unbearably humid afternoon last June, after twelve straight days of rain and periodic flash floods, Alcott Smith and I visited a talus slope at our undisclosed site in the Northeast. Nothing on the rock pile was stable, certainly not the trees rooted in the thin soil at the base of the cliff, many of which had already toppled and lay crisscrossed like pick-up sticks or leaned at precarious angles and seemed ready to fall at the slightest provocation. Walking along the upper talus, a mile from our car, at an incline six times steeper than a ramp for the disabled, Alcott dislodged a rogue boulder the size of a Volkswagen that smashed downhill forty or fifty yards, transmitting a vibratory message to every rattlesnake within the immediate rock field, the very snakes we had hoped not to startle, and sobering us as no snake ever had. Once the boulder settled in place, the thick, stale air reeked, a flinty, gunpowder smell, the by-product of intense frictional heat as rock ground against rock. Now, whenever we cross talus, we go single file like first-graders on a field trip (the only difference is we don't hold hands) ... horizontal baby steps. *Never* vertical.

On talus, I strive to keep my balance and to watch for rattlesnakes, often simultaneously. It's not easy. Imagine a skirt of fractured rocks that begins at the foot of a cliff and slopes to level ground, which may be two hundred yards away if the cliff rises thirteen hundred feet high. A desolation of boulders, millions of rocks that peeled away from the precipice, each breaking apart as it tumbled down. It's all about weathering and gravity. The most massive tend to accumulate farthest from the source, many tons of once out-of-control, spark-spraying boulders, some as big as Buicks, bouncing downhill, that eventually wedged against conspecifics, where they'll remain until the end of time. When a glacier gives birth to an iceberg we call the process "calving," but I know of no comparable term for the birth of a boulder. There ought to be, for the process must be louder, longer, more odiferous, and more destructive (unless you were a passenger aboard the *Titanic*). I've never witnessed a cliff shedding rock, but I imagine it must be formidable. Smaller, lighter rocks collect higher up the slope, having either chipped off tumbling boulders or fallen directly from the cliff, splattering like pieces of broken ceramic. Older, larger rocks, those settled in place for centuries (or longer), are weathered smoother, carpeted with lichen and moss, and stable to walk on, though you wouldn't want to fall into the unforgiving spaces between them; boulders at the bottom of the pile are embedded in the earth, which must dimple under their great weight. More recent additions to the pile are sharp-sided, barren, and unstable to walk on, particularly if the rocks are small. Hiking across small talus is like hiking across a slanted field of seesaws or up a mound of stone boxes, any misstep may trigger an avalanche. Rain, particularly torrential downpours, makes talus walking edgy: rocks are slick and unstable, water having washed away whatever soil helped bind them together.

🐍

The word *talus*, derived from the Latin *talutium*, means a slope with gold. Although some talus may superficially resemble mine tailings, any suggestion of precious metal, except perhaps for colorful October leaves and the odd golden rattlesnake with chocolate crossbands, has disappeared from the more contemporary geomorphologic definition: the accumulation of blocks of rocks broken off a cliff by the inevitable assault of gravity and water—weathering—principally by repeated freezing and thawing, but also by the leaching of heavy rains, as well as wetting and drying. During postglacial times, when freeze-thaw cycles were far more frequent and far more severe than they are today, talus slopes assembled at a more rapid rate.

Every slope has a critical gradient, what geologists refer to as the "angle of repose," determined by a balance between gravity, which attempts to pull the rocks down, and inertia, or "internal friction," a by-product of the angles of the fragments and the grade of the slope. The size and shape of the rocks reflects the spacing of natural joints or fractures in the parent bedrock, which is determined by the chemical composition of the rock itself. Schist, for instance, a metamorphic rock transformed by extreme heat and pressure miles below the surface from previously existing rock that was the under-pinning of some vanished mountain range, fractures more readily than granite and tends to break into platy, less stable fragments. This may explain why I've never seen a rock climber scaling the cliff behind the snake dens. Water, which trickles into cracks and pores

in the cliff, expands nine percent when frozen. Then, the outward pressure of the swelling ice splits off chunks of schist along its natural joints; gravity does the rest, liberating wild blocks, both big and small, that bounce, carom, and ricochet down the pile below, further fracturing along the way. Every talus slope I've ever walked on, even those obscured by forest, has been littered with chips of rock that cause me no little disquiet.

Whenever a slope exceeds the angle of repose, a landslide is possible. The exact angle of failure depends on several features: rock type (granite, marble, shale, schist, gneiss, and so forth), rock size, and moisture content. Dry, homogenous rocks generally experience slope failure when the angle of repose exceeds twenty-three degrees. The angle of failure lowers after rain erodes away whatever meager deposits of earth have built up between the rocks, because the frictional glue that keeps the rocks in place is lost; water also adds to the mass of the entire slope, increasing the impact of gravity. Torrential rain, like those that blanketed the Northeast in June and July of 2013, pour over the cliff and coalesce into intermittent, subterranean streams that flow below the talus, washing away detritus and rock dust, the underpinnings of the slope, loosening both individual rocks and possibly the entire slope. I imagine that this coming winter, rattlesnakes will discover that their ancestral hibernacula, which are deep in the basement of the talus, have been scoured by flowing water.

Wet or dry, ambulating across a talus slope is unpredictable and often treacherous. In the summer of 2013, three teenagers were crushed by a rockslide in Minnesota, and early in the nineteenth century, above the shore of Lake George, a snake hunter died on Tongue Mountain, when the rocks he walked on shifted. In 1983, on an afternoon outing that Randy Stechert astonishingly evokes as if it had occurred yesterday, Bill Brown and Stechert hosted the *National Geographic* photographer Bianca Lavies, on assignment to illustrate Brown's influential timber rattlesnake story. An intrepid Dutch artist with a world of credits on her resume, Lavies worked like Ansel Adams and spoke like Elmer Fudd. Late in the afternoon, higher up the talus, Brown dislodged an eight-foot boulder that caromed downhill, launching a rockslide. Stechert hotfooted out of the way just as the boulder bounced straight up and snapped off a one-foot-diameter tree five feet above the ground. At the base of the talus, Lavies ducked behind a rather inconsequential tree, which she clutched with both hands as though hanging on to the side of a cap-sized boat. Once the rockslide subsided, a cloud of dust hid much of the talus, including Lavies.

"Bianca. Bianca. Bianca, are you alright?"

A thin, unnerved voice answered. "I'm all wight, Wandy. I'm all wight."

🐍

Why do I return to talus? Not because I'm a risk taker—I prefer to be in reasonable control—but because of its accumulated (and odd) natural history, which of course features timber rattlesnakes, the marquee creature of an eastern rockslide. I do not come to handle them, and I do my best not to disturb them, which sometimes is impossible, particularly when the rock I'm standing on begins to move. I come to bear witness to their methodical behavior, which predictably shifts with the temperature and with the season. Watching a rattlesnake emerge from a den is like watching a wildflower unfurl its petals; watching one disappear down a subterranean passage, pulling down the seasonal curtain, is like the faint voice of geese on a moonless night ... and it leaves me with both a sense of loss and a sense of continuity. And also a sense of mystery: what goes on below the rocks?

This is the spring 2013 collective count of timber rattlesnakes from our first two dens heading south: April 29, ten; May 3, twenty-five; May 5, one hundred five; May 10, one hundred eleven; May 17, thirty-two (I only walked a short distance from the first den); May 18, one hundred twenty; June 3, sixty-three; June 18, nineteen. If I had had the inclination to visit all six dens (read agility of a mountain goat and brains of a grasshopper), the count might have doubled. Rattle-snakes love talus, which is the only reason otherwise sane men keep returning. For snakes, it's the perfect place, a thermal Mecca that offers innumerable basking and shedding sites, as well as shady retreats from the midday sun and from aerial predators like red-tailed hawks, which soar above the rocks on the updrafts that deflect off the cliff.

Throughout the Appalachian ranges north of the Great Smokies—the Blue Ridge, the Allegheny, the Pocono, the Kittatinny, the Hudson Highlands, the Catskills, the Taconics, the Berkshires, the Adirondacks, the Blue Hills—and the fractured slopes above East Coast rivers such as the Delaware, the Susquehanna, the Connecticut, the Merrimack, the Hudson, and the Housatonic, talus is the preferred site for denning rattlesnakes, which overwinter in labyrinthine chambers below the surface of the upper slope, where rocks tend to be smaller and more deeply piled. At our location, by mid-October, snakes have crawled down the talus into the detritus layer or entered a subterranean fissure in the bedrock in the root of the cliff, where they knot together in dark, dank corridors surrounded by air that remains nearly fifty degrees Fahrenheit all year, very like a cave. In May, after hours of trekking, I may see more than a hundred snakes in leafless sunshine, grouped together in pods of three, four, five, even ten, each comprised of snakes close to the same size and age. The pods are points on sinuous but traditional lines of egress that lead away both north and south from the den. Keeping track of pods, I connect ophidian dots across my internal map of the snakes' neighborhood, taking note of snakes on rocks, under rocks, and in beds of fallen leaves between rocks. By the time spring grades into summer, most pods have disassembled, rattlesnakes having departed along one of three main routes, visible to us by the presence of the snakes themselves: they either work their way up a chimney to the escarpment and then disperse into the ridgetop woods; or swim across the brook to summer in green, unpeopled terrain; or move north, guided by the cliff, into woods and lowlands along the west side of the brook and associate beaver ponds, where the land is marshy, rocks are in short supply, and voles are abundant. Rattlesnakes that summer in the talus, almost always at traditional sites, are gravid females, which, with a collective anodyne against boredom, hang together like members of a Lamaze class metering sunlight, ripening embryos, and then in late August or September giving birth. Recesses in the rocks, climate-controlled chambers, become semiprivate birthing quarters, for in the Northeast, timber rattlesnakes are wedded to rock, the site of their nativity and the site of their winter dens. Without rocks they'd perish.

On the late morning of June 18, 2013, after weeks of inclement weather—one afternoon, three inches of rain fell in a span of fifteen minutes—and a water table that rose into places I had never seen water before (my basement and the back of my garage, for instance), Alcott and I drove more than three hours through alternating downpours and drizzles to find rattlesnakes, all the while debating whether we should turn around and go home. We arrived at our destination, parked the car, and walked south a mile along the edge of the brook, which had risen nearly to the base of the cliff, flooding out every terrestrial creature that didn't retain a vestige of webbed feet or set of water wings. The air temperature was fifty-four degrees Fahrenheit— not exactly optimal snake weather, the sky was overcast and threatening, and the waterfall, which could be heard a mile away, fell in a wide, tumultuous curtain like a miniature Niagara. Rivulets and cascades poured over and out of the cliff in places I had never seen flowing water before, and an intermittent stream that I had once stepped over was twenty feet wide and had reset its channel. Walking on the talus was perilous. Lichen-covered boulders were greasy, smaller rocks gave way, often in groups, and the steep, muddy trails between rock piles were slippery, forcing us to ascend the slope with our arms as much as with our legs, hand over hand, brachiating from saplings to low-hanging branches like ground-dwelling gibbons. At the first gravid site, seven post-shed snakes, four yellows and three blacks, all fearless and thermally obligated to bask, lay out in the open, coiled or stretched, glistening in the rain, attempting to suck out whatever heat the rocks and air had to offer. A gravid timber rattlesnake attempts to maintain her body temperature between eighty-five and ninety degrees Fahrenheit for at least seven hours a day during the approximately three months of gestation. Every year is different, of course, as different as the weather. First births in 2012 were on August 16. First births in 2013 were on August 22. First births in 2015 were on August 18.

Wherever the snakes had previously loitered on a bed of leaves, either under a rock awning or in the tiny valleys between rocks, leaves were compressed, dents in the forest pillow that became quite obvious once we began to search for them. Not far from the snakes, seven sheds, the older ones tattered by rain, draped the rocks. Slowly, we worked our way from one gravid pod to the next, crossing a stream below a small waterfall, where an updraft hijacked the water before it ever reached the ground and transformed it into an emaciated cloud that drifted north along the cliff for a short distance before being engulfed by drizzle. Next, we passed through oak and hickory woods and over two small talus fields until we reached Sculpin Den, the third hibernaculum in the third major talus slope south from where we had parked. Beyond Sculpin, the rocks were too steep, the slopes too long and too wide to safely walk. Five post-shed snakes, all gravid, three blacks and two yellows, gathered in front of the den crevice. Water droplets glistened on their backs, beaded by a residue of oily secretions that had helped the snakes molt. One of the intact skins that had been divested during the night clung to the rocks like Saran Wrap, fixed in place by the morning downpour.

Late in the afternoon, the sun finally broke through. Two snakes closest to me, a four-foot-long black Methuselah, with an eighteen-segment rattle, which, when held erect, curled backward from the weight of the segments like an upended J (the longest rattle I have ever seen), and her variable tan-colored colleague, a handsome snake with twenty-seven dark-brown bands bordered in yellow that crossed an incomplete orange-tan dorsal stripe, crawled on top of the rocks and stretched out in the sunshine. Eventually, the other three snakes, which were oblivious to our presence, joined them in their vital (and short-lived) quest for heat, positioning and repositioning themselves in the open, tracking the sun as it moved in and out of the bruised cloud bank, always keeping the long axis of their bodies in full light. Never coiling. As the afternoon warmed, threads of steam rose off the rocks, and the heat and the quiescent rattlesnakes lulled us to sleep as well. We dozed off, AARP members napping on warm rocks. Eventually, the sun sank behind the cliff, and we woke up. All five snakes, however, lingered on the rocks for another ten or fifteen minutes and then slowly pulled back, wedging themselves into the talus, a loop or two visible only if we bent down.

For a timber rattlesnake connoisseur, there are two kinds of places: those with snakes and those without. And if you've read this far into the book, you may remember that rattlesnake habitat, particularly in the Northeast, has never been homogeneous, never uniform, never permanent—even at the height of their northeastern distribution a thousand years ago. During the Pleistocene epoch, over the past two million or so years, timber rattlesnakes moved into and out of the Northeast with each of four stupendous glacial invasions and retreats, reappearing most recently five thousand years ago during the worldwide warming spell, the Hypsithermal Interval, when the seasons and climate in Vermont were more similar to those of the piedmont of modern-day Virginia. After the zenith of their distribution, timber rattlesnakes were negatively and unevenly impacted by the "Little Ice Age," a cold snap that lasted about four hundred years, peaking during the seventeenth century, and by the colonial tsunami that began to inundate the eastern seaboard around the same time and still hasn't subsided. According to Randy Stechert, a maven of such matters, of the more than two hundred fifty viable dens in the Northeast that he's personally visited, only thirteen support snakes in anywhere near pre-Columbian numbers, which he places at a hundred or more snakes per den. That's thirteen … as in three more than ten. W. H. Martin believes less than one percent of *all* timber rattlesnake dens, everywhere, support anywhere near historic numbers. All other historic dens, many of which face varying levels of jeopardy, are classified as either medium, small, or critically depleted (think New Hampshire). Most have ceased to exist.

In Vermont, thirty-six of forty-two known dens have been extirpated, several since the middle of the twentieth century. In New Hampshire, twenty-two metapopulations have been whittled down to a single den (Andy's Den) and a handful of snakes, which the state recently purchased from a Boston-based gravel company that had a history of destroying snake-sensitive rock formations. In Connecticut, estimates suggest that the historic statewide population has fallen by eighty-five percent. That timber rattlesnakes persist (though barely) in the Blue Hills, within sight of the Boston skyline, is a testament to both their tenacity and their benign and secretive nature, as well as to the informed ranting of James Condon; that they've all but blinked out in New Hampshire, and would have been gone several years ago without the manic whistle-blowing of Kevin McCurley, is an expression of wanton slaughter, climate change, development, and mining, which collectively contributed to inbreeding depression, which in turn favors the lethal, keratin-eating fungus *Ophidiomyces ophidiicola.*

Back to Vermont. During the last decades of the twentieth century, the increasing volume of traffic on a state highway that parallels four ridgeline dens has become a prime selection factor in the evolution of local rattlesnakes, favoring those snakes that move away from the road after emergence, eastward, up and over the ridge. Roads, even horse and buggy carriage roads, have never been kind to rattlesnakes, which move slowly and coil defensively when startled by the vibrations of an impending vehicle. Virtually every Vermont snake that has dispersed downhill has become a DOR (dead-on-road, the acronym coined by the late Lawrence Klauber) or has been captured (illegally) and rendered into oil, its genes and its pheromone trail lost to the population as decisively as if a predator had taken it out. Absent these snake trails, many ancestral dispersal routes have become forgotten languages (fortunately for the snakes). In central Connecticut, the survival of a stable metapopulation of timber rattlesnakes, circumscribed by state and federal highways (not unlike the Blue Hills), depends on the sympathy of suburban neighbors, who sometimes find snakes mating on their front lawn or migrating along traditional avenues of egress now called Pepitone Road and Larson Avenue. One solicitous man reports that he ushers them off the road with his boy's hockey stick.

Our location is different; it's a world apart, remote and beautiful, a physiographic backwater in the rearview mirror of modern times. Here, rattlesnakes have forgotten the negative side of people and no longer regard us as a threat, even though for many years a local bounty hunter named Sparky Punch raided all six dens (and many, many more). To the snakes, I'm just another rock-side attraction, two-legged and clumsy. Unless I inadvertently startle one or come upon a particularly anxious individual that is easily provoked, they keep to their personal crotaline agendas, whatever they happen to be. I have followed a snake that forded a stream and then stopped to drink in a splash puddle from a waterfall, its head partially submerged, its jowls slowly pulsating, drawing water into its slit of a mouth. One April morning, as Alcott stood as still as stone, a rattlesnake that had recently emerged from the den left its basking rock, crawled down twenty or thirty feet to the brook, drank for fifteen minutes, and then returned to its original rock and soaked up heat all afternoon. In May, perched on an outcrop, I watched a rattlesnake swim across a beaver pond, and in September I watched another swim home, its inflated body buoyantly undulating high on the surface like a piece of driftwood. When the homebound snake passed through a patch of duckweed, its yardstick-straight trail (muskrats zigzag) belied the curvy motion of its swimming, and when it emerged from the water, it was peppered by tiny, green plants, some of which clogged its rattle, deadening the sound.

Our trailless corrugation is forty miles from a town of any relative size, where acres of reeds fill the shallow beaver ponds and cliff-nesting turkey vultures and ravens trail their shadows over a tessellation of forest and open water. In the woods, I find little piles of feathers, which once belonged to ring-billed gulls or crows, the remnants of goshawk kills. Wildness runs as far as the eye can see: pleated hills, row after row, brush the horizon above the flourishing, waterlogged meadows. In May, pickerel spawn in the weedy ponds, their prehistoric, toothy countenance in wide release; water snakes patrol the lowlands, and black rat snakes the uplands. Otter scat and beaver chews are everywhere. An eagle or an osprey may appear above the water like a reassuring dream. On the far shore of the largest pond, one autumn, behind a constellation of geese and ducks and a lone tundra swan, half a dozen great egrets stood as still as pale lawn ornaments. And several years ago, the summer after a bumper crop of acorns and beechnuts littered the Northeast, the woodland rodent population peaked. White-footed mice, chipmunks, and gray squirrels were (almost literally) underfoot. The following fall, after a crash in the mast crop, starving gray squirrels, many suffering from mange, desperately crossed the ponds. One squirrel drowned as I watched, spinning and bobbing in wind-driven chop until, exhausted, it listed to starboard and was swept away.

Talus soil is in short supply. Where it accumulates below the rocks, sunlight can't reach. Therefore, green plants can't grow. In the upper talus, the smaller, more tightly packed rocks may collect traces of soil along their seams, supporting a lattice of vines—Virginia creeper, poison ivy, wild grape—and small stands of stunted, nutrient-starved trees and shrubs like bladdernut, elderberry, mountain ash, scrub oak, sumac, mountain maple, box elder, white oak, and hackberry, and, in season, wildflowers—rock saxifrage, squirrel corn, Dutchman's-breeches, columbine, herb Robert, downy white violet—which gives the gray rocks splashes of color. I frequent the upper talus with all its thermal amenities: the insulating beds of leaves; the detritus; the shady retreats from the sun. Surprisingly, the sun-scorched, often barren rocks, which may reach surface temperatures of more than one hundred thirty degrees Fahrenheit, host a variety of life: five-lined skinks, wolf spiders, the three-inch-long millipedes that flow over the stones, waves of motion passing down their innumerable legs. Garter snakes, black rat snakes, and, less frequently, milk snakes patrol here, and all of them may bed down for the winter with rattlesnakes. One late April morning, I watched a rattlesnake crawl from enduring darkness and into daylight; after the rattlesnake settled on the talus to bask, I turned around to see a rat snake at eye level peering out of a small cavity in a hackberry—three inches of snake, a dark, animated stub, fixed on my position. Porcupines visit the wide mouths of the larger snake dens, which may explain why a quilled snake wound up on Larry Boswell's coffee table. Black bears, coyotes, bobcats, and foxes, both red and gray, also leave their signs across the rocks, messages from other tribes.

Mesmerized, I have watched both the day and the season pass by. One late September afternoon, as darkness congealed over the talus, barred owls and coyotes sent vespers across the river while the stuttering call of an eastern screech owl wafted over the rocks where Alcott and I had napped surrounded by snakes. In early October, if the temperature remains warm, homebound rattlesnakes stay on the talus; if the temperature falls, they go under, braided together like so many twists of licorice.

There's history amid these fields of rock untethered from the hand of man, a phenomenon more commonly experienced in the Desert Southwest. As the day unfolds, the sun rises over the green hills of Massachusetts, heating the talus. Eventually, in accordance with both the temperature of the air and that of the rocks, as well as each snake's personal thermal itinerary, which varies with its state of health, digestion, gestation, and cycle of ecdysis, rattlesnakes either appear on the surface or linger just below. If a snake is cool, it stretches out and basks; if it's too warm, it retreats into the shade of the rocks, of which there are uncountable options, exposing a loop or two to the sun ... or none at all. For a fat-bodied, cold-blooded timber rattlesnake, a talus slope is a geographic heating pad, with a million thermal gradients. There's a rock or a retreat for every occasion and for every snake, and rattlesnakes know the best rocks: thick ones take longer to heat up, but also longer to give up heat; thin ones are less thermally stable, heating up quickly (sometimes too quickly) and cooling off just as quickly. Finding the perfect rock, the rock with just the right thickness, is an evolved art form. Just beyond a major den is a boulder capped by narrow, flat rock. Yearlings love this rock. It's the perfect fit. During ingress and egress, three or four snakelets squeeze under the cap, safely wedged in place. To see them, I bounce sunlight off a mirror to light up the darkness.

Keeping their body temperature within the preferred optimal thermal zone (between sixty-five and ninety degrees Fahrenheit) is the key to a timber rattlesnake's survival in the Northeast, the outskirts of their distribution. As you may have already gathered, timber rattlesnakes are solar collectors. Their dark base color, which is far more prevalent in the Northeast and Appalachian Highlands than on the Piedmont or the Gulf Coast, is an adaptation for solar gain; more than sixty percent of the rattlesnakes at our site are dark-phased, and even higher percentages occur farther north. At some sites, like Merrimack valley (as you may recall), there are *no* yellow morphs. Every snake is as black as a bicycle tire.

Up here, forest frames the talus slopes. Bands of deciduous trees separate four of the eleven rock piles, stunted trees separate the other seven, creating an alternating mix of open rock and woodland that extends for nearly four miles and may support a *large* gathering of timber rattlesnakes, a poikilotherm's nirvana based on rock. The friable wall of schist rises straight up for thirteen hundred feet, more or less paralleling the brook and beaver flowage, before tapering at both ends and fading into forest. Where the cliff becomes impassible foot travel abruptly ends; a hundred-foot-long, lichen-encrusted nylon climbing rope, an artifact left by an old snake hunter, hangs down a chimney in the cliff like a pre-shed snake, a lifeline up and down the precipice, though my boys prefer free climbing. Tied to a large red oak, the rope hangs down the very gully that snakes from a nearby den use during egress and ingress, their ventral scales hugging the slot like wallpaper. Once, a four-foot rattlesnake was on the way up while Alcott began his descent; Smith paused for reflection and the dispersing snake moved right past him and me—I was sitting fifty feet upslope—and disappeared into the next belt of talus. And, last Memorial Day, when my son Jordan, ready for an ascent, reached down and grabbed the rope, a young-of-the-year glided within an inch of his fingers. Unperturbed, both the snake and the boy carried on.

We had discovered the rope late one evening, the year before … and none to soon. On the twenty-sixth of August 2011, having admired the rock-embroidered landscape for decades from afar, I finally visited, harboring a preconceived notion that the rattlesnakes would behave as they do back home, where gravid females gestate along rocks on the lip of the cliff, well above their talus dens, as they've done for millennia. Armed with what turned out to be non-transferable knowledge, Alcott, Paul Jardine, and I climbed up the lower, north end of the cliff and hiked southward for miles, breathless and mindful of the fragile nature of the precipice and the jumble of rockslides, which fell away like a run of avalanches. Shaded outcrops, flanked by tufts of sedge and the exposed ledge, looked promising, but we encountered very few shelter rocks and fewer snakes. One, to be precise, a skittish yellow morph, and one discarded skin. That was it. Far to the south, the cliff assumed knee-knocking height, and directly below it a long, steep bed of talus sloped down into an emerald pond and a phalanx of cottonwoods. Standing on the edge looking down into the valley reminded me of standing on the roof of a Manhattan skyscraper, looking straight down into an elfin world where cars and buses appeared small and delicate and people ant-like. What the cliff and the building had in common, besides their dizzying height, was the chill they generated, which radiated through my body. Standing up here on an escarpment I didn't fully trust, I tried to imagine the landscape as a vulture might see it: the interplay of gray rocks, a ribbon of dark water and green hills. To imagine the view as a rattlesnake might see it, I would have had to convert the charm of the scene into a grossly out-of-focus bloom of color—like a telephoto camera lens used for a close-up picture—a world rendered soft and fuzzy and featureless.

Beyond the high point, the cliff tapered and fractured (rather abruptly) into a series of perpendicular outcrops in a steep, rocky forest. *Very* carefully, unaware that we were fine-tuning a fiasco, we descended a streambed, scrambling over and around boulders, until we reached an old logging road that led us down the valley to the brook. Walking along the brook north, we saw the talus slopes through a web of branches and rattlesnakes, first one then another, which apparently had crossed the water, coiling against driftwood or moving in our direction along the edge of the brook. When the sun vanished behind the cliff, we stopped admiring snakes and force-marched for more than two miles, over and around boulders, through boot-sucking littoral mud, all the while congratulating ourselves for having looped the entire four-mile cliff. Not quite.

In the gloom, we hit that wall, and our forward progress abruptly ended.

"Holy shit. I can't fuckin' believe it!"

"We gotta turn back."

"Turn back? Turn back where? By the time we reach the rocks it'll be pitch dark. That's insane."

"Hey, guys. Whad'ya think? Let's just stay here and slug mosquitoes all night," announced Jardine, with concise amusement, having distilled our choices to *one*. Paul Jardine is thirty-something, single and fit, a crew-cut conservative with a well-seasoned sense of humor, adept at poking fun at his more liberal companions. An EMT at a local hospital and a paramedic in the National Guard, Jardine had been called up for two Middle East tours and, hedging his bets, listed the Vermont Nature Conservancy as the sole beneficiary of his estate, which was designated strictly for rattlesnake protection. We had no light. No matches. No water beyond the stagnant pools. No food. No jackets. No cell reception. And diminishing humor.

"Better find a comfy rock."

"Hey, buddy, that's mine. I'm a federal employee, and I've been eyeing that for a while."

Dismayed, we walked away from the water fifty or so yards and stumbled (literally) on to a chimney with a sixty-degree pitch in an otherwise vertical column. With grim resignation, we gathered around the chimney, stared up at what we hoped would be some sort of physiographic holy grail, and debated whether to risk a climb to the summit, now obscured by a mélange of twilight and woodland shadow. It was then that I noticed the old, knotted, climbing rope.

"Oh my God."

"It's the miracle of the serpents."

"I need a sedative."

We couldn't see what the rope was tied to, but we yanked and yanked and yanked, assuring ourselves that it was secure; at least as secure as spending a sleepless night on a boulder slapping mosquitoes and listening to the subtle sounds of rattlesnakes moving in the dark. And, we hoped, more secure than retracing our route back along the brook, up through the rock-studded woods, and then north along the cliff for more than two miles before descending the last talus slope. After a brief, spirited debate, we decided to climb into the dark.

"Praise the Lord. I'm a believer."

Once we were safely on the summit, Alcott idiosyncratically notified us, "We're right on schedule for birthing," as if announcing a Labor Day mattress sale. "I always figure the last week in August through the middle of September."

Except in this location, either the forest or the talus (sometimes together) runs from the foot of the cliff to the brook or its attendant beaver ponds. But everywhere there are rocks. As I've mentioned, the higher the ledge, the steeper, longer, and more perfidious the rockslide. At the base of the longer talus slopes, hemlocks grow on and between monster boulders, and in the shade of their pagoda-like branches, the deep "caves" between the rocks hold snow and ice into the summer, issuing drafts of chilled air that flow over the rocks as though someone had left the freezer door open. For obvious reasons, rattlesnakes avoid these places.

To repeat, the snakes mostly prefer upper talus, where often smaller, warmer rocks radiate heat like a pizza oven. In fact, in spring, snakes begin shedding here several weeks ahead of snakes from populations several hundred miles farther south. A microclimate based on stone, the exposed talus extends the season of activity in both spring and fall and speeds up the onset of all biological processes.

Between the talus and the cliff, a ribbon of mostly misshapen deciduous trees—red oak, white oak, hackberry, shagbark hickory, basswood, green ash—victims of falling rock, parallels the base of the cliff and extends like woody peninsulas into the upper rocky slopes. Here, snakes bed on insulating cushions of leaves that gather between rocks and just inside stone foyers, and when thermal conditions are to their liking, they surface, stretching out on rocks like so many discarded belts, as they did on that drying afternoon in June when Alcott and I napped with them.

Along the brook and ponds, the trees are decidedly riparian: cottonwood, silver maple, red maple, and swamp white oak, many of which bear the chew scars of beaver. Slightly farther back from the shoreline, a few towering white pines grow out of the rocky earth, their ancient, shallow, serpentine roots looping across the landscape. Closer to the water, in fall, the incandescent red of a cardinal flower interrupts the otherwise monotonous brown mud and pale, withered stalks, offsetting the fetid smell of a half-rotted, unidentified gelatinous mass. Just off shore, broken rocks form shoals on the bottom, while a few big boulders rise out of the ponds like islands. Everything else is a fecund mat of pickerelweed and duckweed, emergent or floating.

※

On the U.S. Geological Survey topographical quadrangle that includes our site, the cliff appears as a six-inch-long smudge, but the aprons of talus bear *no* distinguishing symbol, nothing to either alert a seasoned hiker or obviate disaster for the initiate. The map renders the cliff down to a series of parallel brown contour lines, drawn at intervals representing twenty feet along points of equal height to express the land's three-dimensional ups and downs in two dimensions. The lines are so compressed that even with the aid of a magnifying lens I have trouble teasing out the thicker hundred-foot lines. On a contour map, denser groups of lines represent steepness, and a smudge represents a vertical and potentially fatal plunge. Although depth contours aren't listed for the brook, as they would have been for a deep body of water, marsh symbols, little patches of blue bristles that look like a flotilla of upturned hairbrushes, mark several of the more permanent beaver ponds, reflecting what I see from the dens: mats of duckweed and legions of pickerelweed, strung together by a thread of water. The brook traces the two-hundred-foot contour line, while the cliff peaks at thirteen hundred feet (thirteen hundred twenty-two to be precise). In between is landscape condensation, symbolized by forty (or more) contour lines pinched together into a three-eighth-inch-thick ribbon ... the smudge. The map itself is drawn on a scale of 1:24,000 (America is the only country to use such a quirky ratio; all others employ metric scales of 1:25,000 or 1:50,000); for our purpose, an inch of map equals 24,000 inches of land or, stated differently, a map inch represents approximately a quarter mile of countryside.

From the topo map, I can read both the height and steepness of the longest talus slope: a gain of four hundred fifty feet in over three-quarters of an inch; in other words, the rockslide is approximately a thousand feet long, the equivalent of three sloping football fields of rock. Depending on which direction you go, you either gain or lose nearly a foot in elevation for every two linear feet traveled. The end result: precarious footing. Of course, I didn't need the map to know the landscape is dicey—seeing is believing, and I just needed to conjure the sound of crashing boulders and the pungent smell of burned rock—but this little exercise may help you to more easily visualize the landscape in your mind's eye.

All our snake dens face south-southwest and are found between three hundred twenty-five and six hundred feet above sea level. The lowest-elevation den, which is also the most northern and the most accessible, is buried in rubble below the shortest run of rocks and is unquestionably the largest den, a veritable rattlesnake warren, not only in this complex, but most likely throughout the range of the timber rattlesnake. Depending on whose estimate you go by, the Cash Box, a name redolent of bounty hunting, shelters a boatload of rattlesnakes, some more than four feet long. There's no need to shudder; thankfully for the snakes, most of their facial pits will never relay the flickering, infrared image of a human to their optic tectum. The comings and goings on these rockslides, one of the prime (and primal) seasonal metronomes in eastern North America, represent a gage of yearly variation in degree days and has become another arrow in the quiver of climate-change documentation. A wondrous array of slow-moving color and pattern superimposed on stone, in this talus world toxicity has been rendered a footnote. Although, historically, this not been a universally shared opinion.

According to northeastern New York bounty records, recently sent to me by Bill Brown, between 1968 and 1973, five-dollar bounties were paid on several thousand timber rattlesnakes, of which most were paid to the late Sparky Punch, one of the most prolific and deadly snake hunters that ever lived. Punch, like many people involved with timber rattlesnakes, embodied a dichotomy. Certainly, he killed for cash. Besides collecting rattlesnakes on his own turf, he raided populations elsewhere in New York, as well as in Vermont and Pennsylvania, where he bought rattlesnakes for a dollar apiece at state-sponsored roundups and then, with blithe disregard for the rules of the hunt, bilked various Adirondack counties for bounty payment. In season, Punch would hunt the pleated hills and rock-slides until sunset, and then return home for dinner. Along the way he'd shoot snakes with birdshot fired from one of two pistols, either a .38 or a .22, and then hack the rattles and last few inches of their tails, which he submitted to town clerks for payment. What Punch didn't do, I was told by Randy Stechert during a phone conversation, was butcher seed stock, at least not at *our* isolated gestating sites. Near the waterfall, Punch left the gravids in peace and prosperity, which is one of three reasons we still see so many snakes there today (the other two reasons, to belabor a point, being inaccessibility and hot rocks). Elsewhere, however, Punch did hunt gestating snakes, stripping the population of its future ... wriggling, fully formed embryos, which he tendered for bounty payment along with their mothers.

After the early seventies, when northeastern New York had suspended bounty payments for timber rattlesnakes, Punch began to sell live snakes, more than two thousand of them, to high-paying private collectors. The late Ross Allen, the serpentarium impresario from Silver Springs, Florida, bought snakes from Punch and once accompanied him to the very talus I visit. Allen, upon finding handsome yellow morphs around Cash Box Den, declared them "the golden rattlesnakes" of the Northeast and took several back to Florida with him. In 1983, New York State granted *Crotalus horridus* protection as a threatened species, and Sparky Punch's snake hunting ended. In over fifty years of collecting, he claimed to have removed eighteen thousand timber rattlesnakes from three counties in the foothills of the Adirondacks. An exaggeration? Possibly. But even so, a reflection of Punch's tremendous impact.

A few years earlier, in the late seventies, as if to make amends, but more likely to generate additional income, for a fee of fifty dollars per den Punch led a fledgling Skidmore biologist named William S. Brown to talus dens in upstate New York, an arrangement that eventually jump-started data gathering for the longest capture-recapture study of a metapopulation of reptiles (any animal species, really) on the planet. In the nineties, Punch showed Larry Boswell, the man with the silver rattlesnake inlaid on his eyetooth, how to safely relocate nuisance snakes for the state of New York. It was Boswell, a retired agronomist with the U.S. Department of Agriculture, a Long Island transplant, you may recall, who had summoned Alcott Smith to extract quills from an impaled rattlesnake on a coffee table in a cabin near the threshold of Cash Box Den.

Punch lived in the same county as contemporary folk artist Grandma Moses, though there is no indication that the two ever met. I wish they had. Perhaps then, Moses might have been inspired to scale the rockslides with Punch and paint a landscape with reptiles, which she could have added to her oeuvre of scenes of rural life: sugaring off, Mother's Day, Thanksgiving, Christmas morning, sledding, and so forth. I can imagine that iconic image: *Snake Hunt*. Standing in the foreground would be a lanky and triumphant Sparky Punch holding a blood-stained bag of rattlesnake tails, his hatchet and pistols fixed to his waist, lifeless snakes strewn over the rocks, with the green and gray of outcropped-studded hills rolling away and fading into the distance. Certainly for Sparky Punch snake hunting was unequivocally the premier scene in his own rural life, which he occasionally did right here, right on the slopes where I stand.

Jon Furman, the author of *Timber Rattlesnakes in Vermont & New York*, interviewed and then apotheosized Punch shortly before the bounty hunter died. Punch told Furman that he kept snakes in cages in the basement, outside, even under the bed. Punch had been curious about their thermal ecology before the term was ever coined. He released caged snakes to see how they would respond to thermal gradients. "If it was in the seventies and sunny," wrote Furman, "[Punch] noticed that [rattlesnakes] sometimes headed for shade but often lingered in the high grass.... If it was sunny and the temperatures were in the mid-eighties or higher, they would instantly head for the nearest shade." Punch would induce rattlesnake hibernation in late summer by exposing them to markedly reduced outside temperatures, and in the midsummer by placing them in a cage on the hydrator of his refrigerator. He also suppressed their urge to hibernate by deliberately keeping them warm and well fed all winter, a trick at which NERD's Kevin McCurley later became adept.

Punch also wondered if a timber rattlesnake could regulate its metabolism when faced with a food shortage (a.k.a. starvation). He stopped feeding a snake for two years, providing it only water. Then, satisfied (by what I'm not sure), he fed the emaciated snake six or seven mice over a period of several weeks, which fattened it back up. The snake, he told Furman, appeared to make a full recovery. What did Punch learn from his experiment? He learned, wrote Furman, that timber rattlesnakes can regulate their metabolism in the face of an *acute* food shortage. Punch also studied shedding, biting, feeding, and pheromone tracking (notwithstanding his wife's indelicate response to rattlesnakes under the bed). As Furman noted, Punch might have been a field biologist had he finished high school.

ॐ

That timber rattlesnakes bask in precise, predictable pockets in talus, which conform to both the thermal potential of the rocks and the thermal requirements of the snakes, is an established fact for herpetologists like W. H. Martin, William S. Brown, and Randy Stechert, as well as to snake hunters like the late Sparky Punch. Alcott and I often take a laser thermometer into the field to record digital readings of the rocks and the snakes; the results give us plenty to prognosticate about. We've discovered that if a basking rock is above one hundred degrees Fahrenheit, snakes will remain almost exclusively in the shade. Late one sunny spring morning several years ago, when the air temperature was seventy-six degrees Fahrenheit, the surface temperature of a traditional basking rock had already risen to one hundred sixteen degrees; the undersurface of the rock was twenty-five degrees cooler. Two large pre-shed males, a black morph and a yellow morph, entwined in the shady atrium. The body temperature of the yellow snake was eighty-five degrees; that of the black, ninety-three. Although we enjoy the laser gadget, you really don't need a thermometer to predict rattlesnake behavior. If a rock is too hot for your hand, it's too hot for a snake.

The position of the talus in relation to the direction of the prevailing wind also has a bearing on timber rattlesnake activity. Wind, which lowers a snake's body temperature even on a warm, sunny day, is anathema to rattlesnakes. Gravids avoid a light wind of five miles per hour either by retreating under a shelter rock or by basking in

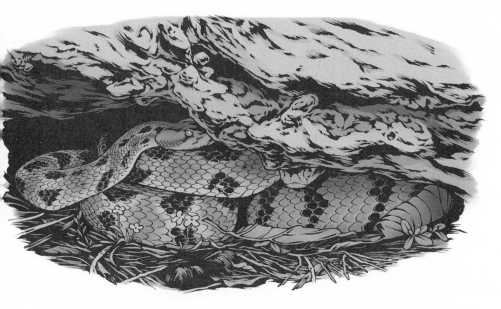

the stillness of a swale, the breeze passing safely overhead. Snakes, even in the slightest dip, soak up the warm air in an undisturbed atmospheric cocoon, still and nourishing. Little in a rattlesnake's life is random.

Depending on the year, snakes reappear on talus between late August and mid-October. Their punctilious timing depends on the nature of the weather—the wind, the sun, the clouds—which translates into how many months and weeks and days and hours their body temperature has approached optimal. Everything about a timber rattlesnake is temperature-driven, and their behavior throughout the season of activity reflects their body temperature, which in turn depends on the temperature of sundry microhabitats, enabling them to maintain whatever delicate thermal balance may be required for the different phases of their life. For instance, the gravid snakes in the talus express a higher and less variable body temperature during pregnancy than either a male or a nonpregnant female. Tiny snakelets heat up more quickly than adults and choose their basking sites accordingly. On a hot day, a timber rattlesnake maintains a body temperature cooler than that of the air; on a cool night, warmer than the air. A snake often accepts a lower-than-optimal body temperature when it is hunting, but it selects a higher temperature when it is digesting.

During a cold night, a migrating rattlesnake may stop en route to its den and tuck under an insulating blanket of leaves, where it may stay until early the following afternoon, when valley fog burns off and the day heats up. After a cold night, Alcott and I arrive at the talus by one o'clock, just as the fog dissipates and the sun reappears. If the temperature is warm, however, say in the mid-fifties or higher, many snakes move all night, funneling along the base of the cliff, one snake after another, hidden below the rock-twisted trees, on a route laid out in pheromones. I once stood on a woodland boulder as an adult glided unerringly toward me, straight as an arrow, over, around, and across fallen limbs and rocks, and then stopped in front of my platform, lifting its head over the rim and tongue-flicking my boot, then moving directly toward another large snake coiled fifteen feet away. It crawled over the snake, and continued north until it reached a crevice in the base of the cliff, where it vanished into the earth, the third snake I had watched disappear down the fissure in two days. On another autumn afternoon, I found three snakelets no more than twenty feet apart, at eye level on a narrow ledge, two coiled in the open and the other wedged into a crack in the cliff like a line of putty.

Although very little is known about the precise nature of the thermal ecology of timber rattlesnakes, it is very likely that every metabolic process has a range of optimal temperatures, as well as minima and maxima, below or above which a snake may be in peril. Every snake has a personal thermal agenda that depends on its size (mass) and duration of inactivity (a week, three weeks, or just waking up from six months of hibernation). For instance, biologists on the Ozark Plateau determined that when a timber rattlesnake first feeds after emerging from hibernation, it uses significantly more energy (and more heat to produce that energy) to digest the meal and rid its body of fecal and nitrogenous waste than it does later in the spring or summer.

In the early eighties, using three snakes from a Sparky Punch den that had been implanted with temperature-sensitive radio transmitters, Bill Brown discovered that although individual snakes migrated home at body temperatures between 55.4 and 64.7 degrees Fahrenheit, there was no difference in their hibernating temperatures. He also noted that during the winter, each snake's body temperature fell at a rate of a little less than one degree Fahrenheit per week until February, when its temperature bottomed out at about forty degrees. Body temperature remained stable in March, Brown discovered. Then, between April and May, it began to rise a little more than one degree Fahrenheit per week. Emergence occurred in mid-May, after each snake reached approximately forty-eight degrees Fahrenheit. Brown noted that over seven months of hibernation, the profile of body temperatures reflected that of both soil and air temperatures. More than thirty years later, on April 18, 2012, a day in the mid-eighties, Smith recorded the first snake up at Cash Box; two days after that, on the third consecutive warm day, we counted fifty-one snakes, which included ten adults that had moved to transient rock sites fifty or more yards from the den, one adult along the brook and two others moving out along the foot of the cliff.

This is the fall 2012 count at Cash Box, a reptilian version of the bell-shaped curve: August 27, three (two neonates); September 15, twelve; September 20, sixteen; September 23, twenty-four; September 25, thirty-nine; October 5, eighty; October 8, ten; October 18, three; and October 26, two. Bird migration is triggered by photoperiod, a universal invariant; snake migration is triggered by photoperiod and apparently influenced by body temperature, which in turn is triggered by the temperature of the air and the ground, a more primitive, deeply personal, and far more mercurial affiliation with the season.

By the time Alcott and I reached the Cash Box Den on April 20, 2012, snakes had already begun to emerge two days before. The temperature was in the low eighties for the third day in a row. Crooked trees were leafless; the den in full spring sunlight. Columbine bloomed, little flecks of red dotting an otherwise somber landscape. Dutchman's-breeches and trillium had begun to pass. Everything was early and warm in the Northeast, including snakes.

Cash Box Den, the main portal in a den complex, which includes that ground-level, horizontal crevice in the cliff and several (many perhaps) pathways through talus into the hibernaculum, is an outcrop of schist that surfaces in the woods thirty feet below the base of the cliff. The entrance is wide enough for a grown man to crawl into—Alcott's gone halfway in to get a digital reading of a snake's temperature—and extends five or six feet, before doglegging right. From there, it appears to constrict to a tubular chamber that veers straight down into the Stygian abyss. Green ash and hackberry grow around the outcrop and on top a veneer of earth supports sedges and a small elderberry. In fall, snakes rarely loiter outside Cash Box. Instead, they bask on the exposed rocks north and south of the den or descend directly into the depths.

During a leafless emergence, I must be vigilant when walking around the den. In April and early May, snakes may be gathered in pods almost anywhere nearby, including inside the mouth of Cash Box. On the twentieth of April, when I peeked into the portal, I saw a huge yellow morph fully stretched inside, nearly five feet of forearm-thick snake with a head the size of a tennis ball. After Alcott had also seen the snake, we left the den and headed south, tallying snakes in pods all the way to the waterfall. Later in the afternoon, we returned to the den. A black male nearly as long had replaced the yellow morph, as if snakes waited below the surface in an orderly line. "OK, Justin's out. It's your turn, Randall. "

I'm a talus Luddite. Always will be. I don't use a GPS to mark dens and basking sites. I memorize the gestalt of the landscape: the pattern of rocks, the kinks in the trees, the angle of the sun. I do get lost, however. I left Cash Box, as buoyant as a balloon, and started down the twilit talus. Then, as I manically admired the last snake of the day, a subadult yellow morph coiled in a rock saddle, gravity and my own questionable balance conspired to topple me and send me ingloriously skidding, ass to talus, until a rock the size of a medicine ball settled on my right ankle.

I limped for two days.

11

Among Rattlesnakes

The den of rattlesnakes ... was found to be an ugly looking place. It is a hillside of tumbled boulders, covered with moss. Here and there a tree shoots up and spreads enough greenery to keep a dim and unreligious light upon the den. You don't see snakes at first glance, but they are there, hundreds of them, under the boulders, or with just the tips of their noses out in the sunshine.

New York Times, South Kent, Connecticut, May 20, 1906

Your second SAT question of the book (I promise there are only two) reads, "ecology is to [blank] as hydrology is to water." The only acceptable answer would be "relationships"; that is, the constellation of relationships between members of the same gene pool and relationships between the various species that live within the same biologic neighborhood. The study of ecology also includes the relationship of each species to the inorganic aspects of the world around them. Because living things have meaning in terms of what they do, I want to know (and, since you've gotten this far into the book, I assume you also want to know), what does a timber rattlesnake *mean*? That well-muscled, four-foot snake the color of fallen leaves and patterned in sunlight and shadow, mobilized in loops and coils, hidden in plain sight, its chin resting on a fallen limb with saintly patience ... a spring-loaded bomb, waiting to be triggered by an infrared signal. What is the meaning of a life lived in the emptiness beyond our senses? In his poem "Among School Children," William

Butler Yeats asked, "How can we know the dancer from the dance?" Had Yeats's poem been titled "Among Rattlesnakes," the text would have taken a different turn, but the essence would have remained the same. A timber rattlesnake and its dance, sometimes gradual, sometimes sudden, are inseparable, a union forged in the fires of natural selection, shaped by the cyclical comings and goings of glaciers that first isolated snakes in shrinking refuges in the Deep South and later liberated them to spread northward into hospitable woodlands, draining and then refilling gene pools; shaped also by the piling of rocks at the foot of exfoliating cliffs, by the nature of seasons and the fickleness of weather, and by the aura of every species that ever lived beside them, from mastodons to meadow voles. What does a timber rattlesnake mean to an eastern deciduous forest or to the talus or to the white-footed mouse or to the hungry red-tailed hawk that hangs on a draft above a gestating site, its wings as level as Nebraska? What precisely does a rattlesnake mean to you? Is it a venomous echo from the Book of Genesis, an unhinged reptile that escaped notice when God breathed life into the charred void? Or is it a traveler, an evolutionary passenger emerging from the mist of deep time, on a singular journey of survival, procreation, and change? Trimmed *in* the crucible of time. And . . . in due time, eliminated.

How would the herpetologist answer the question? The hiker? The turkey hunter? The herpetoculturist? The poacher? The snake handler? The philanthropist who seeks to protect the fracturing habitat of reptiles and amphibians? Or the motorist who runs snakes over, or the outdoorsman who kills them out of ignorance, malice, and spite? How about a westbound Hartford commuter who crosses the Connecticut River on Bulkeley Bridge from his or her residence in Glastonbury, where for perhaps eight thousand years rattlesnakes egressed down a shoulder of a nearby mountain, down nameless creases in the bedrock now called Laurel Grove Road and Bauer Pond Road, where realtors are compelled to tell prospective home buyers that every summer timber rattlesnakes may cross the driveway, bask in the garden, or copulate on the front lawn in an act of self-absorption that can last all day?

ॐ

Incorporated in 1693, Glastonbury, Connecticut, hard on the east bank of the Connecticut River, was originally purchased in 1636 from a Wongunk chieftain for twelve yards of cloth, and has more or less remained an amalgam of farmland, pastoral village, industrial hub, and mountain stronghold, even though the insurance capital of the United States is only ten minutes away. During the American Revolution, Glastonbury produced gunpowder, and for much of the eighteenth and early nineteenth centuries the mountain's white oaks and swift streams supplied lumber and power for mills and boatyards. As if in tribute, the oldest continuously running ferry in the United States still links South Glastonbury to Rocky Hill. In 1847, James Baker Williams, a pharmacist who founded the J. B. Williams Company, built a soap factory in Glastonbury that manufactured the first lather just for shaving mugs. Within sixty years, J. B. Williams had a worldwide reputation for men's toiletries, and on November 17, 1917, filed a trademark with the U.S. Patent and Trademark Office for a line of aftershave lotions called Aqua Velva, which became a staple for soldiers during World War II, not only as an astringent but, because of its high alcohol content, as a beverage. In 1957, J. B. Williams grew too big for Glastonbury and left behind three large brick buildings, which subsequently morphed into the Soap Factory Condominium, one hundred sixteen units, where an eight-hundred-sixty-two-square-foot, two-bedroom, two-bath condo lists online for $134,400.

During the world wars, Glastonbury supplied allied troops with leather and woolen goods. Its factories also milled feldspar, cotton, paper, and silver plates and assembled airplanes. John Howard Hale, a Glastonbury farmer and fruit grower, developed a peach that stood up to New England winter. By 1900, Hale, a fruit-marketing pioneer known as the "Peach King," was growing three hundred fifty thousand peach trees, many on a twelve-hundred-acre Glastonbury orchard, and was shipping fruit throughout the country. In his spare time, Hale helped establish Storrs Agricultural College, now the University of Connecticut. Today, Glastonbury orchards and berry farms grow strawberries, raspberries, blackberries, blueberries, cherries, apples, pears, peaches, apricots, plums, and currants; besides fruit, market stands sell pies, jam, cider, honey, and maple syrup. In June, the town hosts the Under the Strawberry Moon Festival and, in October, the Apple Harvest Festival, a three-day event that features a variety of odd attractions, including carnival rides and politicians.

September 9, 2012, before the day gets too hot and snakes go into hiding, I exit the coagulated traffic on Route 2 and drive into the center of Glastonbury to meet Doug Fraser, a Siena College biology professor I met the previous fall at the annual gathering of the Bashers. (My first impression of Fraser was that he might have been the only Basher who didn't mainline beer.) An Eagle Scout and a lifelong naturalist, Fraser grew up in East Hartford hearing rumors about the nearby rattlesnakes of Glastonbury. In 1957, once he was old enough to drive, Fraser sought out the local rattlesnake oracle for directions to the main den, which he eventually found after traipsing around a remote hillside all day, no small feat for a sixteen-year-old. In the late fifties, while most boys were arguing about whether or not Mickey Mantle was better than Willie Mays, Fraser, who loved snakes more than baseball, shared his bedroom with a small timber rattlesnake, fresh from the Glastonbury highlands. These days, even though Fraser lives in Albany, New York, four hours away, and does formal ecological fieldwork in the jungle streams of Trinidad, he returns to Glastonbury at his own expense many times each year, following a pattern of more than three decades, to monitor rattlesnakes and to educate their immediate human neighbors, not all of whom are receptive. It's Fraser to whom the *Hartford Courant* turns when in need of a quote or a clarification regarding the threat snakes pose to suburban homeowners. Also, the Connecticut Department of Energy and Environmental Protection (DEP) and The Nature Conservancy (TNC) seek out Fraser for advice on land acquisitions that would benefit the state's rattlesnakes. Whenever an application for a building permit that might threaten the rattlesnakes crosses his desk, the Glastonbury town planner alerts Fraser, who holds a research license from the state. Over the years, he has weighed, measured, sexed, aged, marked, and monitored more than six hundred fifty Glastonbury rattlesnakes but has never published his findings even though copious data fill several notebooks. For Doug Fraser, a timber rattlesnake means wilderness, a vestige of primal America that, against all odds, still survives in the corrugated uplands across the river from Hartford, a personal and joyful thread that reaches back to his boyhood.

"My CV doesn't need rattlesnakes. I don't feel compelled for career purposes to publish my Glastonbury results," he told me last autumn over lunch. "Because Bill Brown has published the definitive work on timber rattlesnakes in the Northeast, my paper would only be a small footnote to his body of work. Brown's covered everything. Only if the benefits [to the rattlesnake population] far outweigh the risks would I consider writing." In the thoughtful tradition of the late sea turtle biologist Archie Carr, Fraser devotes a portion of his ecological curiosity to pro bono conservation. "My timber rattlesnake work is for conservation," a clear example of what my nana called a mitzvah.

I first heard of the Glastonbury rattlesnakes in 1987, when the August issue of *Yankee* featured the article "Who Should Live on Kongscut Mountain." Near the end, author Joyce Winslow impersonates an out-of-state house hunter to see what a realtor would tell a prospective client about the possibility of finding a rattlesnake on the front lawn. Winslow tours a home on a road that already has had nine well-publicized sightings that year alone. The homeowner's son, age eleven, shows the author through the house.

His parents were out, but he knew what to say. He did an admirable job until I asked if he'd seen any animals in the creek in his backyard.

"Yes," he beamed. "Deer and raccoon, but no rattlesnakes."

"Rattlesnakes?" I feigned shock.

He reddened and looked down. I turned to the realtor.

"There used to be rattlesnakes years ago, but you never hear of them anymore," she said.

"I have toddlers," I lied.

"There are no snakes here," she said firmly.

Besides being a very short drive from basking rattlesnakes, downtown Glastonbury is also an upscale blend of traditional New England, white clapboards and steeples, and brick-and-glass modernity. Glastonbury's courthouse is gilded, where soft morning light flows like water over every dimple and dent in the leaf. On Main Street there's an alcohol and drug rehab facility that looks like a resort and advertises a five-star chef, luxurious bed linens, and a licensed massage therapist. Also on Main Street: the Connecticut Audubon Center of Glastonbury, Cycling Concepts, Barnes and Noble, the Old Cider Mill, and Daybreak Coffee Roasters. At the corner of Griswold and Main, across from Walgreen's, sits the Griswold Mall, a white, L-shaped complex that claims to be the "Best Shopping Center in Glastonbury." At the mall, you can enjoy a workout, then visit a reflexologist to have the bottoms of your feet kneaded like bread dough; you can dine on gourmet Chinese and Italian food or eat a foot-long sandwich. Then, you can go store to store to buy an ice cream cone, a latte, a rug, a toy, an in-ground pool, jewelry, or real estate among the snakes; you can acquire a cellphone plan, mail a package, or have your nails done or your hair coiffed (two different stores). Three storefronts remain as vacant as the laurel flats of the Glastonbury highlands, which still support one of New England's most robust populations of rattlesnakes.

At some point during their lives, both the gunsmith Samuel Colt and the hockey forward Gordie Howe lived in Glastonbury; Gideon Welles, secretary of the navy during the Civil War, radio personality Mike O'Meara, and Candace Bushnell, author of *Sex and the City*, were born there. Two-time World Wide Wrestling Federation champion Bob Buckland, who refused to sign autographs unless you recited all the U.S. presidents in chronological order, lives there. For a time, Buckland coached varsity wrestling at nearby Rocky Hill

High School, and he ran unsuccessfully for a seat in Congress as a Republican. Timber rattlesnakes didn't have an opinion either way.

Driving down Main Street, I spot an Asian man in a red shirt holding a placard that reads, "Re-elect Srinivasan." By the time I reach Panera Bread, the site of my rendezvous with Fraser, I pass three more copper-colored men in red shirts, each holding the same placard. They all wave to me, white teeth flashing, not the least bit troubled by my conspicuously green out-of-state plates. I wave back.

Energetic and thoughtful, elfin, mid-seventies, balding, with a gray, close-cropped beard, and dressed for the field—khakis, boots, loose-fitting, light-green button-down shirt—Doug Fraser motions me over to his table. As I sit down, his engaging smile dissipates. "There's information you simply *cannot* divulge." Not surprisingly, Glastonbury's rattlesnakes are very accessible. "You could be semi-ambulatory and reach a number of the basking and birthing sites," he laments between bites of scone, crumbs spilling from his fingers. Hartford, weighing in with a population of one hundred twenty-five thousand and having an international airport, is ten minutes away, and in a state of nearly four million, no one lives more than an hour from Glastonbury. Since 2007, the DEP, which Fraser has alerted to both the needs and the vulnerability of the rattlesnakes, provides a conservation officer (CO), a retired detective to patrol the no-trespass basking, birthing, and den sites, and to monitor a number of motion-sensitive surveillance cameras, which, although set high in trees, could just as well be set at eye level; snake hunters don't look up.

Fraser organized a neighborhood watch for residents to report suspicious-looking people and to write down the license plates of out-of-state cars parked adjacent to trailheads that lead into snake zones. Restricted areas are posted with conspicuous red signs that announce, "This area closed to all persons: April 15–October 17." First-time trespassers are warned. A second offense results in arrest and a fine. Although the CO has confiscated cameras and cars and has brought a number of cases before a judge, the rattlesnakes re-main under constant threat from all sorts of people, both poachers and curious amateur naturalists. To make a point, Fraser tells me that the CO arrested the Connecticut "science teacher of the year," along with the owner of a New Haven pet shop that specializes in reptiles and amphibians, in restricted areas, both hunting snakes. Two other men carrying snake hooks, bags, and a hollow boom box, were observed by a third person, and later apprehended. As cryptic as their quarry, the boom box was fashioned to smuggle newborns. An entire family that had recently relocated to Connecticut from Arizona followed printed directions to the den they'd copied off an Internet post on the website called the Field Herp Forum. The father mentioned online that in an effort to get better photographs, he broke his snake hook trying to yank a rattlesnake from under a shelter rock. He was reproached by sympathetic bloggers, who reminded him that timber rattlesnakes are critically endangered in Connecticut, protected by law, and cannot be harassed. "We never heard from him again," says Fraser.

This past summer, the surveillance cameras caught a bare-chested teenager in a bathing suit walking through a no-trespass zone with a snake stick; two other teenagers, one carrying a butterfly net and the other a snake stick; and men with crowbars flipping shelter rocks. The CO stopped two men and a woman from Danbury. All three wore backpacks filled with Tupperware containers. One container held a frog. "*No* frogs in western Connecticut," quips Fraser, polishing off the last of his scone.

Together with his friend, colleague, and esteemed Basher Bob Fritsch, a retired Wethersfield policeman who is one of Connecticut's leading amateur herpetologists and has been closely connected to the uplands since the early sixties, even though he has since retired to Maine, Fraser trains rattlesnake first responders to relocate nuisance snakes from yards and roadways and to retrieve and freeze carcasses. Since this is a suburban population, Fraser emphasizes, "We get called whenever a snake crosses a street or loiters by a stop sign." For every four rattlesnake-related calls, one is for a dead snake, which Fraser refers to as "the one-in-four." In 2013, there were forty-five calls, by far the largest number received in any of the past twenty years; the average is more like twelve or fifteen.

I ask, "Is that because the human population is growing, or have the neighbors become more vigilant?"

"I don't know."

More than half of the town's 52.3 square miles remains undeveloped, hence the persistence of timber rattlesnakes. Beside orchards and farms, rural suburbs, and a pleasant downtown, Glastonbury features a golf course, a private lake, and a mostly remote state forest, punctuated here and there by a few upscale housing developments. The 2010 U.S. census lists the population at twenty-eight thousand seven hundred sixty-two people, of whom more than twenty thousand are over the age of twenty-one. There are six hundred thirty-one blacks and two thousand and three Asians living in Glastonbury; the rest of the population is white. Property taxes support six elementary schools, one junior high school, and a high school, with an enrollment of more than two thousand students. Beginning in the seventies, small clusters of posh developments began to metastasize around the base of the Glastonbury highlands, mostly for commuters taking Route 2 into Hartford's financial district. Apparently, rolling green woodlands look as good to insurance executives as they do to rattlesnakes.

Timber rattlesnakes had the uplands to themselves for perhaps five to eight thousand years, but now run afoul of soccer moms and financial analysts. In 2010, a rattlesnake was severed with a lawn mower, and Fritsch knows a man for whom rattlesnakes mean food. The man told Fritsch that he didn't realize snakes were protected. (If he had purchased a house in the last fifteen years, which he apparently had, his deed would have stated that timber rattlesnakes live in the neighborhood and they are legally protected ... up to a point.)

Not surprisingly, roads delineate the boundaries of the rattlesnakes, which were once part of a much larger genetically articulated metapopulation of five or more dens, which is now divided by Route 2. The uplands support one main den, two lesser dens, and several auxiliary dens, with a combined population that has miraculously remained stable since Fraser began his study. Generally, Fraser keeps the size of the population to himself, because if an unsympathetic homeowner hears the number, one thinks, "My God we're being overrun." But, scattered over seven square miles of corrugated woodlands, the rattlesnakes remain almost as inconspicuous as radio waves … unless they cross a road or a driveway, or you know where to look for them. The snakes might not be obvious to us, but through the endurance of pheromones, says Fraser, "They talk to each other all the time."

Hemmed into their outcropped stronghold, timber rattlesnakes patrol a summer range circumscribed by Larson Street to the west, Pepitone Road to the east, Bauer Lane to the north, and Route 2 to the south. If you punched certain street addresses into a GPS, you'd get directions to within a hundred yards of a basking site.

Although Connecticut listed the timber rattlesnake as an endangered species in 1992, protection of critical habitat leaves much to be desired. "It's a gray area in terms of stopping development," Fraser laments, amid the bustle of Panera Bread, noticeably troubled by the flurry of legal battles over development rights along the periphery of the snakes' summer range. "The State Statute protecting rattlesnakes does not specifically protect habitat, leaving local jurisdictions without explicit guidance and definitions."

I ask, "How do you deal with a proposed subdivision in rattlesnake habitat?" knowing there must be a well thought-out strategy.

Glastonbury has a strict application procedure involving both the planning board and the conservation commission, which has always been sensitive to the rattlesnake. Before the conservation commission agrees to a subdivision, they seek Fraser's advice. Their bottom line is how important is the land to rattlesnakes? On the advice of Fraser, the conservation commission often seeks a compromise, requesting a significant portion of the land be placed in conservation easement, which may result in fewer units. "We stopped one proposal," Fraser says, eyes twinkling, like a mischievous boy in a Norman Rockwell painting. "To avoid litigation, a developer might walk away, but if ground has already been broken without a permit," says Fraser, more soberly, "we're in trouble."

Fraser and the rattlesnakes have an ally in John T. Rook, Glastonbury town planner, whose master's degree project for Antioch University New England, "Habitat Protection/ Preservation for the Timber Rattlesnake Population of Central Connecticut," was written in 1989 and is accessible online. When a proposed subdivision is adjacent to rattlesnake habitat, Rook imposes Glastonbury's strict guidelines, which require a potential developer to hire a herpetologist and to make new homeowners snake savvy. All construction is limited to wintertime, when snakes are hibernating.

Glastonbury wants snakes, which sets it apart from most other communities in Connecticut. On May 8, 2000, reporting on measures to preserve the uplands, the *Hartford Courant* quoted Rook as saying, "The timber rattlesnake is the star of the show." A decade later, announcing a one-hundred-eighty-thousand-dollar grant that the town received from the DEP to buy a fifty-five-acre property in the highlands, Rook told the *Hartford Courant*, "This whole area is prime rattlesnake country. One of the town's goals has been to piece together properties to protect a large, unfragmented area, especially in this part of town." For Rook, the rattlesnake is more than a vehicle for land protection, a means to an end; it's a vestige of wilderness lost, a symbol of ancient instinct colliding with the pressures of the twenty-first century.

Searching the web, I found the minutes from a Glastonbury Conservation Commission meeting held on September 11, 2008, for a proposed three-lot subdivision on a 5.6-acre parcel at 457 Laurel Grove Road. It's typical of the wrangling between developers and commissioners and of the hurdles to development presented by rattlesnakes. A major concern of the commissioners was the presence of snakes relative to the development of lot 3. To provide more open space, the lawyer for the landowner conceded that development on lots 2 and 3 would be moved downslope. Members of the conservation commission, however, remained ambiguous and skeptical.

> Vice Chairman Patrick feels that the previous and current configuration of the houses is still too close in proximity to the rattlesnakes. And that the boundary should be open space so rattlesnake

encounters are limited; he recommends that open space to be given to the Town instead of a proposed conservation easement even if the lot sizes are smaller than the zoning requirement....

Commissioner Huestis asked what is the best way to protect both snakes and people. Mr. Goodin said that they should give education to prospective purchasers and with the deed. Commissioner Schade said that he does not feel that there is a difference if the land is a conservation easement or Town-owned because people tread on it either way....

Secretary McClain wondered if large signs that would warn people about the rattlesnakes could be a requirement of the approval. Commissioner Stern said that she feels signs will disappear and people will not read information in the deeds.... Chairman Harper said that she would like more information provided by a rattlesnake expert to see if this proposal is a significant change and she would like to know if there is any possible rattlesnake mitigation.... Chairman Harper recommended Doug Fraser ...

When money's involved, not everyone heeds Fraser's sage advice. Minutes of the conservation commission meeting on August 25, 2011, address a four-lot subdivision at 479 Laurel Grove Road, where a less-than-sympathetic landowner begs the crucial economic question: whose mountain is this? Excerpts from the minutes read:

Regarding the State Endangered Timber Rattlesnake issue, neither Dr. Douglas Fraser's recommended plan to eliminate two of the rear lots nor Mr. John Rook's alternate plan to eliminate the most northerly lot in an effort to protect the habitat will be implemented....

Secretary McClain commented that Dr. Fraser, a well-known and respected expert, studied the site and determined that, in the best interest of the Timber Rattlesnake, it would be better to reduce the proposal from four lots to two, and Mr. Rook offered a compromise of three lots; both of these suggested plans were rejected. She noted that consideration of this endangered species is required as a part of their deliberations. Attorney [for the property owner] replied that, although Dr. Fraser is an expert, the Endangered Species Act in the State of Connecticut imposes certain requirements, and development of a piece of property for housing is a lawful activity. He doesn't think the Town has the power by virtue of the Zoning Enabling Act to tell his client that she has to eliminate several hundred thousand dollars of value on her property because the Town is concerned about rattlesnakes and Dr. Fraser has written a letter saying there is a den in the area without identifying its exact location. [The attorney] doesn't think anything stated in Dr. Fraser's letter justifies the elimination of half the developable lots in this subdivision application ... and he suggested that the Town purchase the two rear lots if it is concerned about rattlesnakes.

Ever the scientist, Doug Fraser classifies people living in the highlands into four categories: those who relocate nuisance snakes; those who notify the removers when a snake appears under the swing set; those who tolerate snakes and step around them; and those who purposely kill snakes with a blunt instrument or a car—locals called it "Buicking" them to death—of whom there are few. Even though Connecticut has designated the timber rattlesnake an endangered species—penalties for killing one include a fine of up to one thousand dollars and up to six months in jail (an employer who directs an employee to kill a snake is subject to a fine not exceeding ten thousand dollars and a year in jail per violation)—the snakes still don't have a great deal of legal standing. A frightened homeowner can legally kill a snake within five hundred feet of his or her own property. It's not meant to be invitation for mayhem, but you could, more or less, walk down the block and kill one. As Fritsch once pointed out to me, "If you said you were afraid, I don't care where you are in the state, [Connecticut] isn't going to prosecute you." In the early nineties, a resident of Laurel Grove Road bludgeoned a rattlesnake in his garden. The man, who Fraser and Fritsch had counted as an ally, took a hoe to it, claiming that a sudden, irrepressible, deep-seated panic

overwhelmed him. Then, he called Fritsch's cellphone. "I was next door, right next door. Minutes away," Fritsch had told me, extending his arms palms up, as though requesting intervention.

I follow Doug out of Glastonbury center, up winding, wooded back roads, passing an apple orchard, old farmhouses closed in by trees, a cluster of fifties-era ranch houses, each with a wood lot, and several troublesome modern developments, *big* homes punched into rattle-snake summer range, many with shared driveways that curl snakelike back into the woods. Like peeling a potato, each year habitat for foraging gets shaved away, becoming progressively smaller. No one knows where the tipping point lies. Looking west from a roadside promontory, I see Hartford.

Before Perdue chicken and Colonel Sanders, the Saglio brothers of Glastonbury were the premier poultry exporter in the world. In 1948, the grocery chain A&P awarded the Saglio brothers' chickens the coveted title of "Chicken of Tomorrow," and, in 1977, Henry

Saglio, also known as the "Father of the Poultry Business," was inducted into the Poultry Hall of Fame. You can still buy chickens and eggs in Glastonbury, as well as sheep, horses, and goats.

Once a major grower of Connecticut broadleaf tobacco, which was popular for machine-rolled cigars, Glastonbury farms (there are twenty-eight, some registered organic) still supply central Connecticut with almost every imaginable vegetable from arugula to zucchini, herbs and canned goods, including salsa, vinegar, pickles, and pesto. Several farms specialize in annuals and perennials, others in Christmas trees. None sell rattlesnakes.

We're headed to the home of Brett Somers, one of Fraser and Fritsch's minions, a first responder for whom living in proximity to rattlesnakes is enriching. Somers grew up here. His parents still live across the street and own twenty-five acres of prime snake habitat behind Brett's house. At the moment, they have no intention to develop the land or sell it to the state. Somers and his brothers prefer to hunt deer on their own land, unencumbered by the transient public, who have access to Meshomasic State Forest, which surrounds their property, a rattlesnake-inspired addendum to the nine-thousand-acre state forest that sprawls across the nearby towns of Portland, East Hampton, and Marlborough. Somers has two DORs in the freezer, a dark phase and a light phase, and a live snake caged in the backyard.

Brett Somers looks forty-something. He keeps his long, black hair in a ponytail, which exposes ear studs and an anatomically correct timber rattlesnake tattoo that winds up his left arm from wrist to shoulder. As a boy, Somers, who grew up playing in these woods, never saw a living timber rattlesnake, only dead ones displayed by a locally well-known snake-killing neighbor. Now that he knows when and where to look he sometimes sees five or six in a day. "You're lucky if you see them," he says, leaning against his pickup. I can't take my eyes off his tattoo.

The DORs are in Somers's garage freezer, the one his wife won't open. Coiled and frosty in their ziplock overcoats, both snakes appear shorter than his tattoo. Three-year-olds, Fraser guesses. The yellow snake with brown bands was hit across the street, close to his parents' driveway. The dark snake, hit at the intersection of Bauer and Tulip, is lighter than the dark snakes of western Vermont, and appears to be a tribute to Ansel Adams's zone system, a study in gray—pale-gray background, white-rimmed darker-gray bands that grade progressively to black toward the tail. "There are still people who couldn't care less [about timber rattlesnakes] one way or the other," Somers says, handing the frozen snakes to Fraser. Both are unmarked. A neighbor had called another first responder to move the gray snake off the roadside. While hunting for the snake, the responder heard tires squeal. A couple of teenagers driving too fast down Tulip Road had hit the snake, which died the next morning in his home. Several days later, an obituary appeared in the *Hartford Courant*.

A bright mustard-colored adult with yellow-rimmed brown bands idles in a hardware-cloth crate inside the woodshed. As we approach the shed, the snake rattles, a hot-blooded response that renders all other woodland sounds insignificant. Fraser sets up a telescoping net frame, similar to the one used by Bill Brown, couples the nylon net, and then gently teases the snake, still rattling, headfirst down the net, slowly and safely. While the last eight inches of the snake hang over the frame, Fraser spins the net firmly around the tail. Musk fills the woodshed like humidity.

The yellow morph measures sixty-five millimeters from its vent to the base of its rattle, and it is likely a female. To confirm his suspicion, Fraser inserts a probe into the cloaca; the tail is a solid muscular tube; definitely a female, a mature female, a veteran of the housing boom well versed in the cobweb of pheromone trails that lead from bedrock den to basking, birthing, and shelter rocks, to foraging sites in both the woods and the suburbs, where plump white-footed mice and chipmunks come to plunder the backyard birdfeeders. Her broken, four-segment rattle has no taper; from front to back, the segments measure 15.0, 15.0, 15.0, and 14.8 millimeters, respectively. Fraser thinks she's three and a half feet long. Using a marker filled with indelible India ink, Fraser traces a bold line along the crease on the left side of her rattle. If he catches the snake again, he'll determine her shed rate. It's simple math: count the number of unmarked segments, which indicates how many times she has shed since the line was drawn, then divide that number into the lapse of time since she was marked. Presto—the shed rate! This figure tends to be greater for males than for females, who fast during gestating. No food, no growth. No growth, no shed. In the midst of being processed, two rattle segments fall off.

Although the snake still rattles, the sound is muted. I hear crickets again.

"Brett, do you have any crazy glue?" asks Fraser nonchalantly, as if he sought to repair a broken bowl.

"No."

The rattlesnake weighs sixteen hundred twenty-five grams, slightly more than three and a half pounds, equivalent to a full-grown cottontail. Fraser thinks the snake is more than a dozen years old (possibly much older), and that she has bred at least once. Timber rattlesnakes must reproduce on average more than once to keep the population from collapsing—a statement Doug Fraser repeats several times during the morning. Fraser discovered that most female rattlesnakes in the central Connecticut uplands give birth to their first litter around age eight, and that they breed every three years (on average) after that. "If you waved a magic wand and all females bred only once, the population would go extinct," he says again.

You may recall that the population stability of every den relies on the number of mature females, which are the demographic lynchpin of a healthy population. And, as bears repeating, gravid females are vulnerable, particularly to people. In July 1989, the state Department of Transportation (DOT) contracted workers to improve visibility along Route 2. In the process of destroying ledges close to an active den, the men, fearing for their safety, killed five basking rattlesnakes. Four were gravid. The DOT had forgotten to tell them that rattlesnakes were protected in the state.

Doug Fraser, the only person in Connecticut licensed by the DEP to tag rattlesnakes, loads a magnetic chip, about the size of a grain of wild rice, into a syringe attached to a stout hypodermic needle. He injects the chip, called a passive integrated transponder, and commonly referred to by its acronym PIT tag, under the snake's skin just in front of and to the side of the cloaca. The process makes me wince, but the snake doesn't flinch. Each battery-free PIT tag contains a unique code. This rattlesnake is 334-A. When Fraser finds the snake again, he'll be able to identify her by means of a receptor antenna whose electrical field reads the code just as a cashier reads the bar code on your box of cereal. Then comes the science: Fraser plots her movements and calculates her growth and reproductive history. Of course, years may pass before he finds her again, if ever indeed he does.

In the early eighties, modestly funded by the state, Fraser and Fritsch radio-tracked timber rattlesnakes and determined the health of the population. "DEP needed to know what they were up against," Fraser tells me, because some very vocal people on Laurel Grove Road wanted the snakes removed from their neighborhood. Before 2009, he had marked every rattlesnake he found, including the babies. Now, he PIT tags only those snakes recovered by the first responders. "Basically," he says, "I had enough data to hold in front of the state and TNC and say, 'Look, this population is viable. It needs to be protected. It is not going to go away tomorrow. We're not down to the last snake.' That's why I do this science." Fraser continues,

> The state gave us transmitters and receivers. So, the goal of my work has always been for the conservation of the snake, and to this end, I've been determined not to use this information, gotten by [fitting them with transmitters], and molesting them in many other ways to gather data . . . in ways that would harm the snakes, such as revealing the location of the den and population numbers to anyone who didn't need to know. Publishing these data would violate that ethic. It would be very difficult for me to disguise the location of this population. And, I've never felt compelled for career purposes to publish.

By radio-tracking, Fraser and Fritsch were able to pinpoint basking, shedding, and birthing sites, and to follow the snakes' circuitous routes around the highlands. Collectively, the information they gathered determined the age structure of the rattlesnake population and identified habitat critical to the snakes' survival. As a result of Fraser and Fritsch's fieldwork, the state of Connecticut invested millions of

dollars in acquiring land that is now included in Meshomasic State Forest. Although there are still private holdings in the state forest, these landowners cannot develop their property or cut roads across state land to reach them. In fact, the main den is on private property, a mile from the nearest road.

Once Fraser sets the PIT tag, the rattlesnake is ready to be returned to the woods, to be released along the edge of a predetermined— nothing Fraser does in the highlands is random—snake-friendly outcrop. He pours the snake back into the nylon net, which he ties off, and then deposits the net in his day pack. The snake's head bulges out against the net, perilously close to the backpack zipper (more than one herpetologist I know has been nicked through a snake bag). Still rattling, the snake sounds more like a cicada than a venomous reptile, her annoyance stifled by both the shortened rattle and the confines of the backpack.

A trail behind the woodshed winds uphill through Brett Somers's property, a stately forest of tulip trees, sugar maples, and red and white oaks interspersed with a couple of snake-friendly gneiss outcrops that loom out from the hillside like gargoyles. It's warm, mid-seventies. The sky is clear, the air dry. Besides a few crickets and a red-eyed vireo's monotonous song, the woodland equivalent of Muzak, the only other sounds I hear are my own footsteps and a faint rattle, the repetitive sound a constant reminder of her unrest.

We pass by Somers's hunting camp, a sparse affair where rattle-snakes leave their sheds in the woodpile. Entering Meshomasic State Forest, I'm taken by the desolation of the terrain. Regardless of its proximity to Hartford, this woodland supports bobcats and coy-otes, great horned owls and red-tailed hawks. Occasionally a moose passes through, and Fraser recalls a summer when transient black bears looking for ants and yellow jackets overturned rocks, including some favored by rattlesnakes as shelter rocks.

We pause at a spring, where moisture wells and spreads, deep enough to support a pocket wetland with a couple of half-grown green frogs that bolt at my approach. Beneath the sparse shade of black oaks, black birches, and chestnut saplings that rise from ancient rootstocks and then die back before they mature, disorderly patches of scrub oak and huckleberry clog the flats. These woods are home to garter snakes, ribbon snakes, water snakes, milk snakes, hog-nosed snakes (another fat-bodied endangered species), and black racers, which make their own September circuits to rattlesnake birthing sites. Once, Fraser and Fritsch interrupted a ravenous racer that regurgitated a crotaline neonate, and then vanished under a nearby rock. Two other dead snakelets lay nearby. As we've seen earlier, high-strung, long-toothed racers grab baby rattlesnakes behind the head, bite down, and then thrash them to death ... one after another. "A small timber rattlesnake doesn't stand a chance," says Fraser, who, neither reluctantly nor resentfully, bemoans the loss of a rattlesnake to a predator.

Following Fraser's internal compass and an exquisite sense of the central Connecticut uplands—he could assign names to various outcrops—we arrive at a small rock pile, where sunbeams break through a hole in the forest canopy. To me, the rocks look like any of a dozen others we've passed since we entered the woods, but to Fraser and apparently to timber rattlesnakes, these rocks are ideal. "Here," he says, with the instincts of someone who has studied something for a very long time. Fraser opens his pack, removes the net, which he jiggles to keep the contents in the bottom, unties the top, and then spills out the still rattling snake. Promptly, the rattlesnake retreats under precisely the right-sized rocks, thick enough to lose heat slowly, thin enough to heat quickly—not so thin that the rocks get dangerously hot. The perfect place, sheltered from the elements, where a thick-bodied snake (or two or three) could slip below the surface either to moderate its core temperature, or to avoid a harassing predator. Only a black bear or a man with a crowbar could flip these rocks.

That timber rattlesnakes still live in central Connecticut is no surprise. These uplands have never been completely logged or grazed. Except for scattered rattlesnakes that survive on a couple of riverine bumps in Massachusetts, these snakes are the last genetically healthy metapopulation along the Connecticut River, which used to be a major corridor for rattlesnakes' northward dispersal. Formerly, timber rattlesnakes ranged northward along the Connecticut at least as far as Orford, New Hampshire, where Alcott Smith and I spent an afternoon searching in vain for them on Cottonstone Mountain, and across the river in the rockslide below the Fairlee Cliffs. The last of the central Connecticut rattlesnakes, survivors of an eight-thousand-year-old lineage that fell everywhere else before the ax and the crosscut saw, before plows, guns, dynamite, clubs, stones, and, more recently, suburban commuters' cars. Only ghosts mark their passage, echoes that linger in place names, like thorns on trees no longer grazed by extinct ground sloths, which are as definitive of the species' big history as PIT tags are of an individual snake's little history: Rattle Hill, Rattlesnake Swamp, Rattlesnake Brook, Rattlesnake Mountain, Rattlesnake Ledges, Rattlesnake Hill, Rattlesnake Valley, Rattlesnake Den, Rattlesnake Rocks, and (my favorite) Rattling Valley Hill in the town of Deep River, off I-95. Eighty-five percent of Connecticut's historic timber rattlesnake population is gone.

On our way off the mountain, we pass eleven patrol cars parked along side of a paved road, and we see a crowd of uniformed police milling around in the woods. They don't appear to contemplating the joys of living in proximity to rattlesnakes. I wonder what (if anything) a rattlesnake means to them?

In the seventies, when a flurry of cases of juvenile arthritis were detected in Lyme, Connecticut, the disease then known as "Lyme arthritis" was identified as a tick-borne bacterial infection rather than an autoimmune condition. The vector was thought to be *Ixodes scapularis*, the black-legged tick, a widespread arachnid that ranged throughout the eastern half of the United States. In short order, Lyme arthritis became known as "Lyme disease." Numerous cases began to appear along the coasts of southern New England and Long Island. Then, biologists concluded the vector was not the ubiquitous black-legged tick but a similar, undescribed species they christened *I. dammini*, commonly called the deer tick. Although still relentlessly perpetuated by the media, the name "deer tick" fell out of vogue among the arachnid cognoscenti in 1993, when DNA sequencing revealed that the vector for Lyme disease was not two separate species— northern *dammini* and southern *scapularis*—but a single widespread species, the original premise. Since *scapularis* had been classified first, the rules of scientific nomenclature led taxonomists to reinstate that name for the species; in other words, *I. scapularis* subsumed *I. dammini* in the same way that herpetologists had lumped timber rattlesnake and canebrake rattlesnake in a single wide-ranging species; and ornithologists had lumped Audubon's warbler and the myrtle warbler under name the yellow-rumped warbler (much to the consternation of birders, whose life lists decreased by one). Although I know no one who keeps a life list of ticks, taxonomy quickly devolves to a footnote if you're afflicted with Lyme disease.

Let me clear up another misconception, one more germane to our story. Adult ticks *do* mate on deer, and a female, having locked herself in place to feed, is a somnolent and defenseless target for an aroused male. "A whitetail [deer] in the woods of Connecticut, during November," wrote David Quammen, "is like a teeming singles' bar in lower Manhattan on Friday night, crowded with lubricious seekers." However, the number of deer in an area is not a reflection of the population of black-legged ticks, nor is it a predictor of transmission of Lyme disease to humans. Biologists in the Northeast determined that the number of female ticks in an area remains fairly constant whether or not deer are in abundance. (In a study on an island off Cape Cod, state biologists shot seventy percent of the deer herd without affecting the population of black-legged ticks.) In addition—and this is also *very* important—Lyme disease is not inherited. When infected with *Borrelia burgdorferi*, the spirochete bacterium that causes the disease, a mother tick doesn't pass the spirochete vertically to her offspring. Tick larvae hatch disease-free. To summarize, in a given geographic area, say the uplands of Glastonbury, the relationship between the size of a deer herd and the number of future cases of Lyme disease in humans is not linear.

Stick with me. This will relate back to rattlesnakes.

Lyme disease perseveres because black-legged ticks live for two years and have a three-stage life cycle, each stage of which requires a single, separate blood meal from a vertebrate host to fuel both metamorphosis to the next stage and procreation in the case of adults. When feeding, a tick fixes itself to the host's skin for up to a week, ceaselessly imbibing. Larvae, which hatch in midsummer, pick up the spirochete during their first meal when parasitizing small and mid-sized mammals, ground-feeding and ground-nesting birds like hermit thrushes and robins, and lizards, particularly in the South. Collectively, these animals are the reservoir host for the Lyme spirochete. In the uplands along the Connecticut River, white-footed mice, short-tailed shrews, masked shrews, chipmunks, and robins are the most competent reservoirs. *B. burgdorferi* thrives in them.

Engorged, tick larvae drop off their hosts, digest their blood meals, and transform into nymphs, overwintering beneath a blanket of leaves on the forest floor. Meanwhile, the spirochetes multiply within the gastrointestinal tracts of the ticks, penetrate the gut wall, and move into the circulatory system (ticks have an open circulatory system, the GI tract floating in blood) eventually finding their way to the salivary gland. Once the spirochetes penetrate the ticks' salivary glands, they wait. Nymphs reappear the following May or June, scale a blade of grass, a fern, or a tiny woodland flower, seeking new hosts—an activity scientists term "questing," as though the ticks were knights errant in a medieval romance. A nymph infected with *B. burgdorferi* transmits the infection when it feeds on a new host, which is sometimes us. After imbibing, the nymph drops off, molts to an adult, and then seeks a final host in autumn. Adult females overwinter in leaf litter, depositing their eggs in spring. By midsummer, a new generation of black-legged tick larvae hatch, free of the disease. Then, if a questing larva encounters a recently infected host, Lyme disease is passed on.

In the nineties, when everyone blamed deer, deer, deer for the perpetuation of Lyme disease, Doug Fraser took a different approach. Since black-legged ticks have a three-stage life cycle, he wondered how a healthy ecosystem that supported a suite of mesopredators might dampen an outbreak of Lyme disease. Read: did rattlesnakes reduce the transmission of Lyme disease in Glastonbury? Or, in other words, "How can we know the snake from its world?" Drawing on epidemiological data, Fraser examined the Connecticut Department of Public Health's town-by-town Lyme disease frequency, tabulating the number of cases and the rate of infection per one hundred thousand people, a more comparable figure, because not all towns are the same size. Using data gathered between 1984 and 1995, Fraser noted that Glastonbury reported ninety-one cases of Lyme disease, an average of eight cases per year. Colchester, a rattlesnake-free community also on the east side of the Connecticut River, reported three hundred cases, about twenty-nine per year, while snake-free Middletown had one hundred twenty cases, about eleven per year. Between 1984 and 1995, Marlborough, Portland, and East Hampton, rural communities immediately south of Glastonbury that had formerly supported rattlesnakes, sustained an average of two hundred twenty-four, one hundred thirteen, and two hundred thirty four cases of Lyme disease, respectfully, per one hundred thousand population per year, far more than Glastonbury, which had a rate of fewer than thirty cases per one hundred thousand per year.

"To be a strong contributor to total tick infection prevalence," wrote Richard Ostfeld, a senior scientist at the Cary Institute of Eco-system Studies, in Millbrook, New York, in his book *Lyme Disease: The Ecology of a Complex System*, "a host must also feed a substantial proportion of the tick population. Hosts that are abundant (many individuals per unit area) and that are heavily infested with ticks (many ticks per individual) will have the greatest impact on tick infection prevalence in nature." Ostfeld determined that shrews carry the highest tick burdens, but white-footed mice are the most efficient reservoirs for Lyme bacteria. Because they lead short lives rife with danger, white-footed mice invest their caloric resources in reproduction rather than long-term health. Consequently, mice have feeble immune systems incapable of clearing the spirochete. White-footed mice are also pathetic groomers. On them, black-legged ticks picnic virtually uninterrupted in the ears and around the eyes, cheeks, and lips. White-footed mice pass *B. burgdorferi* to more than ninety percent of the ticks that feed on them. (For comparison, on average, chipmunks and shrews pass the infection to approximately fifty percent of the feeding ticks, and opossums, meticulous groom-ers, to only three percent.) I've seen mice that were snap-trapped in my neighbor's barn, their big, naked ears peppered with tick larvae, black dots, no bigger than poppy seeds. For a mouse, which lives fast and dies young, sex trumps cleanliness and health.

In 2009, the Centers for Disease Control reported thirty-eight thousand cases of Lyme disease, the most common vector-borne illness in the United States, but experts agree that the true number may be five or ten times higher. Ostfeld believes the underlying rea-son for the preponderance of Lyme disease is the continuing loss of biodiversity, the chronic simplification of ecosystems. Backyards and flowerbeds favor white-footed mice and shrews over dilution hosts, animals that either remove ticks, clear infection, or both; fastidious groomers with healthy immune systems—grouse, wild turkey, opossum, skunk, and gray squirrel; as well as the litany of mesopredators that dine on small mammals, including large snakes, like our favorite, the timber rattlesnake.

As you may have already surmised, ecology is a convoluted, imprecise science. There may be simple questions, but there are rarely simple answers. Logic often proves wrong—even though adult black-legged ticks parasitize deer, fewer deer do not mean fewer ticks. From the standpoint of complexity, ecology is more like a game of baseball than a game of stickball. For instance, a 2012 article in *Proceedings of the National Academy of Science* suggests that there's a correlation between the rising number of coyotes in the Northeast and the rising number of cases of Lyme disease. This is not because ticks prefer coyotes or because coyotes are slovenly groomers. Here's the less obvious answer: a family of red foxes requires far less territory than a family of coyotes and consumes many more mice, stockpiling what they don't eat. Coyotes eat lots of things from deer to melons, including mice. Coyotes displace foxes. Then, released of predation pressure, mouse populations tend to increase. More mice means more chances for Lyme disease to crossover to people. Now, at night, lulled by the serenade of hillside coyotes, I'm reminded that I haven't seen a red fox in our pastures in more than a year and that Vermont has one of the highest incidences of Lyme disease per capita in the United States.

Borrowing Fraser's data collection, Glastonbury realtors, like the one featured in the *Yankee* article, could have armed themselves with advanced marketing ammunition. For them, rattlesnakes could have assumed a new meaning other than fear, loathing, and lost sales.

In a somewhat mathematical frame of mind, I followed Fraser's lead and went to the Connecticut Department of Public Health website for an update on Lyme disease frequency. Adding, dividing, comparing, I found that from 2001 to 2013, Glastonbury reported one hundred seventy-eight cases of Lyme disease, a thirteen-year average of 13.5 per year, or 47.0 per one hundred thousand per year; for Middletown the numbers were two hundred ten, 17.5, and 50.5, and for Colchester, three hundred nine, 26.0, and 126.0. My stamina waning, I tallied only the average number of cases per one hundred thousand per year for Marlborough, Portland, and East Hampton, and I came up with 119.0, 140.0, and 116.5, respectively, which translates to two and a half to three times the transmission rate of Lyme disease per capita relative to Glastonbury.

Although a quick analysis of that data suggests timber rattlesnakes dampen outbreaks of Lyme disease, Fraser remains equivocal. "I don't know if this is enough to push them below the epidemiological tipping point," he told me in a recent e-mail, dampening my own enthusiasm. He believes that Glastonbury's timber rattlesnakes harvest no more than twenty percent of the mouse population each year. Is that enough to affect Lyme disease transmission? Possibly.

In 2014, a team of biologists from the University of Maryland approached the timber rattlesnake–small mammal–Lyme disease question from another direction. Their results buoyed my spirits ... again. Edward Kabay, who completed his master's degree in biology at the University of Maryland under the guidance of conservation biologist Karen Lips, wanted to determine the role of timber rattlesnakes in regulating Lyme disease in the Northeast. Kabay used published data from Cary Institute on tick infestation rates of small mammals, information from Arkansas on timber rattlesnake energetics, and prey-consumption data from four long-term research sites in New Jersey and Pennsylvania to create a model for the removal of ticks by rattlesnakes based on the snakes' diets in biological communities that vary in both abundance and diversity of small mammals. Rattlesnake predation, as you may recall, is all about chance encounter. Kabay's theoretical model indicates that annually an adult male timber rattlesnake removes approximately four thousand to six thousand black-legged ticks, of which between one thousand and twenty-eight hundred are infected with *B. burgdorferi*, depending on the number of shrews that sashay past a snake's foraging sites.

I tracked down Ed Kabay on the phone last week. He was preparing his thesis for publication and preparing to return to Chapel Hill, North Carolina, where he teaches high school biology. "You get fifteen to twenty more ticks per shrew than per mouse. But the

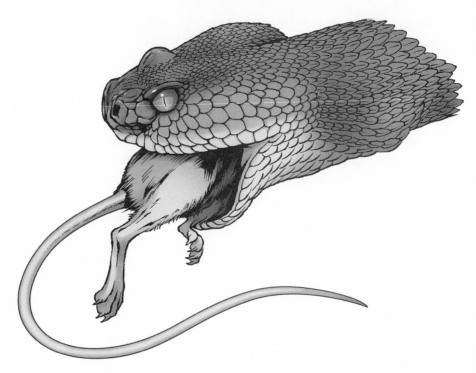

shrews are not conveying the infection to as many of the ticks that feed on them," he said, somewhat surprised by his own findings.

"Are shrews good groomers?" I asked, trying to imagine a tiny, high-strung shrew having the time to do anything but eat and mate.

"Information on shrew [grooming behavior] is lacking because they are next to impossible to keep alive. [Their] humidity and dietary requirements are difficult to maintain in the lab."

Kabay reminds me that the mast crop affects the white-footed mouse population but not that of the shrews, which are strictly insectivorous. To repeat an important point: the most abundant mammal in an oak woodland, the population of white-footed mice may swell from one or two to more than fifty per acre during the year after a big mast crop.

"I don't think rattlesnakes do it alone. Collectively, all the predators are playing a role here. Black rat snakes, copperheads, racers, corn snakes, and milk snakes are efficient mouse predators, too.

Perhaps more efficient. I chose timber rattlesnakes because of their cultural and scientific importance. And because I like them."

ⵣ

After a December 2013 op-ed piece appeared in the *Hartford Courant*, written by William Conway, a senior at Skidmore College, praising the tick consumption of Glastonbury rattlesnakes, a letter to the editor written by a local MD appeared on January 11, 2014.

> As the grandfather of a family with two young children currently building a home in Glastonbury, I was rather astounded to read William Conway's op-ed article, "Leave Rattlesnakes Alone; They Eat Ticks."
>
> As the letter writer states, there were 45 rattlesnake sightings in Glastonbury this past season. A bite from one of these snakes is a potentially lethal event....
>
> In Connecticut, the towns that are Lyme disease–free have controlled deer numbers, nothing else. Mice and other larger animals don't preserve the deer tick population.
>
> If Mr. Conway [the author] wishes to encourage the growth of a rattlesnake population in his backyard, I have little to offer, but please don't suggest that I put my young grandchildren at risk.

Twelve days later, on January 23, an editorial in support of rattlesnakes appeared in the *Glastonbury Citizen*.

> We were bemused by a recent letter to the editor in the *Hartford Courant* demonizing our very own Glastonbury rattlesnakes.
>
> ... Expressing concern over the approximately 45 rattlesnake sightings (as cited by Conway) in town over the past season, he dismissed Conway's observation that timber rattlesnakes are "non-aggressive and rarely attack humans." His thinly concealed outrage suggests that, on the contrary, Glastonbury timber rattlers may be only slightly less hazardous than an infestation of liver-eating zombies.

More people probably die [each year] as the result of sticking a bread knife into a toaster or allergic reaction to hemorrhoid medications.

We would simply point out that 1) killing timber rattlesnakes is prohibited by the Connecticut Endangered Species Act of 1989 and the Federal Endangered Species Act of 1973; 2) the Town of Glastonbury has gone to great lengths to inform and educate residents who decide to move into snake country here (a limited area, by the way); and 3) perhaps the best way to avoid putting your grandchildren at risk from rattlesnakes is to choose a building lot that is not located in snake habitat.

And, if you do choose to build in snake habitat, take the time and trouble to understand and accept the responsibility that entails. After all, the snakes were here first.

Otherwise, welcome to Glastonbury.

May 2, 2013, blue sky, mid-eighties, snake day perfecto. The wind is light, the leaves gently swaying. I rush down I-91, hermetically sealed in my car. Three hours on the road, a descent of five hundred fifty feet in one hundred sixty miles. Eight thousand years ago, rattlesnakes moving northeast out of safe havens among the twisted pines of Long Island, reached the Connecticut River and veered north; pausing here and there to colonize the frost-free chambers inside bedrock outcrops and below talus slopes from Glastonbury to Mount Tom, from Wantastiquet to Cottonstone Mountain and beyond. The trip from the Connecticut River delta north to the Fairlee Palisades in Vermont may have taken snakes five hundred years to complete—rattlesnakes never appear to be in a hurry except when they strike—the landscape constantly modifying their gene pools as the oak-hickory woods dissolved into maple-beech-spruce woods. All that remains of central Connecticut's once robust metapopulation is five dens: two east of Route 2, three to the west, the road's hot, oily patina and the pounding pavement no less a barrier to gene flow than a flaming moat would be. In the end, Route 2 may be the gene pool's final editor, as vehicles eliminate every lusty male that attempts to cross, deleting genetic articulation between the dens... forever.

Doug Fraser, Bob Fritsch, and Dennis Quinn, an independent consultant specializing in amphibians and reptiles, crowd into an outside table at Panera Bread, where I join them for the annual spring rattlesnake survey. We caravan over to Brett Somers's driveway, park our cars, and trek into the greening woods. Everywhere, and gorgeous, are early spring's inescapable rhythms: the flowering of shadbush, the rich euphony of birds—prairie warblers, black-and-white warblers, towhees, solitary vireos, catbirds, juncos, rose-breasted grosbeaks—the edgy cacophony of frogs, peepers, and American toads mostly. Woodland wildflowers. And rattlesnakes, subtle and predictable, basking on the rocks or emerging as though disgorged from the mouth of a den.

Outside a secondary den, a gneiss outcrop with a red maple sentinel and a mouth so wide I could crawl inside, oval pockets of compressed leaves punctuate the ground. No snakes, however. They've slithered up and over the outcrop or down into the wooded valley, segregating themselves into twos and threes amid jumbles of shelter rocks. We'll find twenty-five snakes before the day is over, some of which may have napped where I'm standing.

According to Fraser, the secondary den may have been used for thousands of years as a spillover for snakes that hadn't reached the main den at the time inclement weather arrived. "There is obviously some flexibility in their behavior," he reminds me, as he bends to examine the pillowed leaves. "How else could they have colonized the Northeast?"

We descend the hillside. The snakefest begins.

In the woods well below the secondary den, snake number one: an olive-colored morph with brown bands and rhomboids outlined in yellow slowly withdraws under a rock. Overhead I hear the cascading scream of a red-shouldered hawk, a dark speck in an otherwise deep-blue sky.

On the other side of the rock, a black morph, its head in the sun, its body looped in the shade, watches us pass.

Less than one hundred yards away, amid a larger group of shelter rocks, are a dingy yellow male in need of a shed, a big yellow female with an untapered six-segment tail, and a rubbery, yellowish yearling. Nearby, two black-phase snakes coil in the sun, one totally exposed. Five more yellows and another black. Not exactly a snake warren, but our total hits twelve, eight yellow and four black. This is egress, dispersal from the den into summer habitat—wooded ravines, open forest, wetlands, meadow edges, and a few suburban backyards.

Snakes living in this section of the highlands have recovered after being heavily hunted in the forties, fifties, and sixties. Local police and fire departments used to sponsor a "spring shootout," says Fritsch, who as a young man, once found a snake with its mouth sutured closed. "I took the sutures out. The snake was in terrible condition. It must have been a display snake that people could *safely* handle. I should have put it down."

Over the years, one hunter, a man named Harry Chapman, took hundreds out of the highlands for bounty, for show and tell, for leather, for oil, for meat, for rattle, for whatever a rattlesnake might have offered. Then, in the early seventies, after inadvertently leaving a barrelful of snakes to cook in the sun, killing them all, Chapman transformed from snake slayer to snake champion, a penitent expression ordained by guilt. It was Chapman, back in '57, who pointed a teenage Doug Fraser in the direction of the very den where we're headed now.

But these snakes are still shadowed by Hartford and contend with situations far removed from those facing snakes in more remote corners of the Northeast. Here are snippets of a dialogue I found on the Internet after googling "Connecticut rattlesnakes."

MR. A: Saw a guy last summer throw a handful of M80s at nest of them last year. What a waste. There were some biggies in that splattered batch. Gotta use a 20 gauge bird-shot to get the one u want. Trash the meat for the birds, the skin is where its at to [be] sure.

MR. B: When u are taking a walk in the Glastonbury mtns. Bring a .357–45 magnum pistol. And blow as many as those death tweaks away. Got 29 killed as of 2012. Many more to come in '13. Biggest one I shot was over 4ft. Hope to hit 5 mark this year . . .

MR. C: Timber rattlesnakes are protected in ct there are cameras placed by the ct dep law enforcement at the den sites . . . to protect them from there biggest threat the herp people and snake lovers taking just one or two adult females from a den site has effects on the wild population for years collectors have done more damage than anyone.

MR. D: why not run it over instead?

MR. A: They have no signs, but rattlers are everywhere! . . . Game warden rolled up on me. I'd say it tried to attack my German Shepherd so I put the little bastards out of their misery . . . I've only got three so far, that was with a Bowie knife attached to an oak limb. Taxidermist gonna be mounting me a 5 footer this year.

A mile or so beyond the transition rocks, having passed through stately woods of sugar maple, black birch, tulip poplar, and red oak, we enter a plateau carpeted in lichen. Anemic blueberries and huckleberries, and large, warped mountain laurels rising out of the lichen, beside them a sparse copse of chestnut oaks and pitch pines, their vitality sapped like tomatoes in an infertile garden. White-flowered shadbush lights up the otherwise dun-colored plateau. Boulders rise out of the lichen like islands. A black morph loiters by one rock, a dark-gray snake with a rust-colored dorsal stripe broken by black bands bordered in yellow. Number fifteen. The plateau drops down into a wet, verdant valley, sunlight flickering across tongues of water. Just above the valley floor, a massive gneiss outcrop juts from the slope and turns upward almost to the lip of the plateau. A hemlock perches on the outcrop, a gnarled green advertisement, its exposed roots flowing down both sides.

The den itself is a horizontal crevice midway up the outcrop. Inside, a four-foot-long yellow snake stretches completely out, fitting snugly in the six inches of crawl space between the ceiling and the floor. Head in sunshine, tail invisible in subterranean darkness, the snake is either emerging or meditating. It ignores us. Just below the entrance, another snake, number seventeen, coils in full sunlight, idling on a bed of leaves. Fraser believes these may be the very last snakes to leave the den. "The majority [have] moved on. These snakes may be waiting to shed or may have recently fed and are digesting," a process that may take days or a week or more depending on their core temperature.

Retracing our route back to Somers's driveway, we tally eight more snakes, including another yearling: a total for the day of seven black morphs, eighteen yellow.

Just before I leave, Fraser asks, "Remember all those police cars we saw last summer on the corner of Pepitone?"

"Yeah."

"They were investigating a homicide. A young woman, listed as missing for more than a week, had been murdered and buried in a stream bank by her ex-boyfriend. All that rain we had had washed away the earth. A neighbor walking her dog noticed a hand sticking out of the mud."

12

Into the Abyss

Vermont had decided to kill off all her rattlesnakes, and, with characteristic caution and conservatism, had started the experiment by offering a bounty of one dollar per snake. Of two things you may rest assured:—first, that there are rattlesnakes in Vermont; and second, that there are not very many of them, for otherwise the bounty would have been made smaller.

Even at that price, in a fairly good snake cover, a man could earn better wages than in the hay-field, and if the reward were greater there would be a scarcity of farm hands during the snake season.

ARTHUR F. RICE, *Forest and Stream*, 1895

I imagine that seeing an anesthetized timber rattlesnake must be a little like seeing Pavarotti in pajamas. To refresh your memory (in case you put the book down for a couple of months), in the field, a rattlesnake is the color and pattern of windswept leaves, a virtuoso of camouflage that hides in plain sight, whose form and function are inseparable from deciduous woodland. When startled, a rattlesnake is prone to bluster, coalescing into coils and puffing its single lung with air to exaggerate its size. It may begin rattling at any point in the display ... or not at all. Sometimes the rattling starts before the snake is seen, which triggers a visceral response in the observer. But when anesthetized on a stainless-steel operating table, a rattlesnake simply deflates in plain sight, its heart relaxing to a barely perceptible tick every three or four seconds, like a bobber teased by a nervous fish. If a human had a heartbeat that slow, there'd be a priest in the crowd around the table.

During the spring of 2011, the Vermont Department of Fish and Wildlife, in collaboration with The Nature Conservancy (TNC) and the Orianne Society, a Georgia-based nonprofit and one of several organizations in the world devoted to the conservation of imperiled reptiles and amphibians, began a two-year study of the summer range and status of timber rattlesnakes in western Rutland County. To that end, transmitters were surgically implanted in the body cavities of twenty-two adult snakes, seventeen males and five females, of which two were pregnant. (Another one hundred forty snakes were tagged with passive integrated transponders, PIT tags.) Within the ambit of the study, from late spring through early autumn, the snakes were radio-tracked across rough terrain between the Champlain lowlands and Taconic Mountains. Now that that phase of the project has ended, Scott MacLachlan, the veterinarian who inserted the transmitters, will remove them, and take both skin and blood samples from each snake.

Today, May 8, 2013, I'm MacLachlan's guest. Besides the two of us, there are six others in attendance at his Poultney office. There is Doug Blodgett, a state wildlife biologist and Vermont's rattlesnake recovery leader; herpetologist Chris Jenkins, CEO of the Orianne Society; Heidi Hall, the Orianne PR officer; a fish and wildlife department intern; a college senior from nearby Castleton State; and Mike McBride of the Roger Williams Park Zoo, the veterinarian I interviewed early in 2012 about *Ophidiomyces*, the flesh-eating fungus. McBride drove up from Rhode Island because the zoo had received the grant from the USFWS that had been in the application process when I visited him in Providence. As the coordinator of the interstate investigation of timber rattlesnake pathogens in the Northeast, McBride is in Poultney to demonstrate protocol for blood draws and biopsies.

I have a front-row seat for the procedures, shoehorned into the operating room across the table from MacLachlan and to the left of McBride. I am responsible for holding each snake in place, keeping its head safely inside a Plexiglas tube (the former protective

shipping-case for a fluorescent bulb) while Scott removes the transmitter, draws blood, swabs its cloaca and mouth with long, wooden
cotton swabs, and then cuts a tiny skin sample from both healthy
tissue and, if there *is* any, infected tissue. The previous year, *Ophid-*
iomyces was identified on the belly scales of a male that had died in
captivity, a big, otherwise healthy looking snake, which heightened
the concern that what is now known as the snake fungal disease
(SFD) had somehow arrived in Vermont's rather large but isolated
rattlesnake population.

For the two and a half hours required to work all six snakes through
the operating room, MacLachlan keeps up a jovial commentary
punctuated with self-deprecating humor and bad puns as he talks
about patients, clients, and rattlesnakes he has known. An urban ref-
ugee who spent boyhood summers at venerable Camp Keewaydin,
on the north end of Lake Dunmore, in the Green Mountains—the
camp where Pulitzer Prize–winning author John McPhee learned
to paddle—Scott MacLachlan grew up in a suburb of Hartford,
Connecticut, not far from the central Connecticut rattlesnakes. He
received a master's degree in wildlife biology from the University of
New Hampshire and a DVM from Tufts University, and has been in
practice with his wife Kris since 1986. Although MacLachlan, who's
known as Doc or Dr. Scott to his clients, works primarily with pets
and farm animals, wildlife has always been a part of his practice.
(He's one of several veterinarians on call for the raptor rehabilitation
center at the Vermont Institute of Natural Sciences.) MacLachlan
plumbs McBride's considerable expertise about the least invasive
way to work on timber rattlesnakes, as well as the best restaurants
in Providence. "I'm privileged to work with wildlife," he says, as he
prepares for surgery, shoeless, his feet spread about eighteen inches
apart, his knees slightly bent—a stable athletic position—his white,
high-ankle socks camouflaged against the milk-white floor. He looks
comfortable, as if he's about to stream a movie in his living room
instead of on the verge of pitviper surgery.

In this well-lit rectangular operating room, with a sink in one corner flanked on both sides by a pink Formica countertop, the floor is
plasticized, easily cleanable. There are cupboards, tanks, computers,

and a glass-fronted cabinet filled with a diversity of surgical instruments. The operating table is placed mid-room beneath the outlet of a Y-shaped hose called a partial rebreathing circuit that delivers a mixture of oxygen and measured isoflurane, a gas anesthesia, from a precision vaporizer, which sits on a cart at the head of the table. One prong of the Y is for the anesthesia mix, the other, called the gas scavenger, removes the snake's exhalation so the veterinarian and his helpers don't get dopey and keel over; the bottom of the hose connects to the front of the Plexiglas tube and will convey both the carefully regulated anesthesia to sedate the rattlesnake and the oxygen to bring it back. On the operating table, a heating pad covered by a blue terry-cloth towel will keep the snake's core temperature within the preferred range of eighty-five to ninety degrees Fahrenheit.

While Blodgett and Jenkins decide which snake to bring in, Mc-Bride suggests a game plan based on his experience with Massachusetts and New Hampshire rattlesnakes. First, the cloacal swab, then the blood sample, followed by the surgery. If the caudal vein is difficult to find, blood will be drawn from the heart, a procedure called cardiocentesis. Since the Plexiglas tube shields the heart, a third of the way down the snake's body, the snake must come out. "When it's anesthetized it's perfectly safe," McBride assures me. I secretly hope McBride and MacLachlan find every caudal vein, not because I'm concerned about holding the snake, but because I imagine sticking the heart with a needle would be profoundly disturbing, if not to the snake, then to me. (I turned pale and had to be seated when my wife had amniocentesis.)

Referring to the various ways to draw blood and take biopsies, McBride says, "I want to offer as many options as possible."

"Our only concern is the duration of anesthesia," says MacLachlan, checking both tanks and hoses to make sure everything is in order.

"I'm usually turning off the anesthesia by the time I start the biopsy," McBride counters.

"We have [the timing] down pretty tight. We can [implant] a transmitter in fifteen minutes, less time to pull one out. Do you feel surgery first would affect your samples?"

"We are a little bit concerned," answers McBride, as he assembles needles, syringes, vials, and antiseptic on the back of the operating table. Everything looks small in McBride's hands. He's a *very* big man. "We aren't sure whether cutting into the skin might cause changes in the inflammation. I often draw blood samples while the snake is still going down."

"I like that," says MacLachlan, "but our priority is to get the transmitters out. If you miss a biopsy, it is not as much of a concern to me." If the transmitter's battery dies while the rattlesnake is in the woods, the snake cannot be tracked down by any antenna, and unless it is serendipitously caught, it will bear the hard, red token of twenty-first-century research technology forever.

"We've had snakes under anesthesia for three or four hours [at the zoo], and they've woken up in fine shape," says McBride. But unlike zoo snakes, these rattlesnakes will be returned to the ridge tomorrow to fend for themselves, and they can't bear any impairing traces of anesthesia. "The briefer, the better if [you're] releasing [them] soon after the operation," Scott points out.

After further discussion, MacLachlan and McBride agree on a course of action: blood sample first, then swabs, surgery, and, if all goes well, biopsy. Once the procedures are complete, each snake will be weighed and measured.

Blodgett arrives with the first rattlesnake, a big, brown-banded, mustard-colored male, number 111, its head tightly clamped between his thumb and forefinger. The rest of the snake's body writhes, as though gripped by a red-tailed hawk. Blodgett inserts the snake's head and upper torso into the Plexiglas tube. Obligingly, the rattlesnake crawls forward until plugged. My left hand keeps him there, while my right runs a digital voice recorder. The rattlesnake thrashes like a fish in a boat. Scott attaches the anesthetic induction system to the front of the Plexiglas tube, administers isoflurane, and the snake slowly relaxes, its movements becoming less purposeful.

McBride can't find the caudal vein and decides on a heart stick. He uses a twenty-three-gauge needle but thinks a smaller diameter, twenty-five-gauge, would be better; the twenty-three-gauge needle looks enormous. Even without anesthesia a timber rattlesnake has a rather slow stroke cycle, maybe twenty beats per minute; with anesthesia, it ratchets down so much that I'm not sure I see a pulse. Once McBride finds the heart, he places his thumb and forefinger on either side of it, making each faint beat more pronounced. Now, I see it. Thump ... thump ... thump. Then, shop talk. "You want to sneak [the needle] right into the apex of the ventricle, under the ventral scale behind the heart, more horizontal than vertical," he says.

This is definitely not the sort of needle I'd like somebody sneaking into me.

Although negative pressure helps to fill the syringe more quickly, McBride constantly calibrates his grip on the snake's heart to moderate blood flow. With each vague heartbeat, a small surge of blood enters the syringe. "Since a rattlesnake doesn't have long bones, where is the blood made?" I ask no one in particular.

"Probably in the bone marrow of the vertebrae," answers Scott.

The spleen also makes red blood cells, McBride chimes in. "It's the policeman of the blood. It cleans up old red blood cells," he says while removing 2.5 cubic centimeters of blood, which is the guideline for a rattlesnake that weighs more than five hundred grams. (Number 111 weighs twelve hundred ninety grams, nearly three pounds.) Anticipating my next question, he announces, "A rattlesnake's blood pressure is so low its heart heals quickly." No leakage.

Before Scott sterilizes the site of the incision, I rub a finger over the transmitter, which sits inside the coelomic cavity close to the intestines. Compared to pliable muscles, the transmitter feels like stone. Once the site is disinfected, MacLachlan selects a scalpel, and then opens an inch-long cut. With forceps, he fishes the transmission wire out from under the skin and then gently reels in the red transmitter, the length and width of the first two joints of my pinkie.

While MacLachlan wipes off the transmitter, McBride complains about the four-hour drive from Providence.

"Oh," says Scott, "that's the scenic route."

"I'd like to take a less scenic route home."

"So, we can't expect a return trip any time soon," Blodgett interjects.

"Thanks, but no thanks."

Three blue sutures seal the incision. "The more [surgeries] you do, the smaller the incision," announces MacLachlan, as he knots the last one.

Oxygen replaces isoflurane, hundred percent at first to flush out the gas anesthesia. Three puffs: one ... two ... three. The air sac expands and contracts. As the rattlesnake slowly recovers I lay my fingers across its three-chambered heart, feeling life's faintest echo gradually gathering force, and wonder what its breath against the back of my hand might feel like.

I'm fascinated by the snake's ventral motif, an iteration of three or four cream-white scales followed by four or five heavily speckled scales, each a little map of the night sky ... a linear planetarium. Because a timber rattlesnake hugs the ground, I rarely have the opportunity to examine its linoleum-esque belly scales (unless I'm involved in a necropsy on my dining room table). There are constellations and clusters and supernovas that appear larger and brighter and more misshapen than the spots around them. Of course, such an intricate pattern begs a question. Why? Why bother to evolve a hidden motif? Viewing a rattlesnake's undergarment is a privilege left to mites and springtails and curious naturalists with a lot of time on their hands.

Next patient, rattlesnake number 108, measures forty-nine inches long from snout to vent (the snout-to-vent length, or SVL) and weighs thirteen hundred ten grams, a jet-black morph, an adult male with almost no sign of banding, followed by number 116, a black female, grayer than 108, with an iris as dark as obsidian. Number 116 has a fungal infection along the lesion where the transmitter was inserted a year ago, more visible than the incision itself, a full-blown granuloma, which may or may not be *Ophidiomyces*. (Since most fungi look similar, McBride will send a small sample to a mycologist for identification.) As MacLachlan slits the snake open, a whitish fat deposit oozes out of the incision, a sign that she ate well last summer. MacLachlan is at the top of his game; the transmitter is removed and the snake sewn up in less than a minute.

Because of the granuloma, number 116 is perfect for a punch bi-opsy. McBride grabs the infected skin with forceps, and then, with a tool that looks like a miniature cookie cutter, punches out a very small, circular sample, deep enough to remove dermal tissue. A single suture closes the biopsy.

For a normal tissue sample, MacLachlan removes a healthy scale. Then, under McBride's supervision, Scott hits the jackpot and draws blood from the caudal vein. Timber rattlesnake number 116 is thirty-six inches SVL and weighs nine hundred grams (approximately two pounds), a young adult female that winters in the Crevice Den, a vertical zigzag crack in a bedrock alcove that faces the setting sun. When other dens in the Taconic Mountains are shaded and cooling off, the crevice is still warmed by late afternoon sunlight. And snakes like 116 know it and tend to linger on shelves near the entrance or sus-pend like rolled hose from the narrow interior wall as the sun begins its descent behind the Adirondack Mountains. Unfortunately for Blodgett and Jenkins and their field assistants, the Crevice Den is on private land, and the owners are not sympathetic to the state's purpose. If snake 116 is ever encountered again, it will be because she's foraging in a nearby state park or gestating on the ridgeline owned by TNC. Her PIT tag, 1344677, will identify her. A long shot at best.

In succession, numbers 110, 128, and then 123, a two-pound, yellow-morph female, more than three feet SVL, a postpartum whose loose lateral folds of skin sag like a shirt on a hanger. Isoflurane hushes her rattle, seven untapered, khaki-colored segments (in scientific short-hand, $T = 7$), which fade from agitation to resignation to silence in a matter of seconds. Implanted with a transmitter in June 2012, snake 123 gave birth the following August on a beautiful west-facing ridge that overlooks the Champlain valley.

"Hey, I know your snake," I announce to MacLachlan and Mc-Bride, who busy themselves with the usual pre-op rituals. Last August, in the company of Kiley Briggs, an Orianne field biologist, I had followed 123's radio signals to a shelter rock on a maternity ledge, where several offspring milled around her. Bleached of sensation now, the snake disregards me.

If the weather cooperates, these six snakes will be released tomorrow. Chris Jenkins prefers to release them quickly, but the weather needs to be warm; if not, he'll keep them in a terrarium on a thermal gradient, with plenty of water available. Rattlesnakes need heat to heal, for everything, really. (You knew that didn't you?) After surgery, a rattlesnake will bask for about a week, which is why neither MacLachlan nor McBride implants transmitters in the fall. Should a snake eventually hibernate with an unhealed surgical wound, it will likely perish during the winter from a raging case of septicemia.

Preparing to remove the last of the six transmitters, Scott says, "It is easy to get lost when you do snake after snake," which becomes an issue when handling venomous snakes. Like the carpenter who whacks his thumb with a hammer after pounding too many nails, if you get distracted or tired while handling even the most docile timber rattlesnake, you may inadvertently place you hand where it doesn't belong.

"There are only so many [snakes] I like to do in a session," announces Scott, gliding back from the operating table, his stockings nearly frictionless on the plastic floor.

June 1, 2011. 9:30 a.m. Sitting just off a state highway that threads the Champlain valley south of Burlington, lies a small, weed-clogged pullout for rattlesnake enthusiasts. I've parked here before, many times since the late eighties, but almost always in the company of Alcott Smith. Back in 1987, Smith, who lives two hours away on the New Hampshire side of the Connecticut River, began a second, passion-driven (and uncompensated) career unraveling the natural history of Vermont rattlesnakes. Coincidentally, that same year the state designated timber rattlesnakes an endangered species, which theoretically conferred upon them legal protection from saboteurs and collectors. In fact, though, no one in the parsimonious Fish and Wildlife Department either had time for rattlesnakes or knew much about them, biologists and wardens having relegated snakes in general to a place of less consequence than that of woodchucks. Then, along came Smith.

To my west, farmland and pasture run ten miles gently downslope to Lake Champlain, interrupted here and there by remnant clay-plain forest, where flood-tolerant oaks (several species) mingle with American elm, red maple, and shagbark hickory to form a mesic woodland more closely associated with the rim of the Great Lakes than with the nearby Taconic Mountains (Vermont's less famous range). To my east, a cliff of metamorphic bedrock a half billion years old parallels the highway and forms several miles of the valley's eastern wall. Midway, in a series of short rockslides, rattlesnakes disperse to one of four dens (possibly more), including the crevice, the winter domicile of snake 116.

I'm waiting for Kiley Briggs, the Orianne Society biologist, who several times a week locates and maps the movements of six trans-mittered rattlesnakes and injects a PIT tag, a subcutaneous magnetic chip about the size of a grain of wild rice, into every unmarked snake he can get his hands on (forty-eight at this point in the project), supplying each of them with a bar code for life. It's eighty-seven degrees Fahrenheit, summer humid. American toads croon from

the pond behind the pullout. Mosquitoes and black flies have transformed boot lacing into interval training, their buzzing prompting
spells of compulsory twitching, swatting, and flailing, punctuated by
frequent outbursts of profanity. Briggs arrives in an old blue Nissan,
de rigueur for a recent college graduate. He thinks I'm gesticulating.

After reintroducing ourselves—I met Briggs a couple of months
ago at a public meeting convened to launch the rattlesnake recovery
project—we enter the woods behind the pullout, trace the western
edge of the pond, carefully climb through a barbed wire fence, and
then scale the steep, rock-studded slope. I know the route well. To
our south, hidden below the lip, tucked among embedded talus and
chestnut oaks, is the main den—the one Alcott Smith introduced
me to in the early nineties. Except for the seasonal presence of snakes,
not much recommends the site as Vermont's most productive snake
den, which looks like a hundred other piles of rocks along this west-
facing pitch. Why do we bypass it? Nobody is ever home in June.

Instead, we climb a muddy, leaf-strewn crease in the cliff, a water-
cut wrinkle that Alcott calls "The Chimney." For five thousand years,
The Chimney has been the snakes' umbilical link between den and
forest, their principal route of both egress and ingress, and their
collective busyness has helped to keep the path open. Once we negotiate the rim, we head north by northeast. A quarter of a mile into
the forest, Briggs stops, unlimbers the antenna, wires it to the seven-
hundred-dollar receiver clipped to his waist, and then dials in the
frequency of the first of three snakes we're after, his usual daily quota.
Each rattlesnake broadcasts its presence on a singular waveband, a
series of beeps and squeaks and squelches, an electronic language
programmed by Briggs, which, when heard from behind, sounds
more like cellphone static than a transmission from a snake.

Tall, lanky, with an unconquerable shock of black hair, Kiley Briggs
walks with big strides that eat up the uneven terrain, arms akimbo.
Twenty-four. A graduate of the University of Vermont (UVM to
everyone living in New England), Briggs majored in wildlife biology and understands the odd politics of rattlesnake recovery. He
is loquacious and smiling, a thoroughly committed, sensitive naturalist, friendly, inquisitive, and sharp-witted. "The only thing two

timber rattlesnake biologists can agree on is what the third one is doing wrong," he says with benevolent amusement, a reference to the acrimony that surrounds Vermont's recovery plan. Briggs grew up in nearby Shoreham. During a fifth-grade classroom presentation on reptiles and amphibians, he had an epiphany. "Wow," he thought, "I could make a living doing this," which essentially meant visiting schools with jars of frogs and sacks of snakes.

In the late aughts, on a UVM field trip, Briggs saw his first Vermont timber rattlesnake under the watchful eye of Alcott Smith. With bewildered vehemence, he says, "I wish Alcott would join me. This is *not* how I envisioned the project going forward," referring to Smith's absence from the recovery team. After supplying Vermont Fish and Wildlife with two decades worth of field notes, Smith became an unwitting victim of disingenuous communication within the department, as well as a casualty of Vermont's new endangered-species permitting process, which would have required him to visit the rattlesnakes under the supervision of the very man, whom I'll call "the biologist," he had spent several years tutoring in the ways of rattlesnakes. So the recovery plan moved forward without him. There would be no more spontaneous visits for Smith. No more noninvasive observations. Using twenty-two years worth of Smith's observational data, the state implemented a two-year inquiry into the life of Vermont timber rattlesnakes, a rather Byzantine approach to Smith's contributions that used the media as its unknowing ally by failing to credit the man with the longest, most unselfish history with Vermont rattlesnakes. No one ever sought Smith's opinion about what might be best for Vermont's timber rattlesnakes.

Unsullied by the turmoil, Chris Jenkins, who had unknowingly stepped into a herpetological wormhole, had tried to broker a truce, tried to lure Smith to the project. "After the radios were pulled, I would have liked to hire him to monitor the snakes," he told me in a phone call.

In deference to Smith, Briggs announces, "In order to appreciate timber rattlesnakes, you have to see them for yourself. You have to go to the snakes and watch." I struggle to keep up, battling a maze of dense, pole-sized saplings. "Alcott opened a door into another world to me," says Briggs, a world known to a very select group of Vermont-ers, essentially those people Smith had taken to the ridge: biologists and would-be biologists, teachers, students, TNC staffers, lifelong residents of the Champlain valley. Many of these people were in a good position to protect rattlesnakes and their critical habitat and were respected by their neighbors, even neighbors who were less than compassionate toward snakes, fiercely independent landown-ers who were loath to let state biologists onto their property. There are more than a few such landowners, and included among them were owners of parcels that included two snake dens, the crevice being one of them.

Back in 2010, when Vermont Fish and Wildlife secured both fund-ing and assistance from the Orianne Society, the society hired Kiley Briggs to do the time-consuming fieldwork under the supervision of both "the biologist" and Chris Jenkins. Before Briggs set foot on the ridge, he composed and mailed two hundred fifty packets of infor-mation and permissions to landowners who lived within the likely summer range of Vermont rattlesnakes, the people he most needed as allies. Only twenty bothered to respond. Of those twenty, half refused him. On April 5, 2011, the night of the meeting that publicly inaugurated the rattlesnake recovery program, jointly sponsored by TNC, the Orianne Society, and Vermont Fish and Wildlife and held at a regional high school, the meeting at which I was introduced to Kiley Briggs, most abutting landowners stayed home to watch an off-season football game. Back on the ridge, Briggs peers into unbridled lushness, knowing that beneath webs of green branches rattlesnakes are moving out. Weeks of idleness, of puttering around neighborhood wetlands looking for salamanders and futzing with spring peepers, of learning how to operate the tracking antenna and receiver, of study-ing area topo maps, have given way to daily macro-outbursts of work bushwhacking around the Taconic Mountains: tracking, catching, weighing, measuring, and PIT-tagging rattlesnakes. Kiley Briggs has taken over where Smith left off, and over the next two years he will busy himself with verifying every den, every basking, birthing, and shedding site that Smith had originally documented and with the aid of bar codes and transmitters will confirm every hunch that Smith ever had. Radio-tracking will also unveil mysteries, the summer whereabouts of rattlesnakes, the importance of geography heretofore known only to the snakes themselves.

Last week, when Kiley tracked 007 to the edge of a derelict dam, he discovered a pile of four similarly sized cohorts lounging on a cement embankment, confirming that, as Smith had long suspected, this is an important transient site for annual rattlesnake rendezvous. Today, the only snake on the dam is a frog-stalking ribbon snake, long, thin, and quick as the glint of a sunbeam. Headphones on, Kiley trolls for a signal, slowly turning around and around and around, the antenna held high overhead. If he moved any slower he would segue into a yogic asana: right arm straight up, left hand on hip, body stretched to the limit—rattlesnake biologist's pose. Briggs's antenna delivers static, squelching static. Not a snake.

To pick up a radio signal, you have to stand in the right place at the right time, your antennae facing in the right direction. If you're on a ridge and the snake you're seeking is on a nearby ridge, the broadcast will be deceptively loud, even if you and the snake are three ridges apart; but if you're on a ridge and the snake is in the next hollow, the transmission may be a deceptively faint signal. After traipsing around the woods for half an hour, we pick up 007's signal, a snake Scott MacLachlan surgically implanted nineteen days ago. The day after surgery, 007 traveled north by northeast, four-tenths of a mile from its release point at the main den. This morning, it coils in the shade of a red oak, two hundred feet from where Briggs left it the other day. Juneberry flowers decorate the understory. Finally, resolving the snake in the shadows, I answer Briggs, "Yeah, I can see it," big and dark and piled up like a discarded sweatshirt. We're in the middle of nowhere and if left to my own devices, I'd have to follow the sun to get off the ridge.

A pixilated bumblebee queen lands on the snake's face and goes for a royal stroll. The snake holds tight, impervious (oblivious) to this incidental visit by an insect that would trigger a psychotic reaction in most people.

Next spring (2012), Scott MacLachlan is scheduled to replace 007's battery, which has a life expectancy of eighteen months at seventy degrees Fahrenheit. Warm weather shortens the battery's life; since a basking timber rattlesnake does everything possible to maintain its temperature above seventy degrees—remember that optimal temperature is eighty-five to ninety degrees Fahrenheit—if you wait much longer than a year you run the risk of the battery dying; then, the snake vanishes into a yawning landscape, a calculated risk Jenkins is unwilling to take.

"Why not use a more powerful battery and spare the snake two surgeries in twelve months?" I ask.

"A more powerful battery might disrupt its vital processes," Briggs answers thoughtfully, clearly having weighed the option before.

A distinctly marked black morph, gray base color with thirty-seven white-bordered black bands, like a New Age Ralph Lauren polo shirt, 007 weighs eleven hundred fifty grams (two and a half pounds) and stretches almost four-feet SVL. According to Smith's records, more than ninety-two percent of the snakes that den along this ridge are black morphs. Just a few miles away, on Monday Mountain and across Lake Champlain in the Adirondack foothills, the ratio draws tighter; depending on which den you inventory, the proportion may be thirty to forty-five percent yellow morphs, snakes almost impossible to see against a carpet of freshly fallen leaves. Satisfied that there is nothing here to stuff into her pollen baskets, the bee leaves.

Next rattlesnake up: 001, a black morph, almost as big as 007— eleven hundred grams (2.4 pounds) and forty-six and a half inches SVL. Because of its peripatetic nature—it had cruised nearly a mile within forty-eight hours of release—Kiley has named the rattlesnake Legs. The receiver beeps.

Guided by our ears, we triangulate toward the sound. "It's over here, beyond the beech trees," shouts Briggs, headphones on, unaware he's yelling.

"No. It's over here," he says, self-correcting, his voice drowning out an amorous black-throated blue warbler.

He has *no* clue where the snake is.

The landscape of southwestern Vermont is peculiar. Like a disjunct piece of Indiana, it's all lake and plains along the New York border, then suddenly rises into a bulwark of mountains, into bedrock corrugations, a run of swales—some wide, some narrow—and ridges—some sheer, some gradual—all wooded and pocked with wetlands: streams, springs, beaver meadows, beaver ponds, bogs, seeps, vernal pools, swamps, and small lakes. (When compared to Champlain, virtually everything but the Great Lakes is small.)

Legs blinks in ... then blinks out ... repeatedly. We stop at a high spot. Get a fix. Then, dropping into the neighboring swale, lose it again. Seven hundred feet from 001's last known location; we appear no closer to finding the snake then if we had stayed in the pullout to eat portobello-tomato sandwiches.

There's the signal.

Oops, gone again, engulfed by another bedrock wrinkle.

Mumbling, Kiley reassembles himself into the rattlesnake biologist pose, turns around, fiddling with the knobs on the receiver. Then, he shouts loud enough for someone in Rutland to hear him, "Over here. He's in the swale. Shit, where'd he go? No, he's over here. Damn, the signal keeps shifting." Like a puppet master throwing his voice, the signal comes from everywhere and nowhere.

Just wanting to contribute to the confusion, I say, "Kiley, it sounds louder behind us," having no idea what I'm listening to.

On the next ridge, the signal becomes distinct, seemingly direct and static-free. We drop into the next swale and lose it, again. We're somewhere in the western corner of state-owned land, not far from a town water supply, a mile and a half from the main den portal.

Groping to make sense out of disparate teakettle sounds, suddenly, fortuitously, 001 coalesces out of the leaf litter twenty feet in front of us. No wonder its signal was hard to pinpoint, the snake is moving out, dispersing. Towing its signal behind it. 001 has an agenda. For an enchanted moment, the snake stops, looks around, flicks its tongue. Then, detecting something to its liking, it eases up an oak log. Legs moves along the trunk, hugging the bark, rattle high like a semaphore. Then, the snake pours back onto the forest floor, belly scales gently caressing moldering leaves, a familiar, subtle sound I've heard a thousand times. Jotting down notes, I take my eyes off the snake for a moment. When I look up, it's gone, lost in the shag of the forest floor, a limbless Houdini. But it won't be the last time I see 001.

Looping back toward the ledge, we track fruitlessly after the third snake, 004, but instead find lots of other things: a hermit thrush nest with three blue eggs; a domed-over ovenbird nest crammed with four milk-white eggs with brown, Jackson Pollack speckling; toads and wood frogs and pickerel frogs; a green frog burping on the threshold of a puddle. A throaty-voiced scarlet tanager spills its song from a chestnut oak. Everything but 004. By late afternoon, we give up the search and head back to our cars.

I won't be around tomorrow when Briggs resumes the search.

August 18, 2011, ten o'clock in the morning. Seventy-three degrees Fahrenheit. Big cumulus clouds clot an otherwise blue sky, casting fleeting shadows across the pullout. A west wind, gusting to ten miles per hour, shuffles the clouds and buffets eleven turkey vultures, which hang above the cliff like so many kites; their gray underwings shining brighter than bright. Behind the pullout, a cage full of inflated, football-shaped turkeys, whose white, unkempt, shit-stained feathers represent the quintessential bad-integument day. Standing by his car, Kiley Briggs talks to a mechanic on a cell, a fuel-pump issue.

Last week, says Briggs, a Massachusetts man came off Snake Mountain in Addison and Weybridge carrying a pillowcase and a snake hook, with something large and snakelike squirming inside the pillowcase. Were there records of rattlesnakes that far north on the Vermont side of the Champlain valley, I ask, assuming that was how the mountain got its name?

"There are historic rumors."

Later in the morning, as we enter the woods and head toward the east end of the ridge, across land owned by the town of East Steeple, we pass four men from Brooklyn, who are obviously looking for something on the ground. Kiley asks if they've seen any rattlesnakes.

"If we did see one," answers the spokesman, his volume turned up in that thick, familiar urban accent, words bursting through puckered lips like rifle fire, "I think we'd find dat interesting." Everybody's on his best behavior.

By mid-June, Briggs began finding rattlesnakes hanging around wetlands, beaver ponds in various stages of disrepair, moving back and forth from one saturated hummock to another (much like those in the remote valley Alcott and I monitor). Two weeks after we had tracked down 001—the snake he named Legs—as it moved through the woods southeast of the old dam, the snake arrived at a large, weed-clogged pond. Every day 001 crossed the pond, cutting a trail through the duckweed. Eventually, the snake shed, left the wetland for a month, likely fueled by an urge to mate, and then reappeared. While keeping tabs on the snake, Briggs found six others along the shore of the pond, all males, all opaque; their eyes a clouded baby blue. "Imagine, snakes coming to a wetland to shed," Briggs remarks. One rattlesnake Kiley PIT-tagged disgorged three full-grown meadow voles. By this date in the project, Briggs has tagged sixty-three snakes, of which only two are yellow.

Somewhere in the rumpled landscape between the last ridge and the town water supply, a five-foot-long black rat snake sprawls across our muddy trail. I touch its tail and the rat snake recoils, head up and tail vibrating against the soft ground, a rattlesnake wannabe. Fresh out of nearby streams and puddles, tiny pickerel frogs and American toads, having transformed from tadpoles to froglets, are everywhere under foot. We're seven-tenths of a mile from the pond where, two weeks ago, Briggs found 001 alongside a female by the inlet of a small stream. Six days ago, 001 foraged on a log mid-pond, soaking wet and lime-green with duckweed. Then, three days ago, he idled on a small ridge above the west edge of the pond, which is right where his signal leads us.

The beaver pond is lush and fecund with duckweed. Bouquets of water smartweed and pickerelweed rise out of the green soup. Shaped like an amoeba, a mud dam rims one end of the pond and a long, high stick-strewn dam clogs the stream's outlet, beaver providence. In between, water seeks its own level, saturating the ground beyond, rising when it rains, receding when it's sunny. Trees both living and dead stand in the water. A pair of hairy woodpeckers attend to a dead maple. A number of trails zigzag through the duckweed, none of which I would attribute to snake passage. Muskrat maybe. Or mink. The beaver lodge, safely in mid-pond, sports an outer frock of barkless sticks that gleam in the sunshine.

Although all six radioed rattlesnakes have shed, 001 is opaque and about to shed a second time. He must have eaten well: five, six, seven, maybe eight times so far. A white-footed mouse or a shrew in the gut of a four-foot rattlesnake would not necessarily be obvious, which may be why Briggs never noticed a bolus. Of course, a chipmunk would be a different story. And if a *big* timber rattlesnake swallows a gray squirrel—I've seen this on YouTube clips—the snake appears ready to burst like an overinflated tire, its belly scales agonizingly distended until mid-body scales no longer come close to overlapping. In fact, the Pennsylvania Fish Commission posted a video of a necropsy of a timber snake that had died while swallowing a squirrel, both its esophagus and stomach ruptured. Late this past June, Briggs tracked a snake to the ovenbird nest we had found earlier that month. All four chicks, small and not ready to fledge, were gone; the snake, stretched by the empty nest, showed no distension.

In early July, Kiley painted 001's last rattle segment green. This segment will progress away caudally as, with each shed, a new segment is added. Today, the rattle string has ten segments at full taper (though I'm not sure I see the button), which means 001 is less than ten years old. He looks smaller coiled up than he did fully stretched and on the move. While Briggs takes an external digital temperature reading, his cell rings. It's his ex-girlfriend, a Svengali.

"I have a rattlesnake here and I'm talking to Ted. I'll call you back," he tells her, the cheerful expression draining from his face.

"Please stop calling. I won't pick up," he says brusquely after her eighth call, huffing and puffing up the ridge.

🐍

Some people tend to assume, given the intimate nature of small-town newspapers and Internet postings, that the source and subject of a story, who may be the same person, are credible and that the notion of their authenticity must be indisputable. In the case of timber rattlesnake conservation in the Northeast, this is rarely true ... and if you choose a particular thread of information and follow it long enough, you may find yourself back where you started from. You may also find invention and omission, all of which may cut both ways, either favoring the snake or vilifying it. Timber rattlesnake conservation, as you have already guessed, is filled with dramas imbued with dueling egos. And in some cases it has devolved into a veritable soap opera, with a slow, persistent erosion of optimism about the future of *Crotalus horridus*.

Several years ago, a post appeared in an online version of Vermont's largest newspaper, the *Burlington Free Press*, that randomly lumped together President Obama, the collapsing economy (both national and state), health care, Vermont pioneers, foreign cars, endangered species, transmittered rattlesnakes, and wild leeks. The missive ended with the following frosty lines.

> While I appreciate the work and effort the wildlife groups are putting into saving endangered species in the state, it's time we all stepped back and look at WHY these are endangered species. Wild nature and civilized areas do not mix ... rattlesnakes close

to man = death from rattlesnake bites to children playing in their backyard. There is a REASON these species are endangered ... the smarter, less politically correct residents in this state hunted them out of existence to keep them away from the now civilized areas. Let's not think we are smarter than the previous residents JUST because we haven't lived with these dangerous species (thanks to the previous residents of the state) ...

The sooner these critters are extinct in Vermont the better. They are the reason I alway [*sic*] carry a revolver when I fish over in the East Steeple area.

When I blow the [rattlesnake] to smithereens I'll make sure I destroy the tracking device.

Save the Snake? Sounds like a ... re-election campaign slogan.

Does this mean we will start seeing "SAVE THE SNAKES" bumper stickers next to Vermonters "COEXIST" bumper stickers on their Subarus and Volvos?

I must say [rattlesnakes] are very tasy [*sic*] when grilled with wild leeks and mushrooms.

Other media reports require a more penetrating look and are often negligent based on what they omit rather than what they include. Who, for instance, ought to receive credit for championing a maligned species like the timber rattlesnake, when no one in the state had either the desire or the time? Where is the credit due Alcott Smith?

Fish and Wildlife got involved with rattlesnakes in 2010, twenty-three years after Vermont listed them as an endangered species and thirty-nine years after the bounty on rattlesnakes had been rescinded. Even after rattlesnakes were granted their current "protected" status, for many years, they continued to be slaughtered for oil, hides, and souvenir rattles (which were sometimes pruned from living snakes); they were run over, beaten, crushed, hacked, stoned, shot, trampled, mutilated, decapitated, and collected for the black-market snake trade. In 2014, according to the Vermont Agency of Transportation, on average, fifty-two hundred vehicles pass the pullout each day, a number that has increased yearly. Coming off the ledges and walking along the road back to my car, I've watched vehicles drift across the double yellow line and pass on curves, and I can unequivocally tell you that most cars exceed fifty miles per hour. If drivers are oblivious of pedestrians what chance does a westbound rattlesnake have? Since 2001, the year TNC began keeping DOR records, nine timber rattlesnakes carcasses have been taken off the highway. There were more in the eighties and nineties; in fact, one of the freezer-burned rattlesnakes that Alcott Smith and I dissected had been peeled from the road in front of the weedy pullout. In the early nineties, Alcott heard of a small rattlesnake that had miraculously passed through the gauntlet of cars, only to find itself under siege by a trio of cats in the kitchen of a farmhouse on the west side of the highway, directly across from the Crevice Den.

Time has changed both Vermont's approach to the well-being of rattlesnakes and the behavior of the snakes themselves. Not only has Alcott Smith been excommunicated from the state's rattlesnake recovery team, the snakes themselves have been cut off from the mosaic of farmland, pasture, and swamp that extends west from the cliff base to the eastern shore of Lake Champlain by motorists who exert a modern-day Darwinian pressure, eliminating, either maliciously or accidentally, virtually every snake that attempts to cross the busy two-lane road. Without a pheromone trail to follow, only the odd Samuel de Champlain of rattlesnakes would venture westward into the unknown, braving the scorching, stinking, vibrating macadam. Need I add *oily* and *toxic*? Of the twenty-two snakes implanted with transmitters that Kiley Briggs tracked, most migrated northeast away from the den; a few moved parallel to the highway, but none attempted to cross. By 2014, almost the entire population of timber rattlesnakes from the four-den complex moved northeast, up and over the ridge onto state land, municipal land, TNC land, or land owned by a passel of private citizens with varying degrees of sympathy toward rattlesnakes and state projects.

Between 2011 and 2012, according to the Orianne Society publication *The Ecology of Timber Rattlesnakes (Crotalus horridus) in Vermont: Final Report Submitted to the Vermont Department of Fish and Wildlife*, there were six hundred seventy-two separate GPS locations for the radioed snakes. Of that total, three hundred fifty-seven (fifty-three percent) occurred on protected land—two hundred seventeen on land owned by TCN, one hundred eleven on state land, and twenty-nine on municipal land. The remaining forty-seven percent hung out on private property, land mostly unavailable to state biologists. Briggs learned that foraging timber rattlesnakes like to get their scales wet—their selection of wetlands was disproportionate to the occurrence of such ecosystems within their home ranges—and that all other habitats were frequented according to availability, with the exception of farms and developed areas, which were assiduously avoided.

Kiley's fieldwork also revealed that Vermont rattlesnakes moved farther from their den, in terms of both maximum and mean distance, than did snakes from study sites in Pennsylvania, New Jersey, and West Virginia. Connecting the GPS dots for the movements of each of the twenty-two snakes demonstrated that the average individual home range of a Vermont rattlesnake—determined by local geography and possibly by the size of the population—exceeded that of most Pennsylvania, New Jersey, Missouri, or South Carolina rattlesnakes, which is *not* too surprising when one considers that Vermont is the northeastern outlier in the entire range of the species and that in locations farther south timber rattlesnakes have access to a larger, more densely concentrated prey base and consequently don't require as large a home range.

The movements of Vermont rattlesnakes would more appropriately be compared to those snakes from across Lake Champlain in the foothills of the Adirondacks of northeastern New York, where William S. Brown's thirty-seven-year study remains the hallmark of reptile research, one of the longest, and certainly most detailed single-species capture-recapture investigation undertaken by one person. It could be said with a reasonably straight face that Brown is to timber rattlesnakes what Jane Goodall is to chimpanzees or Dian Fossey is to mountain gorillas (except that he handles his subjects and, when overwhelmed by too many snakes in the field, brings some back to the lab to process); he made timber rattlesnakes seem *real*, animals endowed with a biological and social agenda, more or less limbless, scaled-down version of ourselves: hungry, horny, gregarious. Brown discovered that in northeastern New York, female timber rattlesnakes give birth for the first time at nine and age ten (sometimes later), typically breed every third, fourth, or fifth year and remain amorous until at least age forty-five (a bar that may rise as his research continues). He also showed that these animals are site-tenacious, rarely switching dens, although they do breed with snakes from neighboring dens; that they have rudimentary maternal instincts; that they prefer each other's company; that they are nonaggressive and extremely predictable; and that their home ranges are fixed by certain environmental parameters, including roads. Brown has documented growth rates, shedding rates, and survival rates by age class, average birth numbers, and differences in movement between males and females and between gravid and nonpregnant females.

As I've mentioned before, Brown also documented the "spook factor" (a term coined by fellow Basher Randy Stechert), which he more formally refers to as the "intimidation effect," the collective behavior changes within a population of timber rattlesnakes that is aggressively handled by people, either tormentors or well-meaning biologists commissioned to study them

It's a predator-prey thing. When a biologist grabs a rattlesnake, it responds as though a hawk or an owl had grabbed it. Frequently handled snakes become harder to find, spending more time in seclusion and less in the open; they shift birthing, shedding, and basking sites to less favorable locations; aggregations of snakes are reduced. Less basking means optimal body temperature is less frequently achieved, which alters metabolism, affecting every bodily function from digestion to gestation. In 2005, a paper titled "Scared to Death?" appeared in the prestigious journal *Ecology*. According to lead author David Preisser, the mere threat of predators alters prey behavior. In 2013, David Laundre, an ecologist at the State University of New York at Oswego, noticed that not only did prey behavior change, entire ecosystems changed; through a "landscape of fear," predators exert a greater impact on the prey through intimidation rather than through slaughter. And an agitated snake conveys the message chemically to cohorts. Everyone knows; everyone responds as though they had all been molested.

An enterprising Vermont timber rattlesnake may travel more than three and a half miles from its den and, depending upon your method of connecting the GPS points and calculating the home range, may occupy upward of four square miles, which is a remarkable amount of terrain to patrol for an animal without legs. On the west side of Lake Champlain, questing for food and particularly for females, an obsessive male rattlesnake might roam four and a half miles from its hibernaculum, almost entirely within the inviolate wilds of Adirondack State Park, where its nearly impervious to the shopping and dining habitats of its nearest human neighbors.

Syllogistically, I assumed that based on Brown's study, if the most ambitious timber rattlesnakes moved a maximum of four and a half miles in a straight line from their den, why not measure with a compass a distance equal to that on a topo map? Then, one at a time, stick the compass point in each den and draw a circle with a radius of four and a half miles. All the land inside the circle, particularly to the east of the dens—the land to the west is cut off by the highway—should be considered critical habitat and ought to be protected, one way or another.

Why subject Vermont snakes to "a landscape of fear," when thirty-six years of data gathered on the west side of Lake Champlain already tells you what you need to know? William S. Brown has already done the work for you. Are Vermont timber rattlesnakes *so* different from their neighbors across the lake?

I posed that question over the phone to Chris Jenkins, CEO of the Orianne Society, the biologist who hired Kiley Briggs and then guided the Vermont fieldwork from his base in Clayton, Georgia. "Our goal," said Jenkins, "was to identify critical summer habitat." Using a surgical approach, Orianne's mission was to track twenty-two radioed snakes, determining exactly where on the landscape they spent most of their time. For a timber rattlesnake, there is no politically correct rendering of a landscape; not all parts are equal.

"Why," I ask Jenkins, "was that necessary? Based on Brown's research, you already know, with a *very* high degree of probability, the outer limits of the snakes' movements."

"Within that boundary of critical habitat, what if there are thirty parcels of land for sale and we only have enough funding to buy five? I *need* to know what five pieces to buy," answers Jenkins.

In the winter of 1917, a large, extended Polish family by the name of Galick made up of grandparents, parents, five uncles, and two children, one an infant named Bill, left Schenectady, New York, and bought a farm in Smithville for fifteen hundred dollars, on the habitable edge of a funny little peninsula that is surrounded on three sides by New York and separated from most of the rest of Vermont by the Blatsky River and Lake Champlain's shallow, fecund tail. On a map, the peninsula looks like a flaccid penis, which is why some people refer to it as "the dick of Vermont."

But what a dick it is: remote and majestically pristine. In 1989, TNC shelled out eight hundred seventy-six thousand dollars—at that time, the most money ever spent to conserve a piece of Vermont—to purchase Galick Farm, sixteen hundred acres that included Monday Mountain and three miles of undeveloped shoreline along Lake Champlain. The Galicks, who had lived in splendid isolation from Smithville town hall, once petitioned the state legislature to be their own town. Maybe they could have called it Penisburg.

Except for World War II, Bill and his younger brother Ed stayed on the dairy farm. They shopped in nearby Whitehall, New York—driving across Lake Champlain in winter when the ice was thick enough, boating across in summer. The Galick brothers, who never married, remained remarkably independent of traditional commerce, even on the cusp of the twenty-first century: they hunted for meat and for bounties, trapped, fished, fixed what was broken, grew food and hay, butchered livestock and chickens, raised mink, and retrieved other people's dead cows and horses for their bones and hides and for mink food, living more or less in happy, self-sufficient squalor.

I never met Bill Galick, who could have passed for a character from the HBO series *Deadwood,* or so I presume. Jon Furman met him. When Furman interviewed Galick for his book *Timber Rattlesnakes in Vermont & New York,* for the first hour, "he never had his hand more than six inches away from a loaded revolver, which at first was hard to detect on the liquor-and-pill-bottle-strewn table between us."

Over the years, Bill Galick killed more than two thousand timber rattlesnakes, or so he said, mostly taken from the big den on Monday Mountain, which was the reason TNC bought the farm. His snake-hunting paraphernalia included a .44-Magnum pistol loaded with birdshot, a homemade snake hook, and a small pair of brush cutters for clipping off rattles and snake heads, which were delivered to town clerks in both Vermont and New York. Although most of Galick's rattlesnakes were from Vermont, Washington County, New York, paid five dollars per snake; Rutland County, only a dollar. Guess which town clerk saw most of his kills?

"He knew where on the mountain he could go to shoot snakes. He just didn't know why they were there," Mark Des Meules, a fifty-eight-year-old conservation biologist from Alna, Maine, told me. Beginning in 1982 and for the next fifteen years, while inventorying the flora and fauna of Vermont for TNC's Natural Heritage Program, Des Meules had made it his business to learn what Galick hadn't. As you may have already figured out, Des Meules is a rattlesnake pundit. He had been the first snake warden on Mount Tom in Amherst, Massachusetts, and eventually brokered the deal that made Galick Farm the cynosure of TNC's thirty-two-hundred-acre preserve, now called the Helen W. Buckner Preserve at Monday Mountain, the crown jewel of protected land in Vermont, which in hushed tones is sometimes called the "Galick Farm – Penis of Vermont Preserve." It took years and a lot of liquor for Des Meules to finalize the purchase.

In the late eighties, in the blissful ease of a mid-September afternoon, Des Meules took me to Monday Mountain and showed me a Vermont rattlesnake. Bill and Ed still lived in the ramshackle farmhouse, which was overrun with cats and visited occasionally by black rat snakes that would bask on various west-facing windowsills.

"If you had allergies," Alcott Smith once remarked, "you didn't want to go into that place." Chickens wandered across the yard, around the husks of old machines. Rat snake sheds hung from rafters in a couple of oblique outbuildings like gray crepe paper and a million cobwebs bearing Depression era dust trussed the rafters to the roof. One dimly lit building featured a dangling lightbulb straight out of the basement scene in *Psycho*.

Back in 1982, Des Meules, a lifelong a naturalist with a master of science in biology from Dartmouth College, began searching museum and bounty records for information about Vermont reptiles and amphibians. The trail eventually led him to a retired maintenance worker and amateur naturalist from the Boston Museum of Science named Kinsman Lyons, who lay dying in a suburban hospital. Graciously, Lyons's wife shared her husband's field notes with Des Meules, which provided a treasure trove of reptile and amphibian distribution data gathered from all over New England, including data on a number of species that had not been known to occur in Vermont; one, a cute little talus-loving lizard called a five-lined skink (*Plestiodon fasciatus*), Lyons had collected on Monday Mountain and donated to Harvard University's Museum of Comparative Zoology. Bill Galick had shot the lizard.

Five-lined skinks may have brought Des Meules to Monday Mountain, but rattlesnakes kept him there. Inquiring about the Galicks and their remote farm, Des Meules heard tales that the brothers were rough-and-tumble and not to be trusted. "A lot of people in Smithville thought they were crazy." In the company of friend and rattlesnake mentor (and future Basher) Tom Tyning, at the time a master naturalist with Mass Audubon, Des Meules clandestinely crossed Lake Champlain in a canoe and then snuck up the west side of Monday Mountain, well away from the farmhouse. Combing through the rockslide for reptiles, Des Meules and Tyning noticed that far below, someone on an ATV, serenely impervious to the sound of anything but his own machine, was moving uphill through the woods. Terrified, they ducked behind boulders and followed the progression of the ATV through binoculars. The operator was Bill Galick, replete with pistol, buck knife, and a pair of brush cutters.

"Holy shit. We were in trouble." Trespassing on his property was not the preferred way to get acquainted with Bill Galick. Things became tenser when Galick stopped right below them and idled his engine. "Shut up. Shut up," Tyning had whispered, his admonishing tones barely audible. Minutes seemed to morph into hours, before Galick finally moved on, unaware that two young, charmingly imbecilic naturalists were hiding in the rocks above him. Eventually, they composed themselves and hiked back down the mountain accompanied by the thought that, "Oh my god, we could have gotten killed!"

Des Meules continued. Some weeks later, having steeled his nerve for an inevitable face-to-face meeting, he went through Monday Mountain's "front door" and introduced himself to a queerly suspicious Bill Galick. Des Meules often returned to the farm, and eventually Galick thawed, warmed by Mark's genuine fascination with his hand-hewn knowledge. Together, over the next six years, Mark Des Meules and Bill Galick hunted rattlesnakes, ran trap lines, drank whiskey, ate meals, and documented five-lined skinks scurrying around Monday Mountain. "Galick did not want to eliminate the [rattlesnake] population," Des Meules told me. "They were a *big* part of the family's personality. They brought [Bill] a certain notoriety."

Over breakfast at a recent Bashers' meeting, Des Meules told me, "I did the [original] survey of the Champlain basin historic [rattlesnake] sites and it's incredible that [Monday Mountain] and the ledge [several miles to the east] did not blink out." Everybody collected Vermont snakes, including Sparky Punch, who lived across Lake Champlain, and Rudy Komarek, who lived in his car. For several years after TNC had purchased Galick Farm, both the football coach at Smithville High School and an assemblyman at a door company in Glens Falls, New York, continued to take rattlesnakes off the mountain, which eventually forced the conservancy to post "No Trespassing" signs along the access trails.

"When you're ready, we [TNC] want to buy this place," Des Meules had told Bill Galick over shots of whiskey. Galick deferred his decision for several years, until tired of driving around in a "shitty old pickup," he called Mark at TNC headquarters, in Montpelier, to discuss the terms. Des Meules stopped what he was doing and drove an hour and a half to Galick Farm only to find the brothers in the kitchen, drunk as skunks, pistols on the table. "There would be no discussion," Des Meules told me. "That happened many times." And there had always been the chance (though Des Meules would never mention it) that Bill Galick might take one drink too many.

Then, out of the blue, the Galicks agreed to sell, and it took months for Des Meules to convince Bill Galick that TNC needed the property appraised. Although they wanted a million dollars, the Galicks settled on eight hundred seventy-six thousand dollars and a reserve life estate, which enabled them to remain in the farmhouse among the cats and rat snakes until they died. The two surviving uncles never saw a penny. "You can't advise financially when you [negotiate] a deal like that," said Des Meules. Then, lowering his voice an octave, "Bill ran off to Las Vegas. He blew it all."

Later in the day, over pulled pork and coleslaw, Mark tells me about finding the Patina Den. One mid-September afternoon, while hunting cottontails in the hardwoods above the Blatsky River, on the backside of Monday Mountain, he serendipitously discovered a small snake den that not even inveterate collectors like Galick or Punch knew of. "Holy shit!" No twigs had been broken. No rocks had been moved. Nothing had been overturned. Every lichen, every branch, everything was natural. "And the snakes had a demeanor. You could tell they had not been bothered." Whenever rattlesnakes emerge from the Patina Den, they sport a distinctive dull-orange patina, a residue of iron oxide that comes from a small stream flowing beneath the rocks. I've never seen orange-dusted rattlesnakes anywhere else. It's a singular look. Snakes of Patina are attracted to the subterranean heat and humidity and to the utter remoteness of the place.

"[The Patina Den] was [my] secret spot," said Des Meules, who had been quite ambivalent about divulging the location. "I debated about filling out the [Natural Heritage] forms, as a *good* ecologist would." Of course, he did. "I have to say from what I have seen going on in Vermont — and not just with rattlesnakes — there's something awry in [their] approach to the management of fragile species. I wish I had kept the den to myself."

Patina Den is part of TNC's original Galick Farm purchase. It is considerably smaller than the main den. Des Meules thinks that it may have been the source of replenishment for the overhunted population on the far side of the mountain, a seedbed as it were, a reservoir that dampened overharvesting the main den. Bill Galick, woodsman extraordinaire, purveyor of all things wild, hadn't a clue.

In his spare time, when he's not snake hunting with his wife and grown sons, Mark Des Meules keeps up with the *New Yorker* and *Natural History* magazine and reads books like *Lure of the Labrador Wild*, *Into Thin Air*, and *In the Heart of the Sea*, as well as E. O. Wilson's autobiography *Naturalist* and his novel *Anthill*. There's an obvious theme here: almost all nature, almost all the time. Des Meules, who moved away from Vermont in the late nineties, directs Viles Arboretum, outside Augusta. Pathologically optimistic, he searches for timber rattlesnakes on the warm rocks of southwestern Maine, where they haven't been seen since just after the Civil War. "They're out there. I know it."

"Ironically," Des Meules told me, "I consider researchers almost in the same category as collectors. Certainly invasive researchers." For a combined period covering more than thirty years, Des Meules and Alcott Smith brought people to the ridge dens and to Monday Mountain. "Alcott's changed attitudes, he's enriched lives," Mark says, a tinge of sadness in his voice, since he too has been ignored by Vermont's timber rattlesnake team.

"Issues like this, [rattlesnake] turf wars," Chris Jenkins had said in our recent phone conversation, "never come up in North Carolina," where the Orianne Society recently began a study of timber rattlesnakes in the southern Appalachians.

Exercising rhetorical powers, Des Meules asks, "What do the snakes need? Number one is habitat. Number two is threats taken away. We can learn all we need to know about [the ridge] site by watching. You have your binoculars and you spend all day looking. Alcott gathered data by observational ecology. He handled snakes, but he didn't take them home. No one ever availed himself of [Alcott's] information. And [when] I read that [Vermont Fish and Wildlife] is trying to document where the [Monday Mountain gestating] sites are, 'Well, give me a call!'"

In an effort at détente, Chris Jenkins mentioned to me that Vermont Fish and Wildlife is drafting a recovery plan for the timber rattlesnake, an umbrella document that will combine all of Alcott Smith's observational data with the results of the Orianne's two-year tracking study. But no one in Vermont has been in contact with Des Meules, a rather major oversight given that his involvement with rattlesnakes began five years before the state formally recognized that they were endangered.

On three different occasions during his tenure at TNC, Des Meules escorted Vermont reigning governors—Philip Hoff, Howard Dean, and Madeleine Kunin—to Monday Mountain, to see the rattlesnakes. Kunin referred to Galick Farm in her memoir as Eden. And Des Meules agrees. After a *long* Vermont winter, "You sit on that ledge and everything is greening up. And it's like—*holy shit* this is amazing," he says, lost for a moment in the memory of a place that he's no longer allowed to visit.

I vouch for the view, which I've seen on more than a few occasions with Alcott Smith who once chaperoned eight Bashers up the mountain, tenured herpetologists from across the range of the snake. Stunned by the jaw-dropping view, some found it hard to believe they were in Vermont, where from a promontory overlooking the great green sweep of the Champlain lowlands, all the way to the twin geographic humps called The Saddles and the rest of the eastern foothills of the Adirondacks, and beyond to the cold granite domes of larger, more distant high peaks like Algonquin and Marcy. From the summit, you can also listen to gossiping ravens, see a peregrine tear the sky or a red-tailed hawk or a vulture float past, and look down on wild turkeys in a green meadow, heads bobbing above the grasses. And on the summit, you can bask in autumn sunshine on the rim of a world that features rattlesnakes and little lizards with bright blue tails, while daydreaming of an age not so distant, when the sweep of life was more in sync with the "Age of Bill Galick" than with the "Age of handheld technology."

I ask Mark, "What do you think the formula for protecting timber rattlesnakes in Vermont should be?"

"Protect habitat—it's pretty simple—and then stand back. Eliminate the obvious threats. Employ passive observation and gather data. Buy habitat. I don't go for conservation easements. I would rather own the land," answers Des Meules, who was responsible for most of TNC's land purchases in the eighties and nineties. Now, with data generated by the telemetry study, the state or the Orianne Society will be able to do that, earmarking the most critical parcels, which is always important when funding is limited, particularly for an animal of limited political and social capital. (Timber rattlesnakes may be charismatic, but they are not fuzzy.)

Mark began his rattlesnake research in Vermont with an inventory. He'd identify historic sites and current sites, check old museum records and newspaper accounts, and then follow up on sightings, both old and new. Because of his breadth of knowledge, Des Meules was asked to write the Vermont chapter for *Timber Rattlesnake: Life History, Distribution, Status, and Conservation Action Plan*, a thirty-state summary of the biology and conservation needs of timber rattlesnakes sponsored by the U.S. Fish and Wildlife Service, a tome twenty years in the making. Des Meules tracked down twenty-five historic dens, all in rocky, low-elevation slopes adjacent either to the Connecticut River and its major tributaries or to Lake Champlain, the two warmest parts of the state. At one site, the rockslide below the Fairlee Cliffs, which is a few miles from my home, I eat blueberry pancakes at the Fairlee Diner and gaze longingly across Route 5 at snake-free talus.

With unimpeachable gravitas, Des Meules invested hope in every rattlesnake rumor, even the most outlandish. In 1983, he followed up a rattlesnake sighting in Barton, way up in the Northeast Kingdom— the refrigerator of Vermont—a distributional no-way, a sighting that would raise suspicions in any herpetologist. Although the director of a local science museum had endorsed the sighting and a local newspaper had reported it as *fact*, the observer admitted to Mark that the rattlesnake had been an invention to stop a neighbor from developing the adjoining lot. Who wants to live in the company of rattlesnakes? "Well, we got to the end of that one. Not a sighting," Des Meules says with uncontained mirth.

And then, taking a deep breath, he unleashes a pent-up cataract of historical footnotes and grievances aimed at the Vermont rattlesnake tribunal.

> Timber rattlesnakes are a species at risk, in fact, a species at great risk. When I began visiting Monday Mountain, I reached out to the expert, Tom Tyning. I didn't say, "Fuck this, I'm going to do this myself." I reached out to Tom. He was my go-to person. I'd call Tyning all the time. I didn't feel any degree of competition with him. Are you kidding me?
>
> [The state] never called me, not even once. And clearly they knew about all my work at Monday Mountain. I didn't have any skin in the game. I would have gladly come and said, "Hey, I'll tell you what I think." [The biologist is] making the decisions now. [Ours] is a very small field of expertise. It's not like there [are] a hundred [biologists] out there and [Vermont Fish and Wildlife] is saying, "We can't get in touch with all of them." There's like *two ... three*. If the [Vermont] Endangered Species Committee were given charge and responsibility to sign off on the [timber rattlesnake] research proposal, I'd have to conclude that they were not doing their job. Why didn't they say, "Are there experts out there, who we need to be hearing from before we sign off on this [permit]?"

When I electronically posed my questions to "the biologist," he smugly wrote back that he had "neither the desire or the time to engage in debate."

"No one but the people involved bear the responsibility for so many questions," says Des Meules, simmering. "I find it quite incredible that an individual who professes to have conservation as [his] banner, cannot defend [his] decisions or data in a court of peers. Worst of it all is that the general public trusts that the very best conservation decisions are being made on ... behalf [of the animals]. I think [the biologist's] response was not only pathetic but very sad ... very sad ... [as well as] rude and empty. At the same time, it certainly indicates to me that this boy has something to hide. The fact that he did not respond to your questions suggests there is a keen awareness and nervousness about what has been going on." When I mentioned Jenkins's comment about "rattlesnake turf wars" in a recent e-mail to Des Meules, Mark replied, "I suspect [Orianne] had no idea about the situation or the circumstances. The bigger question for me goes to the Endangered Species Committee, the Herp Sub-Committee, and to [the biologist], as he knowingly navigated [rattlesnake recovery] around *big* chunks of experience and data."

William S. Brown, who has a record of pulling good herpetologists into his gravitational field, told me in an e-mail, "I don't totally dismiss the actual findings.... What I do question is why they chose not to recognize Alcott's survey results and den counts and why they gave no recognition at all to the existing Vermont Recovery Plan that did have the good sense to utilize Alcott's work. Dismissing Alcott and the recovery plan are a serious error of omission and commission."

Any study that involves a charismatic species, especially a venomous snake, often adds to their mystique, which confers a sort of showman or macho-man aura to those involved in doing fieldwork. Newspapers love it. "I recall," Brown continued, "many years ago, when I was avidly reading my first technical book on amphibians and reptiles, that [Roger] Conant warned his young readers ... saying not to brag about or use snakes to gain an advantage in the social arena."

After he read *The Ecology of Timber Rattlesnakes (Crotalus horridus) in Vermont,* Brown echoed Chris Jenkins's point of view: "their study did accomplish an objective of quantifying the movements and habitat use of the snakes, and therefore it will probably prove useful to land managers, but [it] only peripherally covered population aspects of the snakes." Which, of course, is understandable: it's tough to replicate in two years' work to which someone else devoted thirty-six.

Deliberately and without malice, Brown continues, "[The report's] main deficiency is not taking into account the prior extensive surface counts compiled by Alcott for many years during which very few humans, other than Alcott himself, were visiting the dens. Thus, they should have used his data as prior work done and they should have compared his results with theirs.... Not only didn't they include Alcott's data, they don't even have the common courtesy of recognizing his work." This is data, I relay, Chris Jenkins assured me will be used in Vermont's forthcoming recovery plan.

Based on the need to have movement and habitat data specific to the west face of the northern Taconic Mountains, I recognize the value of the state's two-year study. Who owns the lands the snakes use? Knowing the answer is useful, which is the project's strength. Its weakness is the apparent selective overlooking of Alcott's and Mark's previous work, which is inexcusable. Why would the state ignore the two men who brought rattlesnakes to their attention?

September 12, 2012, ten fifteen in the morning. Sixty-seven degrees Fahrenheit. Lightly overcast, with cloud shreds screening the sun. After yesterday's thunderstorm, the woods behind the weedy pull-out are slick, and every oak and maple and hickory leaf droops with rainwater, showering us as we pass through the forest.

Four months ago, Scott MacLachlan implanted transmitters in six-teen additional snakes, twelve males and four females, of which two are pregnant. On the ledge, Kiley and I visit the red cedar rookery, a squat, gnarly cedar growing between two shelter rocks that Alcott had taken me to in the early nineties, likely in continuous occupation since the last wildfire scorched the ridge more than a century ago. On either end of the first flat-topped shelter rock, two litters entwine in each foyer, gray-pink snakelets as thick as my thumb, milling to-gether, becoming forever acquainted. One pod has an adult female in attendance. Like a little strip of gray cellophane, a recent oily shed sticks to the underside of the rock. A baby yawns.

I ask, "Why are you tracking different snakes this year?"

Raising his voice to be heard over the ridgetop wind, he explained to me, "Tracking the same snakes wouldn't have given us any new information, since they go to the same spots over and over again."

Yesterday afternoon, concerned about lightning, Kiley had ditched the aluminum tracking antenna before fleeing the woods. He had marked the site on a GPS, and after fussing over the snakelets, we bushwhack a mile and a half into the woods, which gives me a vague notion of what it must be like to be a homeward-bound timber rattlesnake tracing a pheromone trail through a trackless forest. Guided by our own brand of planetary magic, we walk directly to the antenna.

The population of timber rattlesnakes along the ledge is apparently large, but given that there's no one but each other to breed with, cut off as they are from the rest of the snake world by the two-lane highway, the cabal most familiar with Vermont rattlesnakes—Briggs, Brown, Jenkins, Blodgett, Smith, Des Meules, the biologist—hopes there is enough genetic diversity to stem the cascade of consequences that are usually brought on by inbreeding. Several months ago, a large male died in captivity; when *Ophidiomyces ophiodiicola*, the flesh-eating fungus, was found on its ventral scales, everyone's optimism began to erode.

We're after a rattlesnake Briggs named Secretariat, which had moved 0.3 miles to the northeast before any of the other radioed snakes had left the vicinity of the den. Secretariat then moved a mile in a single day, perhaps an ophidian cross-country record, before setting up shop along the shore of Maris Lake.

"I lost him for a few days and then he turned up at the lake and stayed put," says Briggs, flexing the antenna.

I follow behind, listening to the ring tone of Kiley's cell and then to the robotic list of messages floating back like something out of *Star Wars*: several friends, a colleague, the guy from the local garage. Merrily, he plunges deeper into the woods. He won't pick up.

We find Secretariat in a tree-ambush pose, a few body coils pressed against the trunk of a red oak, head and S-shaped neck flush to the bark pointing up, waiting for a hapless chipmunk or flying squirrel or a white-footed mouse to descend.

"He's one of the ridge's more spectacular snakes," says Briggs, with all the enthusiasm of a Little League coach, "particularly after shedding." Secretariat is dark gray with brown-bordered black bands, a wraith among shadows. Our elevation is approximately eight hundred feet, close to the height of land. The dens are a good deal lower. Elsewhere at our latitude, however, timber rattlesnakes den up to thirteen hundred feet. Vermont snakes don't have that option;

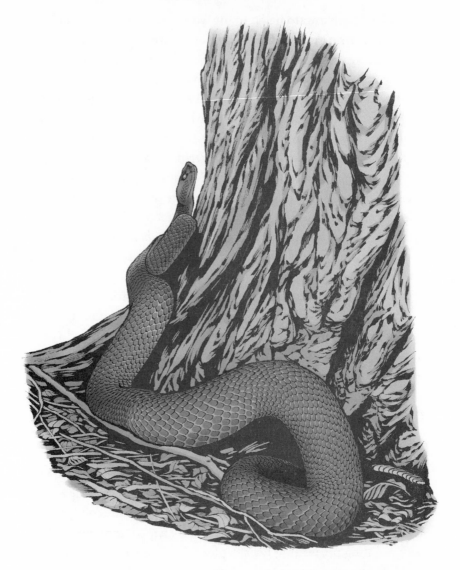

thirteen hundred feet would put them in the clouds. And three hundred fifty miles south, timber rattlesnakes in the mountains of West Virginia den up to thirty-six hundred feet; five hundred fifty miles south, in North Carolina they den up to forty-eight hundred feet.

One adventurous snake led Briggs over the length of the eastern slope from stem to stern, and back again. "He's like a rich tourist. He just ups and goes wherever he wants."

Walking back toward the dens, we locate a radioed female named Pretty Girl, a smoky-gray snake with coal-black bands, each with a yellow border. A trace of brick-red dorsal stripe separates each band. Coiled, recumbent after a summer of debauchery, Pretty Girl is an ophidian trollop that Briggs has seen in the company of a number of cavorting males, a number of times. "She was more than a mile and a half from here last week."

Then, serendipitously, while leaving the woods for the open ledge, we almost step on a PIT-tagged female, a black morph wedged in the groin of two small outcrops: number 1348018.

Not far from the rim, a postpartum female, number 123, chin against a log in traditional ambush pose, awaits whatever small mammal happens to wander by. Several of the neonates we saw earlier in the morning were hers. (Eight months from now, I'll hold 123 in a Plexiglas tube as Scott MacLachlan surgically removes her transmitter.)

Out of the blue, Kiley says, "Remember my ex-girlfriend, who wouldn't stop calling me last year, when we were on the ridge," as if I could possibly forget that she had called eight times, harassing him and disrupting the wild emptiness with space-age ring tones.

"Yes."

"Well, several days later, she abducted me at gunpoint. She held me hostage for a week. Tied me up and fed me toast. She sabotaged my cellphone," Briggs continues as though he's recalling the high-lights of a soccer game.

"She hit me with a baseball. My head was swollen."

She called Kiley's phone contacts and left disturbing messages, including to Chris Jenkins in Georgia, whom she told that Briggs quit Orianne and would no longer be monitoring rattlesnakes.

"How did you escape?" I ask.

While his succubus was on the phone, Kiley knocked her down, got to his car, and, after a chase around the western slope of the Taconic Mountains, escaped to a UVM professor's home, where he hid for several days. Kiley got a restraining order, returned to rattle-snakes until the last snake went down for the season, then drove to Georgia to work on gopher tortoises and indigo snakes.

Later that fall, Briggs abruptly returned to Vermont. He had been charged with abuse.

13

The *Very* Last Rattlesnake

Forgetting is another kind of extinction.

TODD MCGRAIN, sculptor

I would hop the wobbly, chain-link fence that separated affluent members of the South Side Sportsmen's Club from Sunrise Highway riffraff and disappear into the crooked pines to fulfill every frontier fantasy I ever had. As Davy Crockett, I'd hack my way through the Alabama wilderness. As an Algonquin warrior, I'd spy on colonial outposts. Some days I was Natty Bumppo or Chingachgook (my fantasies were unbiased) or Daniel Boone, even a Rocky Mountain fur trapper, for which my imagination assumed a bigger than usual role since the South Shore of Long Island is nearly flat and where I roamed was nowhere more than twenty or thirty feet above sea level.

Once over the fence and beyond sight of the preferred entrance, my plans were as shapeless as water, my head swirling with television tunes, particularly the theme song from Walt Disney's *Davy Crockett* series. It was irrelevant that Long Island lacked a "mountain top" or was light-years away from having been the greenest anything "in the land of free." Although bordered on all sides by development, my immediate surroundings were neither inhabited nor inhibited by humans, which mattered most to me; my imagination took care of the rest. On a whim, I'd sneak around a large impoundment on Connetquot Brook and head west to the edge of a trout hatchery, keeping well away from the three-story wooden clubhouse and assorted outbuildings; other times I'd head into the heart of the pine barrens, ripping through the loops of briars that ensnared trees and shrubs, giving weight to the word "bushwhacking." Whenever I reached the edge of the brook, whose name I learned years later was Rattlesnake (for a very good reason), I'd walk the stream bank northeast, gorging on highbush blueberries and huckleberries, which grew in profusion. When I wasn't preoccupied watching something (or pretending I was), I tried to avoid poison ivy, which also grew in profusion as both a vine and a shrub and had been called a "vegetable rattlesnake" by the *Brooklyn Daily Eagle*. Spellbound, I might stare at water striders as they trolled for hapless insects on the surface of a quiet pool, towing their balloonish shadows along the shallow, sandy bottom. If I saw a caretaker or a watchman, either of whom made any outing more thrilling, I'd hide behind a fat oak tree, stifling my breath, until my racing heart was still. I was never caught, never pursued. Mostly I had the barrens to myself.

On its way to join Connetquot Brook, Rattlesnake Brook flows for a couple of miles southwest out of coalescing seeps in Bohemia County Park, into and out of two small manmade ponds, before annexing two smaller streams, West Branch and Middle Branch, and then disappearing into a culvert under Sunrise Highway. The diminutive watershed appears on a topo map as three wavy blue lines that collectively drain a green swath, representing several thousand acres of forest and swamp. Nowhere along Rattlesnake Brook are the woods too dark or the trees taller than sixty feet. Just below the leaf litter is sand, sand, and more sand all the way down to Long Island's geologic basement, several hundred feet below the surface. I'd roam across carpets of oak leaves and pine needles and into and out of muck-filled depressions, sinking to my ankles among the skunk cabbage. In patches of open sand, I'd find arenicolous insects: solitary wasps digging burrows, where they would stuff hapless caterpillars, spiders, and beetles; and ant lion larvae digging weird conical pits and hiding just below the surface, all head and vise-grip mandibles, lying in wait for wayward ants, which I obligingly dropped in.

One morning, I saw an osprey—a very rare bird in the early sixties. Always I'd see Fowler's toads, sometimes garter snakes, sometimes red-backed salamanders, which I'd find by overturning logs. There might be a box turtle out for a stroll or a recumbent water snake on a partially submerged limb or a muskrat or a black duck or an antediluvian snapping turtle, its dark shell rising out of the water like an igneous island. Once, I chased a black racer up a tree. Towhees, prairie warblers, brown thrashers, and catbirds were everywhere and noisy, singing in the pines, nesting in the thickets, and every now and then, I'd catch a glimpse of a red-tailed hawk or hear its razor-edged scream. The notes of a hermit thrush were reason to pause. At twilight, the voices of screech owls, great horned owls, and whip-poor-wills haunted the pine barrens, transforming it into what seemed like a landscape without borders. Cottontails, gray squirrels, and chipmunks busied themselves in the undergrowth, and on more than one occasion I spotted a deer or a shrew or a plump little mouse (which I now assume was a pine vole). Raccoon and opossum transformed muddy shorelines into wondrous tableaus, and there were long-tailed weasel, mink, red fox, and perhaps skunk and gray fox, but I never saw any of them. Although an East Coast race of greater prairie chicken called the heath hen had vanished from the Long Island Pine Barrens by the time of the Civil War and from the planet by 1934, there were still plenty of ruffed grouse and quail. I knew that Long Island had long ago been home to wild turkey, bobcat, mountain lion, wolf, black bear, otter, and elk, but I had no idea in the early sixties that these airy, sunlit woods had once been a rattlesnake stronghold . . . or that in 1962, on July 20, less than four months before my fourteenth birthday, Roy Latham, an Orient Point farmer and legendary naturalist, whose published reports on the nature of the island were as prolific as they were diverse, saw a rattlesnake in East Moriches, the *last* rattlesnake reported from Long Island.

I was born in 1948. My parents moved from Manhattan to Long Island in 1952. In the late fifties, my father, a garment district executive, joined a modest country club in the rural hamlet of Sayville. To get there from our home in south-central Nassau County, we drove east on Southern State Parkway into Suffolk County, and then took Sunrise Highway further east. By the time we reached Oakdale, properties were marked by the acre, not the square foot, and there were stunning mansions and farmhouses. Oakdale is level, laced with streams, and wooded, mostly pitch pine and a variety of oaks, some no thicker than my pinkie and no taller than a couple of feet. South Side Sportsmen's Club, which owned nearly thirty-five hundred acres and leased an additional twenty-three hundred more, ran from the north side of Sunrise Highway all the way to the south side of Veterans Highway and was the largest, wildest, most mysterious piece of undeveloped land left on Long Island. As you might imagine, the hundred-year-old club was the exclusive playground for New York's gold-coast aristocracy. It boasted among its membership, past and present, many of whom had summer homes in Oakdale, William Vanderbilt, Charles Louis Tiffany, William Bayard Cutting, August Belmont, Henry Sturgis Morgan, and restaurateur Lorenzo Delmonico, owner of the first and most upscale restaurant in the Western Hemisphere, which pioneered Delmonico steak (a personal favorite of mine), Delmonico potatoes, lobster Newberg, chicken â la King, eggs Benedict, baked Alaska, and Manhattan clam chowder (the restaurant even served passenger pigeon pie, for which Delmonico himself selected the pigeons at the Fulton Market). J. P. Morgan, Andrew Carnegie, Henry Clay, Daniel Webster, William "Tecumseh" Sherman, the Prince of Wales, and the chancellor of Germany had been guests, as had five presidents—Ulysses S. Grant, Chester A. Arthur, Grover Cleveland, Theodore Roosevelt, and Herbert Hoover—and two-time presidential hopeful Adlai Stevenson. South Side Sportsmen's Club was decidedly not the natural habitat of a middle-class, prepubescent Jewish boy.

I had been aware of the woods since my parents joined the golf club. How could I have missed it? How could anyone have missed it? Development along the north side of Sunrise Highway stopped at Connetquot Avenue and resumed again at Pond Street, three miles to the east. If you included adjoining Bohemia County Park (one hundred seventy-two acres), Bayard Cutting Arboretum State Park just across Sunrise Highway (six hundred ninety-one acres), and nearby Heckscher State Park (sixteen hundred acres), there were more than nine square miles of primeval Long Island. These wild, lonely barrens were still visited by wildfire, which would loosen plumes of summer smoke into an otherwise cerulean sky. Once, while standing on the ninth green in the company of my parents, I could smell smoldering wood. Only the ocean seemed more desolate to me than the South Side Sportsmen's Club.

When Henry Hudson first sailed into New York Harbor, early in the seventeenth century, rattlesnakes had free reign of Manhattan, Brooklyn, and Queens and Long Island; besides Oakdale's there are records from Freeport, Centre Islip, East Islip, Great River, Great Neck, Commack, Sayville, Yaphank, Smithtown, Wyandanch, Sag Harbor, Patchogue, Shirley, the Hamptons, Moriches, Bayport, Babylon, and Bay Shore. In addition to Oakdale's Rattlesnake Brook, Long Island has three other rattlesnake place names: Rattlesnake Stream in Sag Harbor, Rattlesnake Swamp in Yaphank, and a second Rattlesnake Swamp in Sunken Meadow State Park. The last timber rattlesnake in New York City was killed around the time of the Spanish-American War, in Seton Falls Park, in the Eastchester section of the Bronx. Long Island rattlesnakes made a prolonged last stand in the pine barrens, where between 1869 and 1899, well over fifty were dispatched at the South Side Sportsmen's Club, thirteen in one year, and twenty-nine by the game watchman during the 1890s.

In the 1880s, Edward Knapp, Jr., in a self-published family history, wrote that in front of the clubhouse, "It wasn't uncommon to see ten or a dozen snakeskins stretched out on boards, drying in the sun." As late as the turn of the century, young boys were not allowed to wander alone along the banks of Rattlesnake Brook. "It was infested." Even seasoned quail hunters stayed away. "It was a bit dangerous for the hunter," recalled Knapp, "and very dangerous for his dog."

During my furtive treks into the pines, I never saw bedrock. Compared to the mainland, say the Hudson Highlands, the surface of Long Island is geologically new, *quite* new—post-Pleistocene, less than seventeen thousand years old—and has been fashioned by ice and surf. There are no outcrops. No cliffs or ledges or caves. No rockslides. No shelter rocks. Nothing really to suggest that Oakdale once supported a thriving population of timber rattlesnakes, yet for a century, the employees of the South Side Sportsmen's Club did-in every one they saw, encouraged by the Club's longstanding payment of a five dollar gold piece for every dead snake. In 1924, a former superintendent told the *Suffolk County News* that the last rattlesnake to be bountied was so small the employee who turned it in "admitted that he had been overpaid."

For many years, rattlesnakes on Long Island had been newsworthy—their presence, their absence, and the hunt for them. Apparently, the length of the snake and the number of rattle segments, which were simply called "rattles," were more important to note than either color or pattern, which I assume was as variable as (though somewhat different from) those of the upland timber rattlesnakes I have described. On August 30, 1884, the *Brooklyn Daily Eagle*, once the most popular afternoon newspaper in the United States (and edited by Walt Whitman from 1846 to 1848), reported that a five-foot rattlesnake with twelve rattles was killed in Oakdale near the clubhouse at the South Side Sportsmen's Club. On September 15, 1888, the *South Side Signal* detailed the afternoon killing of four rattlesnakes near the clubhouse by a wood-gatherer named Thompson. (The newspaper called the snakes "beautiful," all black and yellow and boldly marked, and reported that Thompson would wait for winter before he would gather more wood.) If you planned ahead, the preferred instrument of slaughter was a shotgun or a bullwhip; if the encounter was spontaneous, a stick.

On September 2, 1898, the *Suffolk County News* mentioned that an Oakdale cyclist slew a rattlesnake and "retained as a highly prized relic the snake's tail, which was ornamented with eight rattles and a button." Another incident involving a cyclist and an unfortunate Oakdale rattlesnake appeared on the third page of the *Eagle* on August 8, 1902.

> Bay Shore, L.I. August 8. H. D. Burd of this village, while riding his bicycle along the south road near the South Side Sportsmen's Club, at Oakdale, yesterday, saw a large rattlesnake. Dismounting and securing a heavy stick Mr. Burd killed the reptile, which when measured was found to be three feet ten inches in length. It had eight rattles. Mr. Burd has preserved the skin as a souvenir.
>
> So far as is known the swamp at that point is the only place on

Long Island where rattlesnakes are ever found. An average of one per year is killed there. This is the second one killed this summer.

The *Suffolk County News* added an odd detail: Mr. Burd had been an upholsterer.

In 1963, when their taxes had become too great a burden, the club sold their land and buildings to the state of New York and then leased them back for the next ten years. In 1973, the state kept the trout hatchery going and designated the property Connetquot River State Park, a crown jewel among Long Island's parks. While leafing through the 1976 summer edition of *Comments on Connetquot*, the park's quarterly newsletter, I found a story of the killing of one of the last rattlesnakes of Rattlesnake Brook. Again, a bicycle was involved, only this time it was the murder weapon. One morning in 1913, Bertram Smith and his young nephew Harvey pedaled to work at the Bayard Cutting estate. En route, Uncle Bert noticed a rattlesnake basking along the roadside and promptly rode over it. According to Harvey, who survived whatever emotional scars were associated with the incident to become the director of the Suffolk County Marine Museum, the snake was three feet long and had three rattles.

Early in the fall of 1880, the *Suffolk County News* reported that an agitated rattlesnake awakened an Oakdale deer hunter who had fallen asleep at the foot of a small tree. The hunter shot the snake (surprise, surprise), which was more than four feet long and had eleven rattles. Fearing some form of serpentine retribution, the hunter scurried up a tree, which was where his companions found him.

And then there are the legends. On July 23, 1970, the *Islip Bulletin* reprinted a "Historic Long Island" column, written by a historian named Paul Bailey and selected and supplemented by someone named Carl A. Starace. (I have no idea when or where the original column first appeared.) Bailey's story about the death of Lorenzo Delmonico, the famed restaurateur, opened with, "A stream on the grounds of the South Side Sportsmen's Club ... was named Rattlesnake Brook because of the abundance of those reptiles." Bailey wrote that Delmonico had died deer hunting in the pines and then recounted two versions of his untimely death. In the first, Delmonico died of twin puncture wounds in his leg, the aftermath of a fatal snakebite; in the second, he died of a heart attack, his heart having given out at the excitement of seeing his first deer, what fellow club members referred to as "buck fever." To get to the truth of his death, I googled Delmonico's obituary in the *New York Times* and discovered that neither of Bailey's versions was correct. Lorenzo Delmonico died of dropsy at his home in Sharon Springs—he wasn't even *on* Long Island. Then, I discovered that it was his uncle, Giovanni, the founder of Delmonico's, who had died deer hunting in Oakdale. Giovanni had accidentally shot himself. There was no mention of a rattlesnake.

Although in the early sixties I wasn't aware of Long Island rattlesnake lore, I was acutely aware that compared to my hometown, Oakdale was untethered; once on the other side of the fence, with little effort and little direction, I could encounter wildlife crawling, slithering, scurrying, bounding, running, climbing, swimming, floating, digging, flying, sleeping, ambushing, feeding, or singing. Had I ever suspected that rattlesnakes once lived here or, in the case of not-too-far-away East Moriches, were still living among the pines I would have been ecstatic.

⚬

Long Island is one hundred twenty-three miles long, twenty-five or so miles wide at its widest point, and shaped like a fish—perhaps a sucker, with Jamaica Bay in the far southwest as its roundish, downward-pointing mouth—a geographic detail Walt Whitman had made popular without the advantage of air travel. On a map, the island's fishiness is quite apparent—something I had been made aware of in grade school—but from a plane window, seeing it is always a revelatory confirmation of what I already know intellectually.

Pre-Cambrian bedrock lies hundreds of feet below the surface of Long Island and has been subsequently covered by both Carboniferous and dinosaur-era debris; upon all that lies the more familiar unconsolidated rubble of the Ice Age, the island's thick and porous integument—stones, gravel, sand, silt, and clay—which sustains both a high water table and deeper, layered aquifers. Two long, linear, glacially pushed-up hills, Ronkonkoma Moraine and Harbor Hill Moraine, extend west from each fork of the fish's tail and mark the southern limit of the last two glacial advances. (A third moraine lies offshore, on the continental shelf.) Older and longer, Ronkonkoma stretches from Montauk Point (the tip of the south fork) to Brooklyn and supports much of the Long Island Expressway. Erosion of the moraines gave rise to Long Island's outwash plain, the nearly level South Shore, my boyhood landscape: barrier islands and their beaches and dunes; the salt marshes and lagoons, bays and inlets; the narrow greenbelts hemmed by malls and office buildings, which pass quickly into swirls of track houses that radiate outward from improbable centers; truck farms; and the relict pine barrens and cedar-tupelo swamps of central Long Island. The point is that Long Island is one big sandbar, more or less devoid of boulders (at least on the outwash plain), and where I grew up along the South Shore is not my idea of a *Crotalus* paradise.

In a gap in the Ronkonkoma Moraine, just below exit 58 on the Long Island Expressway, the water table reaches the surface in a series of seeps and springs that merge into Connetquot Brook (Secatogue for "Great River"), the premier aqueous lifeline across the Oakdale barrens. Sustained by rain and snowmelt and by rising groundwater, Connetquot Brook drains twenty-four square miles of pinewoods and swamps before emptying its freight into Great South Bay, seven miles to the south. Historically, both Connetquot Brook and Rattlesnake Brook, which merge south of Sunrise Highway to become Connetquot River, were two-way highways for spawning fish. In May, thousands of silvery, shimmering alewives would leave the bay to spawn upstream beneath the pines, and in November, hundreds of sea-run brook trout would head in the opposite direction.

Both brooks and their tributaries mark the deeper outwash streambeds that had flushed glacial meltwater to the sea; shallower meltwater beds became swamps. Then, between four and eight thousand years ago, during the same prolonged drought and global warming that ushered rattlesnakes into Vermont, much of Long Island's outwash plain became pine barrens and a tongue of tallgrass prairie (the Hempstead Plains), both plant communities favored by wildfire and supported by coarse, sandy, porous, acidic, nutrient-poor soil. Without fire there would have been no barrens, no prairies, and possibly no rattlesnakes.

As I've mentioned earlier in the book (several times I bet; after all it is a long book), biologists currently lump *all* timber rattlesnakes into a single species, *Crotalus horridus*, with no recognized subspecies. W. H. Martin, one of the editors and authors of *The Timber Rattlesnake: Life History, Distribution, Status, and Conservation Action Plan,* divides the snakes into five separate geographic units, referred to as "eco-regions." Within each region, snakes are similar in both appearance and behavior, and are at least subtly different from snakes that live in other regions, having been influenced by the strictures of the regional landscape and climate, which is why the base color of a timber rattlesnake from Vermont differs from one from Iowa ... or, for that matter, from Long Island. Martin further subdivides the five units into twenty-eight subunits. Long Island rattlesnakes form a discrete subunit of the Mid-Atlantic Unit, which also includes the New Jersey Pine Barrens, the Delmarva Peninsula, and the western shore of Chesapeake Bay. This is the big-picture family tree of our favorite species, one any genealogist could admire, like a map of the bloodlines of European monarchs, except that timber rattlesnake biologists have found that the randy, roving nature of male timber rattlesnakes helps reduce the fallout from inbreeding.

Here's how Long Island fits into Martin's system. Over the past two-and-a-half million years, during each glacial advance (there were four big ones), the level of the ocean dropped two or three hundred feet, at times lapping the rim of the continental shelf. There were a number of isolated glacial refuges in the Southeast where timber rattlesnakes survived the Ice Age; one was probably off the modern coast of South Carolina and Georgia, where Martin believes snakes lived in wooded stream corridors much as they do nowadays in the New Jersey Pine Barrens (or once did on Long Island). Then, around six to seven thousand years ago, as the level of the sea rose during the peak of the last interglacial warming period, rattlesnakes moved inland out of the continental shelf refuge, up the coastal plain to Long

Island and across the piedmont into the Appalachian Mountains, where they continued northward, eventually reaching outposts in southwestern Maine and the Champlain valley of southern Quebec. Three thousand years later, around the time a pair of stone tablets was inscribed with rabbinical Laws (which had shortsightedly left off an ecologic ethic as in thou shall not lay waste to thy planet or its creatures), the rising Atlantic Ocean cutoff Long Island from the coastal plain of New Jersey, but not quite from the nearby Hudson Highlands. Around seven hundred years ago, continues Martin, the beginning of a period known as the "Little Ice Age," due primarily to climatic factors, Long Island's rattlesnake population probably became isolated from those populations in the Hudson Highlands, became insular without an ophidian version of the Throgs Neck, Whitestone, and Triborough bridges to the mainland.

On Long Island, rattlesnakes probably denned in swamps, like those along Rattlesnake Brook, alone or in pairs, in tunnels left by decaying roots or by busy rodents, close to the water table, where the groundwater kept them hydrated and relatively warm in winter. Nowhere is there evidence that they ever gathered by the hundreds or even by the dozens ... Long Island's landscape couldn't accommodate such congresses. Written in a rather different key, it is very likely that Long Island rattlesnakes behaved much as they do nowadays in the New Jersey Pine Barrens, denning in swamps, hunting in pines, loitering in sunny, woodland openings, which eventually included roadway shoulders and railroad tracks. How else could you explain death by bicycle ... or death by train? In 1915, the journal *Copeia* noted that Long Island rattlesnakes "acquired the fatal habit of sunning themselves on the railroad embankments, and of lying across the heated rails."

On July 22, 1882, the *South Side Signal*, a Babylon weekly, reported that a trackwalker employed by the Long Island Railroad had killed a rattlesnake near the Centre Islip train station. The snake was reported to be seven feet long (ha, ha, ha) with thirteen rattles. Later the same year, on September 9, the *Signal* proclaimed that a train in Yaphank severed a large rattlesnake, which when reassembled measured four feet long and nearly eight inches around, and had eleven

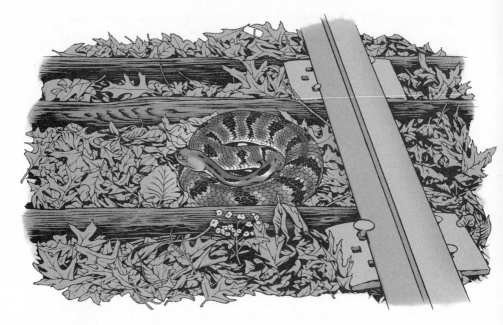

rattles. A decade later, *Port Jefferson Echo* posted a note that three miles west of the South Side Sportsmen's Club, a four-foot snake with seven rattles was killed near the Sayville train depot; the paper noted that this was the first snake ever to have been discovered in that location, which must have made the neighbors *very* pleased.

Politically speaking, Long Island's crusade against timber rattlesnakes was unorganized, unsubsidized, and unrelenting. Local historian Richard Welch wrote, "Authorities probably assumed that Long Islanders would kill every snake they found. They were right." Because Oakdale rattlesnakes tended to den alone, dine alone, and die alone, Long Island never produced the snake-killing analog either of Rudy Komarek or Sparky Punch. Or a Peter Gruber, a.k.a. Rattlesnake Pete, an ophidiopath, who wiped out every snake in three dens on the west side of a Finger Lake called Honeoye and then paraded around Canandaigua, New York, in a three-piece suit made entirely of seven hundred rattlesnakes. His ensemble included snakeskin coat, vest, shirt, pants, hat, shoes, tie, and gloves; rattle segments for buttons; vertebrae for watch chain and charm; and a tie pin fashioned from a gold-plated fang. On Long Island, rattlesnake hunts were happenstance, which resulted in their slow ebb over centuries. It takes a long time to overcome inertia.

Recently, I returned to the South Side Sportsmen's Club, now Connetquot River State Park. Rather than hop the chain-link fence, which at my present age would require more of a Herculean effort, I drove through the main entrance station, off the westbound lane on Sunrise Highway. I parked by the stables and walked to the old clubhouse, the wood-shingled three-story building that I used to avoid. Annie Macintyre, an environmental educator with the state parks department, led me past her downstairs office and the dining rooms (men and women ate separately; if a club member wanted to eat with his wife, he ate in the ladies' dining room) and down a long hallway into the billiard room, an unheated recreation room at the far end of the clubhouse. The room was crammed with noteworthy clutter, a century of artifacts, such as a billiard table and rack of cues, a Franklin stove circa 1830, narrow wooden lockers to fit liquor bottles, duck boats, antique fly rods and fishing tackle and trout flies arranged in glass cases, a series of one-hundred-year-old shorebird and duck decoys, portraits of woodcock, portraits of hunting dogs, ice saws, a map of Islip, a wood and leather rocking chair, a stuffed raccoon lying crosswise over a branch, mounted wild turkeys— strutting, flying, feeding—stuffed owls of numerous species and a huge whistling swan, cabinets of mounted birds (mostly ducks and geese) and a large, freeze-dried snapping turtle that weighed less than an afternoon newspaper.

The billiard room also housed the fishing records, entered into enormous red logbooks, which listed date, location, and species caught (mostly one of three species of trout), size of the fish, and name of the fisherman who caught it. In addition, the logbooks listed who, where, and when someone killed a rattlesnake, the length of the snake, and the number of rattles. On August 12, 1887, for instance, listed among rainbow trout stats was a note that Thomas Reilly, head caretaker and manager of the hatchery, bludgeoned a three-foot-long rattlesnake with five rattles between the trout ponds and clubhouse. The following day, in the same general location, Reilly killed a four-foot-nine-inch snake adorned with six rattles. On July 8, 1898, Reilly killed two rattlesnakes in the woods along the West Branch of Rattlesnake Brook, very likely a conjugal pair.

Later that morning, I took off for the pinewoods, which as far as I could tell were in need of a good wildfire; there were very few young pines and pitch pine woods were ceding their domination to black oak and red oak. Southern pine beetles, a recent invader on Long Island, had lain waste to almost all of the mature pines, whose skeletal branches stood bleak against the cold, gray December sky. After more than an hour of following a xeroxed map in every direction but the right one, I found Middle Branch and West Branch and eventually worked my way to Rattlesnake Brook. Wherever sunlight broke through tributaries, loops and lines of greenbriers made walking all but impossible, much as I remembered from years ago. Floodplain swamps supported red maple and tupelo and in the open skunk cabbage; woods had an understory of highbush blueberry, huckleberry, and sweet pepperbush. I saw few Atlantic white cedars. Rattlesnake Brook flowed in and out of patches of sunlight, around blowdowns, and through swamps, appearing as wild and clean as I remembered it.

I could well imagine a hundred years ago, as dawn broke across the South Side Sportsmen's Club, night coming to an end and the glow of the Morning Star, ten degrees above the western horizon, beginning to fade by the minute. Five miles to the south, in a radically different world, gray chop rolls to shore, foaming water against a pale beach, sugar-white sand under an incandescent sky. Inexorably, morning breaks lumen by lumen, citrus colors bloom in the southeast, and a male grouse begins to drum on a pine log, his beat the audible pulse of the barrens. In my mind's ear, I can almost hear a whip-poor-will calling for the thousandth and final time before settling down for the day, and from a pool within a tupelo swamp, choruses of spadefoot toads taper off, the last two or three sounding like nestling fish crows, full-throated and twangy. Swaddled in shadow, I imagine a rattlesnake unwinding under an oak log and then moving out slowly, easily, as though time were irrelevant. Urged by tendencies best understood by rattlesnakes (and by Rulon Clark), it glides along the rim of a wetland—just like the one I'm standing in—past crooked pines, and then threads through a jungle of scrub oak, the sylvan version of a bad hair day. Gently, the snake brushes the shell of a torpid box turtle. A towhee hurls invective. I imagine the snake, resolute and inscrutable, loosely coiled, a muscular spring whose chin rests against an oak log, waiting . . . some rodent's nightmare. It sucks to think that such a scene is no more.

Raymond L. Ditmars, the legendary curator of reptiles and chief of the Department of Mammals at the Bronx Zoo, author or coauthor of more than a dozen popular books, including *Snake-Hunters' Holiday: Tropical Adventures in Search of Bats and the Bushmaster* (1935), had among his fans the *New Yorker*'s Roger Angell—at age eleven Angell considered Ditmars as important an author as Dickens and Conan Doyle—and Newt Gingrich, who at age thirteen wrote Ditmars a letter and received a loving reply from the Bronx Zoo that the great naturalist had died in 1942, fourteen years earlier, the year before Gingrich was born. During the nineteen-twenties and thirties, Ditmars wanted to know whether rattlesnakes still lived on Long Island. Ditmars thought a live Long Island timber rattlesnake would be the "most interesting accomplishment on zoological records in several years." In 1921, he told the *Brooklyn Daily Eagle* that over the years, employees on the Vanderbilt estate in Oakdale had regularly reported seeing rattlesnakes. Although rapidly suburbanizing, Ditmars believed Long Island still had numerous wild nooks and crannies, "enough to enable the snakes to live comfortably." The two most likely places to look, he thought, were the pine barrens of Oakdale and Yaphank's remote Rattlesnake Swamp. "I am planning to visit the [Vanderbilt] place next spring to ascertain whether the snakes observed are rattlers or not," he said.

Although I spent hours online searching archived newspapers from Suffolk County and from metropolitan New York, I could never determine whether Ditmars had actually followed through and hunted for rattlesnakes in Oakdale. Likely, his obligations at the zoo, his neotropical collecting expeditions, and his writing projects kept him too busy. Instead, he encouraged enthusiasts to make the trip, asking hunters to deliver any suspected rattlesnakes directly to him at the Bronx Zoo. Obligingly, hunters shipped to Ditmars a treasure trove of Long Island snakes, which included water snakes, black racers, hognose snakes, and the charred, severed head of a milk snake, but ... not a single rattlesnake.

One of Ditmar's acolytes, Edwin A. Osborne, took on the challenge. A native of Richmond Hills, Osborne was an accomplished rattlesnake hunter, a writer, and Long Island commissioner of the Reptile Study Society of America—an organization that included among its stated goals to assist "timorous young ladies" in the proper capture of black racers. Not exactly David Livingston searching for the herpetological equivalent of the Nile, Osborne's intended pursuit of a rattlesnake on Long Island nevertheless generated press coverage. On March 21, 1923, the *Eagle,* which through the years would publish a number of his snake stories, announced that Osborne, whom the newspaper considered more "discriminating than St. Patrick," would lead a rattlesnake hunt into the hinterlands of Oakdale, Ronkonkoma, and East Moriches. The *New York Times* mentioned the forthcoming trip as well, noting that Osborne would be joined by a handful of Boy Scouts from Queens who had studied "snakeology" the previous summer at the Boy Scout Museum in the Ramapo Mountains, where Osborne had gained notoriety as a rattlesnake hunter.

Did Osborne find a Long Island rattlesnake? I have no idea. Once again, after many pleasant but fruitless hours searching online for the results of his trip, I decided that the notion of a Long Island rattlesnake hunt was more newsworthy than the results of a Long Island rattlesnake hunt, which likely reflects one of two possibilities: either Osborne and his entourage never went on the expedition or they did go and didn't catch anything.

Although Lawrence Klauber never mentioned Long Island rattlesnakes in his epic book, my online surfing did yield curious information. The *New York Times* apparently had limited interest in the presence or absence of rattlesnakes on Long Island; nonetheless the

Brooklyn Daily Eagle was all over the subject. In 1910, an anonymous "special" to the *Eagle*, titled "Saw Big Rattlesnake," contained no useful information other than that a man from Sayville saw a snake in Oakdale. In 1921, the *Eagle* reported that Dr. Gilbert J. Raynor, a Brooklyn high school principal who summered in East Moriches, had been out picking huckleberries on Raven Point when he "heard a noise as much like the rattling of dry beans in the pod." The snake was small, twenty-four to thirty inches long. There was no mention of the number of rattles, but the story added that a woman picking arbutus in Wyandanch also saw a rattlesnake. There were other interesting, but less relevant, titles, such as "Wife Thrust in Den of Rattlesnakes, Then Drowned by Mate, Is Charged" and "Orchestra of Rattlesnakes at Bronx Zoo." On December 16, 1923, the *Eagle* published Osborne's own article "Rattlesnake Has Brains, or, at Least, Appears to Have, Long Island Expert Believes." Although the story was based on the study of snakes in his menagerie, I'd like to think that Oakdale's last wild rattlesnakes had outwitted Osborne and the Boy Scouts, thus prompting the storyline.

For a couple of years during the late twenties, the *Brooklyn Daily Eagle* sponsored a weekly exhibit in a special booth on the third floor of the Brooklyn Museum, which at the time was a repository for both works of art and works of nature. (It's entirely an art museum now.) In a prominent newspaper article, the *Eagle* dubbed the exhibit "Hidden Treasure of the Brooklyn Museum," and the museum provided both a public and members-only lecture to accompany each "treasure." Among these treasures were a five-thousand-year-old Egyptian boat; an eighteenth-century picture plate of Harvard College, when the school was housed in one building; a thousand-year-old figurine of a Greek girl taking a bath; a fresco of a woman dug from the ruins of Pompeii; a three-hundred-year-old bullet-damaged statue of Buddha, shot as it was being stolen from an Indian temple by a Pennsylvania Dutch farmer; a Gauguin study of a Tahitian girl; a set of sixteenth-century Venetian glassware; and ... a pickled Long Island rattlesnake, which the *Eagle* claimed was "the rarest specimen of rattlesnake in the world."

On December 18, 1927, the *Eagle* revealed that week's new hidden treasure with the headline "Extinct Rattlesnake Formerly Plentiful Over Long Island: Exterminated by Railroad, Engines Caught Serpents Napping on Rails," claiming that its specimen was the best preserved anywhere. Actually, the museum had two specimens in its collection; one, killed and mounted in 1772, was pictured in the newspaper slouched against an upright limb, looking somewhat like a deflated bicycle tire. The lumpy mount appeared to have been the work of "affordable taxidermy." The curator told the *Eagle* that the rattlesnake "would hardly be recognized in the best reptile circles as a rattlesnake at all." The "treasure snake" was tightly coiled in a jar of alcohol and was listed as fifty inches long and six inches around. It was adorned with nine rattles and lacked a button. Without unwinding the snake, I'm not quiet sure how the *Eagle* knew it was fifty inches, a length that doesn't jibe with the snake's reported age of four or five years. Short of training with Alex Rodriquez at Biogenesis or being fed all winter by Kevin McCurley at his New Hampshire snake-breeding facility, a Long Island rattlesnake would be unlikely to reach fifty inches long in less than twenty years. The *Eagle* also mentioned that the snake had probably been caught and strangled in Islip in 1882, by Dr. A. G. Thompson, who gifted it to the Long Island Historical Society, which regifted it to the Brooklyn Museum.

I want to see a specimen of a rattlesnake from Long Island, either a mount or a skin or a spirit submerged in alcohol or formalin. Although I'm not too particular, my own quest has proven almost as difficult (though certainly not as newsworthy) as Ditmars's and Osborne's was ninety years ago. A filmy memory has kept me looking.

When I was in grade school I recall seeing a dusty, faded rattlesnake skin tacked on the wall of an old Sunrise Highway roadhouse in the village of Seaford. In lieu of delinquent taxes, Nassau County had acquired the tavern and eighty-four acres in 1938 and then transformed it into Tackapausha Museum and Preserve, a county institution devoted to Long Island natural history. In addition to the snakeskin, the museum was crammed with a hodgepodge of mounted birds and mammals. There were a few living reptiles and amphibians in terrariums, as well as wall charts, maps, pressed plants, and drunken piles of books and magazines and handouts (which I collected) about various aspects of the nature of Long Island—geology, trees, insects, saltwater fishes, and so forth. I have no proof that rattlesnakes ever lived in Seaford (or next door in my hometown of Wantagh). I assume they may have lived around here in the seventeenth and eighteenth centuries because back then, much of southern Nassau County was still a mix of Atlantic white cedar swamps and maritime forests, and there are records of rattlesnakes from Freeport, ten miles west of Seaford, and from Babylon, ten miles east. By the time I arrived on the Island in 1952, the south shore was houses, houses, and more houses, dwindling greenbelts (except for Tackapausha), and the remnants of a few farms, including one active dairy farm and a commercial nursery (where I could still find black racers, garter snakes, and little brown snakes). In the early fifties, what few rattlesnakes existed on Long Island were sequestered in Oakdale, Yaphank, or East Moriches, which was fortunate for my father, whose natural history ambitions were centered around eradicating Japanese beetles on my mother's roses and dandelions on the lawn; I'm not sure he had the time to deal with rattlesnakes.

During the creation of this book, whenever I began an inquiry into an obscure tidbit of timber rattlesnake information, very often the first person I'd contact would be W. H. Martin, an avuncular presence for any rattlesnake neophyte. A deep-thinking, self-employed herpetologist, Marty finds value in everyone's point of view and has become the conduit for everything to do with timber rattlesnakes. Within his circle are biologists, enthusiasts, herpetoculturists, and even notorious renegades like the late Rudy Komarek, who called Martin at all hours until his wife had Komarek's number blocked. Marty put me in touch with Allan Lindberg, a wildlife biologist with the Nassau County Department of Parks, Recreation, and Museums, which maintains Tackapausha. Lindberg said no record existed of the Tackapausha snakeskin, speculating that it might have been de-accessioned to the trash bin in the late sixties, when the museum relocated away from Sunrise Highway. Then, at Lindberg's request, a number of other Long Island naturalists and historians joined the effort, but the results were the same as at Tackapausha. There were no extant specimens either in the collections of the Long Island Historical Society or in any of the various local historic societies, nor were there any at the William Floyd Estate (which is part of Fire Island National Seashore) or in the collections of the State University of New York at Stony Brook. Neither were there any bounty records for Suffolk County municipalities. One can only imagine what might have happened to the Brooklyn Museum's "hidden treasure."

Then, by enlisting the support of Al Breisch, Martin spared me the tedium of placing ads in newspapers or of going door to door to check old Suffolk County farmhouses for a specimen. A stalwart Basher and retired herpetologist with the New York DEC, Breisch plugged into the country's largest, most prestigious reptile-collection databases, but unfortunately his results were no different than Lindberg's. No rattlesnakes from Long Island were housed in the collections at the American Museum of Natural History, the Smithsonian Institution, the Chicago Field Museum, the Carnegie Museum, or the California Academy of Sciences. Neither were there any among the skins and spirits at Harvard, Yale, or Cornell. Then, right on his doorstep, Breisch turned up a lead.

Although the New York State Museum in Albany didn't list a rattlesnake from Long Island among its reptile collections, a friend of Al Breisch named Joseph Bopp, a curator of mammals and birds, uncovered an old handwritten note in a file box, which indicated that the museum has (or had) a mounted specimen ... somewhere. Bopp thought the snake might have been used in an exhibit and not returned to the proper specimen collection. After additional curatorial detective work, he found the stuffed rattlesnake in a cardboard box in a cabinet subsumed under assorted study skins of Long Island birds and mammals. The specimen tag read "Oakdale, Suffolk Co., NY," and the specimen card, which Bopp later found across the street in an index at the fish lab, noted that the snake, which was labeled *canebrake* rattlesnake, had been collected in 1915 and donated in 1959, a gift from the South Side Sportsmen's Club. Although Bopp questioned whether the specimen was actually from Long Island or killed elsewhere and then given to the Sportsmen's Club, Lindberg e-mailed Breisch, "I think I recall in my recent research of material for Ted that some LI records (possibly in the *Suffolk County News*) mentioned LI TRs as Canebreaks [*sic*]. So a listing of a Canebreak [*sic*] in this case might not be a deal breaker."

It wasn't. Records of the South Side Sportsmen's Club, which are on file at Connetquot River State Park, indicated that a "yellow phase" rattlesnake had been killed on the grounds in 1915, mounted and exhibited, and eventually donated to the New York State Museum.

After making arrangements to see the snake, I drove over to the museum on a cold, clear morning just before Christmas. Bopp met me in the lobby, registered me with the security guard, and then escorted me to the third floor. At the far end of a narrow hallway, inside a heavy metal door was a warehouse-like storage space lined with cabinets, each teeming with miscellaneous bird and mammal specimens, mostly stuffed. One cabinet held small flocks of extinct North American birds, one species per shelf: passenger pigeon, Eskimo curlew, Carolina parakeet, Labrador duck, great auk, ivory-billed woodpecker, and heath hen. In the middle of the room were metal shelving units that supported fossil bones and tusks of mastodons and wooly mammoths, and assorted gleaming, white molars the size of soup bowls. On an upper shelf that extended across the room was the disarticulated skeleton of a fin whale that had been scavenged on a Nantucket beach. The collector was Tom French, Massachusetts Division of Fisheries and Wildlife's director of the Natural Heritage and Endangered Species Program, the man charged with keeping timber rattlesnakes in Massachusetts and James Condon's supervisor in the Blue Hills. According to Bopp, French collects and processes beached marine mammals as a pastime, preparing the skeletons, which he donates to educational institutions like the New York State Museum. Along with the whalebones were a series of photographs of French at work, trucking the carcass home to central Massachusetts, burying it in the yard, and then covering the mound with a huge tarp. French then waits, sometimes for months, as insects and bacteria digest the rancid blubber and putrid meat. Before being taken to the museum, French brought the bones to the DEC's Delmar lab, where they were buried under four truckloads of horse manure from Al Breisch's farm to remove the oily patina and pungent smell.

Specimen number 324 was forlorn, so dingy that it looked more like a black morph than a yellow morph as it idled forever coiled in a shallow cardboard box packed with graying cotton. Its dusty gray body had dark brown bands bordered by time-faded tan. Someone had painted five white Vs on the side of the snake; almost mimicking its chevrons, each a discrete right-angled row of dots that pointed toward the head. The mouth was open, one fang missing. Scales littered the bottom of the box. I stood fingering the stuffed rattlesnake and gazing into vacant glass eyes, a decrepit specimen whose vitality nevertheless endured as a symbol of something lost ... my youth, its essence, the wildness of an island I once called home.

Epilogue
The Ambiguous World of the Timber Rattlesnake

You can't study the darkness by flooding it with light.

EDWARD ABBEY

May 3, 2015. 10:00 a.m. Rattlesnakes are abroad ... *finally* ... two full weeks later than in 2012, when they first appeared on April 19. It's sixty-four degrees Fahrenheit, discreetly overcast. A light south wind creates the illusion that the brook and its attendant beaver ponds flow north, an echo of the waterway's natal direction. The temperature inside the mouth of the den is 45.2 degrees; an outside wall is seventy-five degrees. Two big male snakes, the keepers of the gate, their colors and patterns dulled by winter, coil just outside the entrance on a bed of leaves, side by side, touching: one snake is yellow; the other, black. I'm three feet away, pressed against a stout, horizontal limb. The snakes couldn't care less.

When the black rattlesnake's external temperature reaches eighty-seven degrees, he unspools, withdraws to the den, where he sprawls in the lobby, head down the hallway and leaves just a single posterior curl, a dark question mark, outside. A few moments later, in recognition of his neighbor's absence, the yellow snake lifts his head (slightly), flicks his tongue. Neither seems in a hurry. Are they ever?

The valley is on the verge of spring: peepers chorus in woodland pools; wood frogs have already spawned; great blue herons and kingfishers patrol the rim of the ponds. Cottonwood and black birch catkins jiggle in the breeze, and except for floodplain green ash and swamp white oak, which are bud-swollen but leafless, the hillside is tinted green, a splendid mix of baby leaves that are too small for the wind to notice; spotty and subdued virescence, more pastel than crayon: gray-green, red-green, yellow-green, lime-green—a chromatic inverse of autumn. Amid the roots and rocks, just below the base of the cliff, sprays of Dutchman's-breeches, rock saxifrage, and bloodroot punctuate the loam. I love these days, standing here on the threshold of emergence, when sunshine bathes the whole world, heating the rocks and the forest floor; for now, the slope belongs to vagrant rattlesnakes, which can be found anywhere, moving in any direction. In less than a week, however, as the forest canopy draws closed, most snakes will begin to avoid the shaded upper slope and will stick to the more exposed vine-laced talus, where they'll assemble in leafy hollows between the rocks or sprawl on the rocks themselves. Later in the day, as the talus heats up, they'll withdraw into the recesses of the rocks, immersed in Earth's cool breath. To find them, I'll bend over and comb the nooks and crannies with my LED beam or reflected sunlight off a mirror. Some snakes will be visible, a loop here, a loop there, a rattle, a head, perhaps an unsettling buzz, but most will remain hidden in passages just beyond the flashlight like the knowledge that they belong here in the Northeast, living lives uninterrupted as apex predators.

Mid-afternoon. Standing on a horizontal oak that grows out of the base of the cliff, a foot and a half off the ground, somewhere south of the main den, I hear the light sibilance of a snake gliding through dead leaves. A big, black morph, a male, passes directly underneath me; slow enough that I count the rattle segments . . . four, broken, no taper. He's moving out. And I move with him.

For the next half hour, I shadow the snake, staying twenty paces behind, but just close enough to chart his progress. The snake slithers over the sun-mottled ground, crossing logs; crawling over, under, and around rocks and boulders; into piles like stone tents and out the other side, head periscoping above the leaves. On and on, on an ever-southward trajectory.

A second snake, a yellow male, arrives from the north and embarks on the very same path as the black male, stone by stone, following the irrepressible lure of pheromones. After about thirty yards, for reasons beyond my understanding the yellow snake turns back, and then, several hundred yards farther down the line, the black rattlesnake enters a grotto and doesn't emerge. Although I search in vain with my flashlight and can't relocate the snake, I couldn't be happier. I've followed him for a several hundred yards through the upper, wooded talus. It's Hanukkah without candles.

February had been a long, challenging month, a throwback to the mid-seventies. Halfway through, immured by another deep freeze, I yearned for emergence, looked forward to my favorite den's hemorrhaging rattlesnakes — despite the lingering ice in my veins . . . despite the wind, the snow, the forty-three consecutive days of below freezing temperatures. I imagined half a dozen snakes woven together in the courtyard of the den, basking in leaf-filtered spring sunshine, the image a lifeline out of my winter doldrums. Back in mid-March, when I wanted to believe we were bearing down on rattlesnake emergence, one twinkling night I heard the call of a barred owl, the cry shattering in the cold. It was minus thirteen.

February had been memorable for all the wrong reasons. The Hudson Highlands had never had a colder month. In southwestern Maine, where rattlesnakes had lived until just after the Civil War, the average temperature was fourteen degrees Fahrenheit; in west-central Vermont, where they still live, it was 5.2 degrees; and for ten nights in Glastonbury, Connecticut, the temperature never breached zero. New York's Finger Lakes district had fourteen days at zero or below; Ithaca had a monthly average temperature of 10.2, one degree warmer than Syracuse. Nearly six feet of snow fell on the Blue Hills, breaking the previous one-month record by almost two feet. By March first, Worcester, Massachusetts, had had more than nine feet of snow; normally snowfall barely breaks four feet. No wonder rattlesnakes were late to the surface in 2015. Wouldn't you have stayed in bed?

Unequivocally, on May 3, the most enthralling aspect of the talus was not the snakes, but the rockslide itself. All winter long, meltwater had seeped through a network of fractures in the cliff, repeatedly freezing and thawing, expanding and contracting, until a piece of ledge twice the size of a Silverado, more than two hundred cubic feet of rock (a *very* rough estimate on my part), broke off, shattering on impact, which leveled full-grown trees, redrawing the northern perimeter of the slide. One chunk, a rectangular block, ten feet by four feet by three feet, plowed a groove in the rockslide down to the brook's edge, reconfiguring the talus, creating a stone swale through the rattlesnakes' route of egress. Several transitory basking sites were so altered I momentarily became disoriented, reminiscent of my first visit to South Florida three days after the passage of Hurricane Andrew. At one transition site, where I had been accustomed to seeing ten or more snakes, I saw a singleton and then never more than three during subsequent visits. Sharp-edged boulders, busted rocks, stone chips, and dust were everywhere, which made the slide dangerously unstable. Until mid-May, when leaves finally hid the gray, weathered cliff, I could see the lighter scar of exfoliation, an enormous section of wall exposed to the elements for the first time in more than half a billion years.

Sooner rather than later, rattlesnakes will figure out the thermal qualities of their jiggered neighborhood—its their heritage to read the rocks; very likely they'll assemble at new basking and birthing and shedding sites, setting up new traditions on the rockslide, however transitory. Then, in years or decades or eons, the cliff will give up more of its face, and they'll have to do it all over again.

May 5, 2015. 8:00 a.m. Last night, the temperature never fell below fifty-five degrees. This morning it's sixty-three, breezeless, and thickly overcast. At the moment there are no snakes in front of the den. But eight adults, the vanguard, have reached the north side of the waterfall, a quarter mile south, and sprawl across the rocks. I see the large black morph I escorted two days ago, the one with the busted four-segment rattle, and an equally large black cohort with two vertical, mid-body scars, parallel and well-healed, gruesome dents in otherwise smooth lateral symmetry. The scars resemble a pair of botched appendectomies, two inches long and three-eighths of an inch wide, likely the result of an attack by a red-tailed hawk or a great horned owl, both of which hunt rattlesnakes on the talus.

Like a big, furry moth, a red bat flutters by. A redtail screams. A raven passes overhead with an egg in its mouth. High on the horizontal limb of a red oak, a porcupine naps, legs splayed across the branch, a recumbent pincushion.

Late in the afternoon, once the cloud-clotted sky clears and the day heats up, I work my way back to the main den. A large male, sulfur yellow with white-edged brown bands, basks near the entrance. When my laser-spot, handheld, infrared thermometer tells me his external body temperature has reached eighty-eight degrees, the snake crawls back inside, brushing against an emerging black female, whose temperature is sixty-seven. Within half an hour, the male cools down to seventy-five, and the female warms up to seventy-three. Then, he

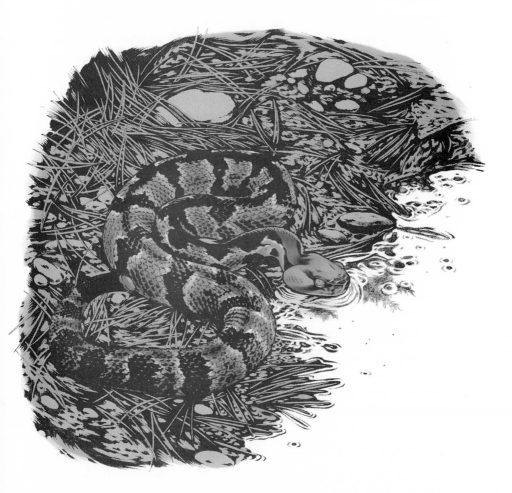

decamps, returning to the sunshine, and entwines with his neighbor ... yellow, black, yellow, black. Another light-colored morph,
more brown than yellow appears from the north and joins the first
two, a third strand in the ophidian weave.

May 17, 2015. I'm with Alcott Smith, my inveterate snake-watcher
buddy. It's 10:00 a.m., seventy-eight degrees. The vernal clock has
been recalibrated; the season is back on track. Columbine, which I
first noticed on May 8, has gone to seed. Dutchman's-breeches have
died back, and fringed polygala has flowered. Trees rooted in the
upper talus are in full leaf, shading the den, which rattlesnakes now
avoid. Mourning clock butterflies and tiger swallowtails are out. A
gray tree frog calls.

Wherever the sunshine smears the most perfect rocks, which also
block the wind, snakes bunch together, a knot here, a knot there, all
the way to the waterfall and beyond—scenes so transfixing you have
to remind yourself that you're standing on treacherously loose rock.
They have begun to ascend the slide along the far side of the plunge
pool, joining forces with snakes from a smaller den to the south, and
together they scale a cleft in the ledge, rockslide by rockslide, up,
over, and out into the forested world.

Once you get to know timber rattlesnakes, it's hard to remain neutral;
they're venomous and potentially deadly, no question, but they're
also beautiful, helpful, long-lived, social, mellow ... and predictable,
which is why, among thoughtful herpetologists and enthusiasts, if
the conversation turns to den location and population size there's a
binding code of secrecy. To pinpoint the rattlesnakes for the whole
world to find would expose the snakes to poaching and slaughter as
much as it would help curious people appreciate them. It's a balancing act. Paradoxically, a timber rattlesnake's emotional impact can be
so profound, so encompassing, that people have crafted careers as
their champion or as their bane. There's still a lucrative black market
for rattlesnakes in New York City's Chinatown, where gallbladders
are in demand. "Wall off the public," emoted a prominent seventy-

four-year-old professor emeritus at a recent Basher gathering. "They are our enemy with their Internet communication bullshit. I hate that shit."

Not long ago, I fielded a series of e-mails from an "all-star team" of Bashers, three biologists who between them have nearly one hundred fifty years of experience in the field, as well as on the frontlines of the rattlesnake wars. William S. Brown, as you may recall, conducts sophisticated scientific research on his home ground in the foothills of the Adirondacks, while W. H. Martin and Randy Stechert do fieldwork throughout the Northeast and Central Appalachians, a pair of gypsy biologists who monitor timber rattlesnake populations that are threatened by sundry developments from gas and oil pipelines, hydrofracking, transmission towers, and wind farms to condominiums, resorts, golf courses, roads, parking lots, malls, and pig farms. Collectively and routinely, the trio work with federal, state, and local governments and author and coauthor numerous peer-reviewed, scientific papers that appear in prestigious worldwide journals. They have contributed chapters to dense, multiauthored books with titles such as *Biology of the Vipers*, *Biology of the Pitvipers*, and *The Biology of Rattlesnakes*, and have battled some of the country's most nefarious poachers like the late Rudy Komarek. They also testify on behalf of the timber rattlesnake in court and lobby state officials for the snake's continued protection. Brown, Martin, and Stechert have been indispensable in the writing of this book, having generously shared with me their knowledge and their study sites, fielded innumerable phone calls, and welcomed me into their exclusive, yet informal club, the Bashers. Now, however, they worry that I may have overstepped explicit (and mobile) boundaries they have set by revealing *too* much surreptitious information, which is a recurring dilemma when one works with timber rattlesnakes.

Back in the spring of 2011, as I stood on a teetering rock, Brown assessed my literary intentions, interrogating me as though I were an indicted felon and he a prosecuting attorney. On that warm morning, in the foothills of the Adirondacks, Bill made it perfectly clear that to proceed any further with him, I could not say where I was. Anything geographically revealing was verboten. I could not mention the names of roads, rivers, lakes, valleys, mountains, cities, villages, or geologic formations. (I was grateful Bill allowed the planet to still be in play.) On the other hand, snake natural history, snake lore, and the interpretation of his thirty-six years of research were admissible. The *number* of snakes he has captured, marked, and recaptured—the lynchpin of his accumulated knowledge, the data source for virtually every journal article he has published—was out.

Early in February 2015, once I completed the first full draft of this book, Bill Brown sent me a cautionary e-mail, a message that was far more magnanimous than was our first meeting on the wobbling rocks—after all, we did have more than four years to get to know each other—emphatically extending his prophylactic concerns over the geographic secrecy of his snake dens and numbers to dens and numbers everywhere, including the ones Alcott and I monitor.

> Over the years, in my public lectures and demonstrations, I've deliberately shied away from mentioning two things: numbers and locations. I decided to do this because I realized that I could not "win" with either. That is, the public would be split on the first issue with half saying it wasn't enough [snakes] and half saying it was way too many, and on the second issue ... revealing where the sites are located invited unwanted visitors; just like a gold miner isn't about to tell his competitors where the main lode is, or a trout fisherman isn't about to tell other fishermen where the best fishing holes are, a rattlesnake hunter isn't about to tell other hunters where the best basking areas or dens are (this last example is actually what Sparky Punch himself told me was his general rule of thumb, as it was with Francis Wilbur, the most infamous bounty hunter beside Punch that operated in these parts). Aren't these the two basic problems you're dealing with?

Absolutely.

Two weeks later, on a bitter morning in mid-February, I joined Bill at his comparative anatomy laboratory at the State University of New York at Albany, where in retirement from Skidmore he teaches a class each semester. I had volunteered for his "preservation party," a once-in-a-blue-moon event in which Bill pickles all the dead timber rattlesnakes that have accumulated in his freezer, and then donates the specimens to the New York State Museum's research collection, where they'll be available to biologists—like books in a library—in perpetuity, a vital linage that began well before Darwin. Today, a total of thirty, one per ziplock bag, of which twenty-six had died crossing a road, mostly males. (That comes as no surprise. Right?) An ignoramus bludgeoned another in a state park; two were found listless in the woods; one died in captivity. Several rattlesnakes were so badly damaged we trashed them. Bill wore a thigh-length, immaculate white lab coat and looked as though he were about to fill a cavity or a prescription. Matt Simon, his long-time field assistant, and I wore waist-length, robin's-egg-blue lab coats, as thin as tissue. Goggles shielded our eyes from formalin. My job was to rinse gore off the thawed snakes and to record data that Matt read off each snake's waterlogged card, items that included sex, color morph, general size (average, small, large), date killed, date collected, collecting site, collector, and any cool miscellaneous facts; one three-foot-long rattlesnake, for instance, had a chipmunk in its throat—I could feel the chipmunk's rock-hard skull; another had an enormous fecal deposit in its large intestine, longer and thicker than my thumb, which was crammed with small-mammal fur, not unlike an owl pellet. If a snake had been marked after being captured alive in the field, part of Brown's long-term study, either Bill or Matt called out its scale-clip number. I wrote it down. For inclusion in the New York State Museum's collection, Bill assigned each snake a unique catalog number. Afterward, he injected formalin into the rattlesnake's tail, cut vertical slits from its cloaca to its head to ensure that the preservative entered the coelomic cavity, and then wound it into one of several formalin-filled five-gallon jars, snake on top of snake, where they remained for a week like a stack of curly poker chips. Once the rattlesnakes were fixed, Brown flushed out the formalin, which would have eventually destroyed bones and rendered DNA unreadable, and then transferred each one to a separate jar of ethyl alcohol, its personal eternal bath.

Number 206, marked in 1981, when it was at least ten years old, had been found DOR in 2006. "Holy shit," Brown blurted out, still aroused after all these years. "This is amazing. He should have been an athlete." He had recaptured the snake in 1987 and 1988, and again in 1999, recording a progressive increase in its length and weight. All afternoon, we preserved and dissected snakes (some were in excellent shape), opening and closing windows in a futile attempt to straddle a *very* narrow zone between succumbing to formalin fumes or to hypothermia from the windy, single-digit temperatures outside. Our conversations always returned to my obligations to the rattlesnakes and to the people who shared their careers with me. Said Bill, repeating his axiom in the sternest tone he could muster while inhaling carcinogenic vapors in a draft cold enough that we could see our breath:

> And the real test is this: if you had mentioned your own favorite area, you too would have felt betrayed and insulted by your very own words because you would be fearful that your area would be subjected to invasion by poachers and "field herpers." These are the feelings of each and every one of us. The only way we can overcome the threats to the timber rattlesnake caused by the Internet and, unfortunately, by your book is to rise above the past indiscretions of Ditmars and Kauffeld who, in their naïveté, were responsible for much of the destruction of the areas they revealed.
>
> My very strong advice to you is don't repeat their mistakes. Again, please, don't repeat their mistakes.

In the process of researching this book I became the custodian of privileged information tasked with keeping secrets that biologists willingly and, in a few cases, grudgingly shared with me. I was taken to dens I could not write about, saw snakes I could not describe, heard rants I could not repeat. Recently, when I broadcast an e-mail to the Bashers that detailed the location and number of snakes I had seen at my favorite site, I unintentionally instigated a thread in which the authors, many of whom I am indebted to for sharing with me their *Crotalus* concerns, demonstrate, with more than a tincture of paternal rebuke, just how complicated and convoluted it is to protect a wondrous and venomous endangered species. This thread circulated for several days, having taken on a life of its own.

For example, Illinois biologist Brian Bielema lamented that, in the years he received state non-game funds to study timber rattlesnakes on a Department of Natural Resources (DNR) nature preserve, he had been commissioned by the DNR to produce reports that were *not* considered sacrosanct, a serious transgression of research etiquette. According to Bielema, anybody could have located his study site in the busted limestone country above the Mississippi floodplain from the state's website, which had uploaded his reports, or with a simple Google search that include the words "rattlesnake study Illinois." In addition, administrators of the fund chastised Bielema for not sending press releases to local media.

> With some horror a friend of mine pointed out that several of my [timber rattlesnake] reports to the Illinois DNR were available from their documents online. Early ones had maps that I had included in my reports thinking that they would only be seen by DNR biologists. It turns out that when I applied and received small non-game wildlife grants . . . the resulting reports were available for the public.

Needless to say, Bielema now bears the expense of his own fieldwork. Responding to his quandary, retired New York State herpetologist Al Breisch wrote that, having acquired an exemption to the state's Freedom of Information Law, New York's Department of Environmental Conservation takes a more enlightened approach to sensitive field data and allows biologists to withhold site-specific information from public disclosure, which, of course, includes website posts. Various federal agencies that deal with endangered and threatened wildlife, however, are not exempt from the federal Freedom of Information Act. Cautioning his audience, Breisch wrote, "Any of your reports that get to a federal agency are essentially public documents."

W. H. Martin's fifty-eight-year career as an itinerant rattlesnake biologist, which is still going strong, trumps everyone else's. Marty

Martin has been hired to survey and inventory rattlesnakes throughout the eastern United States and has visited more than five hundred
dens. Thirty-six years ago, he helped Bill Brown mark snakes; he's
also an original Basher. In 2008, when I first thought seriously about
this project, I called Cornell herpetologist Harry Greene for advice.
The first contact Greene recommended was Marty, who promptly
invited me to join him on a two-day rattlesnake survey along the
Blue Ridge, just over the Virginia line.

As part of the e-mail thread, Martin mentioned that a couple of
computer-savvy colleagues had hacked the "secure" file of the Pennsylvania Fish and Boat Commission just to prove that it was *not* difficult to gain access to the state's rattlesnake files, where nearly twelve
hundred GPS coordinates were stored. Like the old E. F. Hutton
television commercials, when W. H. Martin talks, Bashers listen; and
when W. H. Martin posts, Bashers don't delete. "I have said most of
this before, but in case somebody didn't get it ... once the cat is let
out of the bag, it is out of the bag. No data is 100% secure. It is helpful
to keep data (without GPS coordinates) and maps in separate files."

For similar reasons, Martin sent the Bashers, all pedigreed rattlesnake biologists and caretakers, another e-mail in the thread that concluded with "In the interest of security my future group e-mailings
will lack specific numbers or locations. Details may be exchanged on
a case-by-case basis by request."

As those reprimanding e-mails piled up, I tried to assure my
correspondents that my ambition was not to be a pariah. I had no
intention to subject timber rattlesnakes to additional vandalism, to
encourage further illegal capture, to foster more political ill will or
worse no will at all. Rest assured, I insisted, I certainly don't want
to raise a cry that, "Oh my God, there's venomous snakes in New
Hampshire and the state's encouraging them" ... or that rattlesnakes
live within a *very* leisurely drive from Boston or New York City. I did
not intend to replicate what Carl Kauffeld and Raymond Ditmars
accidentally did in the mid-twentieth century, when their books un-
leashed a torrent of snake hunters and poachers into the American
outback. It is an especially important concern in the twenty-first
century, in our age of instantaneous digital communication, when

both field herping and hiking websites post photographs of timber rattlesnakes, often with GPS coordinates affixed to the picture, and chat rooms, blogs, Facebook announcements, tweets, and threads from group e-mails (including the Bashers) reach unintended eyes.

After all, colorful, boldly patterned timber rattlesnakes still sell for over one hundred dollars apiece on the Eurasian black market, legal protection be damned. In early summer 2014, as many as twenty gestating snakes were stoned to death in north-central Pennsylvania, blood staining their basking rocks, a second-degree misdemeanor punishable by a fine of up to five thousand dollars and two years in jail. No one was caught. And in August 2015, seven teenagers camping in a Maryland state forest were arrested and charged with grilling a timber rattlesnake, a state endangered species.

As I mentioned in the prologue, to take precautions to protect the snakes, I've changed a person's name or two, invented names for several recognizable roads, physical and manmade features, and, in some instances, have been deliberately vague, both geographically and statistically … but not always. Although I remain true to the nature of the snake and to the nature of the people devoted to its well-being, I also remain true to my own nature as a curiously stubborn naturalist who wants nothing less than to encourage people to regard the timber rattlesnake (all species of snakes, really) as not just a species that deserves protection, admiration, and—unequivocally—survival, but as an animal whose history converges with our own (if Congress were to choose a "National Reptile" what better choice than *Crotalus horridus*?) … and most of all as a fascinating, functioning member of North America's eastern deciduous woodlands. From the Amazon to the Antarctic, the timber rattlesnake is on Earth's short list of species that boast more than a few surviving members born during the summer of Woodstock, animals that still consort with the opposite sex and still produce viable young. A deep understanding and appreciation of the nuances of rattlesnakes is much like understanding the influence and expression of *terroir* on wine. Just as local geology, geography, climate, and subtle landscape elements make wines distinctive, the expression of these aspects also makes timber rattlesnakes from the Blue Hills different from snakes from the Hudson Highlands or from Nebraska ... or, once upon a time, from the Pine Barrens of Oakdale, Long Island.

No precaution is too trivial to safeguard timber rattlesnakes, even among the chosen few, who often bicker with one another about how a biologist ought to conduct timber rattlesnake research. Discussing the possible downside of attaching either an internal or an external transmitter to a snake in order to find its den, one Basher had this to say:

> My objection to gluing the transmitter onto either the rattle or the body of the snake is that it is liable to get caught in the tight confines of a hibernaculum or a shelter and either get pulled off or worse cause the snake to be entrapped.
>
> I think there needs to be a clear conservation goal for doing any kind of invasive study. There is likely to be some mortality and you will alter the behavior of the snake by surgically implanting a transmitter. If you are not competent to find a den without putting a transmitter in a snake, maybe you should be doing something else.

Cornell's Harry Greene considers invasive procedures in terms of the cost to the population being studied (if the cost of research is high at a given location, don't do the research there) versus the benefits to the species in the broader context. "Done as best we know how," Greene e-mailed me, "radio telemetry is a powerful tool for understanding snake biology, and I'll argue we rarely can know in advance that some new finding will not, someday, have an important conservation impact."

To look at it another way, conservationists have come far in changing our views of and biases toward native wildlife. We've turned wolves from a menace to a symbol of wilderness; we've turned bison, which may have less cerebral capacity than rattlesnakes, into a symbol of the Great Plains; we've turned great white sharks from a cold-blooded terror into a symbol of globe-trotting, oceanic benevolence; we've reshaped our opinion of raptors from a source of consternation to a source of glory, gathering at sites where hawks migrate, watching eagle-cams on our laptops. We've gone from the greatest whale-killing nation on Earth to the greatest whale-conservation nation. Is it possible to extend timber rattlesnakes that same courtesy?

Apparently, you *can* teach old snakes new tricks. No matter how predictable timber rattlesnakes are, they would not have survived as a species for two million years if collectively they were an ophidian version of Reb Tevye from *Fiddler on the Roof*. Over deep ecological time rattlesnakes shift around, ranges expand and contract, damning tradition. Is there any other way to interpret their reappearance in the glaciated Northeast and Upper Midwest? If timber rattlesnakes didn't have a bit of Magellan in their DNA, they'd be out of luck whenever trees and shrubs shaded over a basking site or a cliff loosed rock and altered the landscape, like the rockslide Alcott and I visit. And since the eighties (possibly earlier), rattlesnakes have begun to colonize a number of rural, sun-heated human refuse heaps, what ecologist refer to as "edificarian" environments: rock walls, a tire graveyard, a pile of highway rubble, quarry slag.

Snakes that relocate along natural gas pipelines face a perilous situation. In the short term, their predictability makes them easy targets for vandals and poachers. In the long term, gas lines are ephemeral, bound to the whims of the economy. An active natural gas pipeline maintains a constant temperature of forty-five degrees all winter, which raises the immediate frost line two-feet closer to the surface, providing rattlesnakes with an alternative overwintering site, both shallower and warmer than their ancestral one. If the utility abandons the line and gas ceases to flow, the site might abruptly switch from luxury accommodations to ecological death trap, a sepulchral monument.

On a warm, overcast Tuesday morning in early July 2015, I meet Richard Thorp at a Starbucks in an undisclosed location in the Northeast, and we drive nineteen miles to a secluded trailhead within a very prominent watershed. A pair of small rivers frame the ridge with its half-dozen snake dens before joining the mother flow a few dozen miles to the south. Part of the ridge runs northeast-southwest and follows the drainage of the larger of the two rivers. East of the ridge, a fertile valley; west of the ridge, rolling wooded hills, one after the other like ripples in a lake. Within the timber rattlesnake's entire range, this is unquestionably an isolated population, but close to the epicenter. It is also close to the epicenter for natural gas pipelines in the Northeast. Clearings in the woods mark where bedrock was blasted and dredged and pipes were buried. An avenue of rubble parallels each pipeline, and intermittent rocky knolls interrupt each brush-hogged right-of-way.

How long did it take timber rattlesnakes to discover this man-made, sun-blessed nirvana? Apparently, not long.

That snakes are attracted to sunny rock piles is no secret. It's their equivalent to morning coffee; it's what gets them going and keeps them going. Finding the rocks might be random; staying there is not. In a nutshell, here's how it might have happened: snake finds rocks and basks; other snakes, following pheromone trail of first snake, find the rocks and join in. Then, snakes shed on site. Males include the rocks in their search for prospective mates. Females abandon traditional rookeries for the "new" rocks and stay behind to gestate and give birth. Neonates shed and stick around and evidently follow a thermal gradient through the rocks toward the warm pipe, which in the Northeast is approximately midway between the surface and the frost line. These snakes grow up, the pipeline their all-purpose habitat. Through the years, a few adults and juveniles join them. Eventually, within a decade or two, as the pipeline becomes the indispensable social center, a demographic sink, serviceable basking, shedding, and birthing sites have been abandoned, forest traditions lost (or at least forgotten); the ancestral den, lifeline of survival, attracts fewer snakes.

"I've seen basking, shedding, courting, male-to-male combat, and birthing along the line. Some snakes overwinter there. It's become such a predictable, vibrant rattlesnake habitat that black racers come in from the woods in mid-August to hunt neonates along the pipelines," Thorp tells me, both hands on the wheel, as he negotiates a rutted logging road. "There is no safe period when rattlesnakes are absent, which places a burden on the gas company whenever it wants to work on the pipeline. Winter is the only safe time to brush hog."

Once we leave the car in a remote pullout at an unmarked trailhead, Richard takes off. He walks at a brisk pace, long flowing strides over a network of crisscrossed trails, through rushing brooks, up a hemlock ravine, and onto a pipeline right-of-way. Having spent years on Vermont ridges chasing after Alcott, I am prepared to keep up. Richard Thorp is a tall man, at six two, with skin and hair as pale as marble and a head the size of a muskmelon. He's a lifelong amateur herpetologist to whom timber rattlesnakes are catnip, a retired policeman who has busted a number of poachers during a distinguished forty-five-year career in law enforcement. Everybody trusts Richard, and Richard trusts me.

Half an hour later, we reach our destination, a long line of jumbled rocks that undulates through the right-of-way like the arched back of a snake. Richard takes me directly to a large, flat rock he calls the "Mary Rock" after the remains of an inscription that reads "Mary Shimansky I Love You," which was painted across the top sometime in the late seventies, shortly after the pipe was laid.

Three gravid rattlesnakes, all yellow morphs, idle together in front of the Mary Rock, as vulnerable as if they had been crossing a road, which was why twenty (or more) were slaughtered at a Pennsylvania gestating site in the summer of 2014. "I've seen as many as twenty-six snakes on this pipeline alone, all females. In late August, babies pile up," Richard said, a tone of resignation in his voice. "We can't afford to have a snake warden patrol the pipelines. We can't afford to fence out the public. What good are surveillance cameras if you don't have anyone to monitor them?"

Clearly, these rattlesnakes are under siege and something must be done, or the poaching and vandalism will go unabated.

"I'm running out of patience," Richard laments, having rounded into the homestretch of lifetime of rattlesnake work. "I don't care if they put me in jail. I'm seventy-four years old. I don't have much time left."

Thursday, April 20, 2012. 11:50 a.m. Low seventies, a strong, warm wind blowing up from the south. Ribbons of cool air that waft out of icy caverns around the base of the rockslide create a microclimate in to which no snakes will be found. Higher up the talus, the rattle-snake zone, already columbine and rock saxifrage bloom in warps and cracks in the cliff, and in the loam around the base Dutchman's-breeches, hepatica, and bloodroot flowers wilt. It's the Arctic on one end of this sloping mass of cliff fragments, Virginia on the other.

I'm with Alcott Smith (of course). We've tasked ourselves with locating the main den, south of Larry Boswell's cabin, where Alcott had rescued the quilled snake. By mid-afternoon we find it, a cavity in an outcrop, more than a mile from where we parked. Three feet inside, an entrance large enough to crawl into, is a mound of white, chalky powder, an unexpected island rising out of a bed of leaves. Lines of rattlesnake spoor inscribe the mound, squiggly notations marking their progress. While we speculate about the nature of the mound, a snake nearly five feet long emerges out of a passageway at the back of the den, slowly crosses into daylight, and then pauses mid-mound, head up, tongue out, the largest timber rattlesnake I've ever seen, an almost patternless male, his yellow base color faded like an overexposed sunrise, his head the size of a tennis ball, white powder dusting his flanks.

All afternoon the den disgorges snakes. All afternoon Alcott and I discuss the odd mound. Having never peered into the foyer of such a large snake den before, I have no clue where the powder came from. A pile of uric acid, the millennial accumulation of snake wastes? A pile of calcium carbonate, a precipitate leached from the ceiling? A million dollars worth of abandoned cocaine that the snakes, by their presence guard like the Gila monster in *The Treasure of the Sierra Madre*? Pixilated, neither of us is particularly committed to our guesses, and we truly don't believe the third possibility; we're just fishing like an osprey above a beaver pond. Just before dusk, once the entrance has cleared of rattlesnakes, Alcott collects a handful of powder, which we place in a labeled ziplock bag. I'll deliver the sample for analysis at the U.S. Department of Agriculture office in White River Junction, Vermont.

Monday, April 24, 2012. 7:50 a.m. One of the joys of being a naturalist is the thrill of self-discovery. Ziplock bag in hand, as I prepare to leave for White River Junction, the phone rings. It's Randy Stechert. He wants to know if we found the den. I review the day, the den, and the snakes. Eventually, I mention the mound and our speculations. If anyone would have an educated guess what the powder might be, it would be Randy, the rattlesnake savant, who has peered into the doorway of more than three hundred fifty dens, including this one.

"Return it," he says, pausing to laugh. "It's not in your best interest to have anyone examine it, particularly someone employed by the federal government. It's Larry Boswell's ashes."

He sleeps with the snakes.

Acknowledgments

This would have been an entirely different book without the insight, encouragement, recommendations, introductions, and company of Alcott Smith, to whom I owe an immeasurable debt. Of course, there were other key actors that helped me in many ways, many of whom Alcott introduced me to, including Kevin McCurley, Randy Stechert, Bill Brown, and Marty Martin. I did stumble on to James Condon and Chris Jenkins on my own and I already knew Mark DesMeules and Tom Tyning. Thank you all and to all Bashers: past and present, holo and hemi, and wannabe.

Many scientists (herpetologists or otherwise), historians, and rattlesnake enthusiasts fielded queries, shared their research results, and provided perspective. Thank you to Bob Aldridge, Brian Bielema, Dick Birnie, Doug Blodgett, Joseph Bopp, Al Breisch, Kiley Briggs, Jonathon Campbell, Jim Chestney, Rulon Clark, Brendan Clifford, Norm Conners, Lou-Anne Conroy, Lee Fitzgerald, Doug Fraser, Tom French, Bob Fritsch, Bob Giffen, Jeff Gray, Harry Greene, Tom Henderson, Chris Jenkins and the generous staff of Orianne Society, Dan Keyler, Allan Lindberg, Annie Macintyre, Mike McBride, Ed McGowen, Scott MacLachlan, Michael Marchand, Zach Orr, Lou Perrotti, Joe Racette, John Rook, Andy Sabin, Andy Soha, Brett Somers, David Scopaz, Anne Stengle, Les Tyrala, Richard Welch, Tom Wideman, and Christopher Wren.

A host of generous people read or listened to portions of the manuscript, and then provided constructive critics on matters of rattlesnake biology and economics, conservation history, and human decency. Thank you to Bob Aldridge, Brian Bielema, Al Breisch, Bill Brown, Mark DesMeules, Doug Fraser, Chris Jenkins, Ed Kabay, Dan Keyler, Scott MacLachlan, Kevin McCurley, Jed Merrow, George Pisani, and Randy Stechert. Others were literary sounding boards: Frank Asch, Jeannie Kornfeld, Casey Levin, Jordan Levin, Sandra Miller, Ruthie Sproull, Steve Rashkin, John Tiholiz, Susan Tiholiz, and Annie Weeden. A few did both. Marty Martin, Alcott Smith, Harry Greene, and an anonymous reader reviewed the whole thing and provided indispensible recommendations on current biology, anatomy, and taxonomy, as well as on the imperative to more carefully veil my snake-watching locations.

Several research librarians unknotted a sheepshank of obscure literature searches: Nancy Mason at the University of New Hampshire Special Collections, Amy Witzel at Dartmouth College's Social Science and Humanities Library, Connie Rinaldo, the librarian at Harvard University's Museum of Comparative Zoology, and the public relations director at Audubon Society of Rhode Island.

I am indebted to editors, who worked with me on various rattlesnake stories that appear here in different form. At *Audubon*, David Seideman and Jerry Goodbody; at the Boston Globe's *Globe Magazine*, Veronica Chao; at *Northern Woodlands*, David Mance III; at *OnEarth*, Laura Wright, George Black, and Douglas Barasch; and at Vermont Public Radio, Betty Smith.

My agent, Russ Galen, prodded me to go well beyond my initial proposal, and never waivered in his commitment to a book about rattlesnakes. Marian Cawley, my friend and literary muse, provided editorial guidance, made many constructive suggestions, kept my baseball metaphors to a minimum, and had an uncanny knack for understanding what it was that I meant to say. At the University of Chicago Press, Christie Henry provided unflagging support and encouragement, and deft editorial suggestions. Her colleague Mark Reschke was a surgically precise copyeditor, and Logan Smith kept track of all of us.

Any errors that remain are mine alone.

This book is immeasurably enriched by the drawings of Alexandra Westrich, who benefitted from the unselfish provision of reference photographs to guide her drawings. Thank you Polly Smith-Blackwell, Matt Simon, Kevin Stohlgren, James Condon, Nicole Corman, Bob Fritsch, John Mizel, Casey Levin, and Jordan Levin for sharing your digital images.

For the past thirty years, I've watched timber rattlesnakes in the company of many people, all of who added to my knowledge and enjoyment of *Crotalus horridus*. Some were guides, others guests. These include Steve Andreski, Kevin Andreski, Ginny Barlow, Charlie Berger, Brian Bielema, Doug Blodgett, Kiley Briggs, Bill Brown, Marian Cawley, Rulon Clark, Brendan Clifford, James Condon, Nicole Corman, Sherry Crawford, Mark DesMeules, Bret Engstrom, Doug Fraser, Bob Fritsch, Jeff Gray, Bill Hoffman, Paul Jardine, Chris Jenkins, Dan Keyler, Erik Kilburn, Amos Kornfeld, Jacob Kornfeld, Jeannie Kornfeld, Brian Lombardo, Casey Levin, Jordan Levin, Linny Levin, Marty Martin, Kevin McCurley, Murray McHugh, Sandra Miller, Tom Palmer, Bob Pitler, Dennis Quinn, Steve Rashkin, Andy Sabin, Carl Safina, Matt Simon, William Smart, Alcott Smith, Avery Smith, Polly Smith-Blackwell, David Scopaz, Randy Stechert, Timmy Tattoo, and Joan Waltermire.

I am deeply and humbly grateful for all my friends and colleagues, for years worth of intellectual support and morale-boosting. Not the least of whom is Hannah Kornfeld; one of the key ingredients in my ongoing efforts was her delectable chocolate chip cookie bars. If I have omitted anyone in these acknowledgments, I am truly sorry.

Bibliography

Books

Adams, E. Clark, and John K. Thomas. *Texas Rattlesnake Roundups*. College Station: Texas A&M University Press, 2008.

Audubon, John James. *Birds of America, and Interspersed with Delineations of American Scenery and Manners*. 5 volumes. Edinburgh, Scotland: Adam Black, 1831–1839.

Brennan, C. E. *Rattler Tales from Northcentral Pennsylvania*. Pittsburgh: University of Pittsburgh Press, 1995.

Brown, Fred, and Jeanne McDonald. *The Serpent Handlers: Three Families and Their Faith*. Winston-Salem, NC: John F. Blair, Publisher, 2000.

Brown, Marvin L., Jr. *Baroness von Riedesel and the American Revolution: Journal and Correspondence of a Tour of Duty, 1776–1783*. Chapel Hill: University of North Carolina Press, 1965.

Brown, William F. *Rattlesnake Tales: Fact and Fiction about Texas Diamondback Rattlers*. Beaumont, TX: WFB Enterprises, 1999.

Brown, William S. *Biology, Status, and Management of the Timber Rattlesnake (Crotalus horridus): A Guide for Conservation*. Society for the Study of Amphibians and Reptiles: Herpetological Circular No. 22, 1996.

Burton, Thomas. *Serpent-Handling Believers*. Knoxville: University of Tennessee Press, 1993.

Campbell, J. A., and E. D. Brodie, Jr., eds. *Biology of Pitvipers*. Tyler, TX: Selva Press, 1992.

Clamp, Heyward, Jr. *Adventures of a Carolina Snake Hunter: Tales of Tails*. Sanibel Island, FL: Ralph Curtis Publishing, Inc., 2009.

Collins, Joseph T., and James L. Knight. *Catalogue of American Amphibians and Reptiles*. Salt Lake City, UT: Society for the Study of Amphibians and Reptiles, 1980.

Conant, Roger. *A Field Guide to the Life and Times of Roger Conant*. Provo, UT: Canyonlands Publishing Group, 1997.

Curran, C. H., and Carl Kauffeld. *Snakes and Their Ways*. New York: Harper and Brothers Publishers, 1937.

Darwin, Charles. *Expression of the Emotions in Man and Animals*. London: John Murray, 1872.

Ditmars, Raymond L. *Snakes of the World*. New York: Macmillan Company, 1931.

Dobie, J. Frank. *Rattlesnakes*. Austin: University of Texas Press, 1982.

Ernst, Carl H. *Venomous Reptiles of North America*. Washington, DC: Smithsonian Institution, 1992.

Ernst, Carl H., and Evelyn M. Ernst. *Venomous Reptiles of the United States, Canada, and Northern Mexico*. Volumes 1 and 2. Baltimore: Johns Hopkins University Press, 2012.

Furman, Jon. *Timber Rattlesnakes in Vermont & New York: Biology, History, and the Fate of an Endangered Species*. Hanover, NH: University of New England Press, 2007.

Gibbs, James P., Alvin R. Breisch, Peter K. Ducey, Glenn Johnson, John L. Behler, and Richard C. Bothner. *The Amphibians and Reptiles of New York State*. New York: Oxford University Press, 2007.

Gloyd, Howard K. *The Rattlesnakes, Genera Sistrurus and Crotalus: A Study in Zoogeography and Evolution*. Chicago: Chicago Academy of Sciences, Special Publication Number 4, 1940.

Gotch, A. F. *Latin Names Explained: A Guide to the Scientific Classification of Reptiles, Birds, and Mammals*. New York: Facts on File, Inc., 1995.

Greene, Harry W. *Snakes: The Evolution of Mystery in Nature*. Berkeley: University of California Press, 1997.

Greene, Harry W. *Tracks and Shadows: Field Biology as Art*. Berkeley: University of California Press, 2013.

Gutkind, Lee. *The People of Penn's Woods West*. Pittsburgh: University of Pittsburgh Press, 1984.

Hadden, James M., Lieut. *Hadden's Journal and Orderly Book: A Journal Kept in Canada and Upon Burgoyne's Campaign in 1776 and 1777*. Albany, NY: Joel Munsell's and Sons, 1884.

Hayes, Lyman Simpson. *History of the Town of Rockingham including Bellows Falls*. Published by the town of Bellows Falls, VT, 1907.

Hayes, W. K., K. R. Beaman, M. D. Cardwell, and S. P. Bush, eds. *The Biology of Rattlesnakes*. Loma Linda, CA: Loma Linda University Press, 2008.

Hood, Ralph W., Jr., and W. Paul Williamson. *Them That Believe*. Berkeley: University of California Press, 2008.

Hylander, C. J. *Adventures with Reptiles: The Story of Ross Allen*. New York: Julian Messner, a Division of Pocket Books, Inc., 1951.

Jensen, John B., Carlos D. Camp, Whit Gibbons, and Matt J. Elliot. *Amphibians and Reptiles of Georgia*. Athens: University of Georgia Press, 2008.

Kauffeld, Carl. *Snakes: The Keeper and the Kept*. Garden City, NY: Double & Company, Inc., 1969.

Kauffeld, Carl. *Snakes and Snake Hunting*. Garden City, NY: Hanover House, 1957.

Kemnitzer, John W., Jr. *Rattlesnake Adventures: Hunting with the Oldtimers*. Malabar, FL: Krieger Publishing Company, 2006.

Kemnitzer, John W., Jr. *Snake Hunting the Carolina Tin Fields*. Stanfordville, NY: Wild Side Press, 2004.

Kimball, Douglas W., ed. *The Timber Rattlesnake in New England—A Symposium*. Springfield: Western Massachusetts Herpetological Society, 1978.

Kimbrough, David. *Taking Up Serpents: Snake Handlers of Eastern Kentucky*. Macon, GA: Mercer University Press, 2002.

Klauber, Lawrence M. *Rattlesnakes: Their Habits, Life Histories, and Influence on Mankind*. 2nd edition. Volumes 1 and 2. Berkeley: University of California Press, 1997.

Lazell, James D., Jr. *This Broken Archipelago: Cape Cod and the Islands, Amphibians and Reptiles*. New York: Quadrangle/ New York Times Book Company, 1976.

Linzey, Donald W., and Michael J. Clifford. *Snakes of Virginia*. Charlottesville: University Press of Virginia, 1981.

Mattison, Chris. *Rattle! A Natural History of Rattlesnakes*. London: Blandford Publishing, 1996.

Minton, Sherman A., Jr., and Madge Rutherford Minton. *Venomous Reptiles*. New York: Charles Scribner's Sons, 1969.

Morris, Ramona and Desmond. *Men & Snakes*. New York: McGraw Hill Book Company, 1965.

The New Encyclopedia of Snakes. Princeton, NJ: Princeton University Press, 2007.

New York–New Jersey Trail Conference, Inc., and the American Geographical Society. *New York Walk Book*. 4th edition. Garden City, NY: Doubleday/ Natural History Press, 1971.

Ostfeld, Richard F. *Lyme Disease: Ecology of a Complex System*. Oxford: Oxford University Press, 2011.

Palmer, Thomas. *Landscape with Reptile: Rattlesnakes in an Urban World*. New York: Tichnor and Fields, 1992.

Pope, Clifford H. *The Giant Snakes: The Natural History of the Boa Constrictor, the Anaconda, and the Largest Pythons, including Comparative Facts about Other Snakes and Basic Information on Reptiles in General*. New York: Alfred A. Knopf, 1980.

Pope, Clifford H. *Snakes Alive and How They Live*. New York: Viking Press, 1949.

Quammen, David. *Spillover*. W. W. Norton & Company, 2012.

Romer, Alfred Sherwood. *The Vertebrate Body*. Philadelphia: W. B. Saunders Company, 1956.

Rubio, Manny. *Rattlesnakes: Portrait of a Predator*. Washington, DC: Smithsonian Institution Press, 1998.

Seigel, Richard A., and Joseph T. Collins, eds. *Snakes: Ecology and Behavior.* New York: McGraw Hill, Inc., 1993.

Seigel, Richard A., and Joseph T. Collins, eds. *Snakes: Ecology and Evolutionary Biology.* New York: Macmillan Publishing Company, 1987.

Smith, Hobart M. *Evolution of Chordate Structure: An Introduction to Comparative Anatomy.* New York: Holt, Rinehart, and Winston, Inc., 1960.

Souder, William. *Under a Wild Sky: John James Audubon and the Making of The Birds of America.* New York: North Point Press, 2004.

Stafford, Peter. *Snakes.* Washington, DC: Smithsonian Institution Press, 2000.

Stone, William. *The Campaign of Lieut. Gen. John Burgoyne, and The Expedition of Lieut. Col. Barry St. Leger.* Albany, NY: Joel Munsell, 1877.

Tennant, Alan. *A Field Guide to Texas Snakes.* Lanham, MD: Gulf Publishing, 1998.

Thatcher, James, MD, Surgeon, Continental Army. *Military Journal of the American Revolution, 1775–1783.* Gansevoort, NY: Corner House, 1998.

Thompson, Elizabeth H., and Eric R. Sorenson. *Wetland, Woodland, Wildland: A Guide to the Natural Communities of Vermont.* Hanover, NH: University Press of New England, 2000.

Timmerman, Walter W., and William H. Martin. *Conservation Guide to the Eastern Diamondback Rattlesnake Crotalus adamanteus.* Society for the Study of Amphibians and Reptiles: Herpetological Circular Number 32, 2003.

Tyning, Thomas F., ed. *Conservation of the Timber Rattlesnake in the Northeast.* Lincoln: Massachusetts Audubon Society, 1992.

Tyning, Thomas F. *Stokes Nature Guides: A Guide to Amphibians and Reptiles.* Boston: Little, Brown and Company, 1990.

Wasmus, J. F. *An Eyewitness Account of the American Revolution, and New England Life: The Journal of J. F. Wasmus, German Company Surgeon, 1776–1783.* New York: Greenwood Press, 1990.

Wheeler, Donald G. *Tales from the Golden Age of Rattlesnake Hunting.* Lansing, MI: E.C.O., 2001.

Wideman, Tom. *Texas Rattlesnake Tales.* Abilene, TX: State House Press, 2006.

Williams, Samuel, LL.D. *The Natural and Civil History of Vermont.* Walpole, NH: Isaiah Thomas and David Carlisle, 1794.

Magazines

Abele, Ralph W. "Sacking Contests and Resource Management." *Pennsylvania Angler,* November 1984.

Adams, James T. "In Search of the Pine Barrens Timber Rattlesnake." *Reptile and Amphibian Magazine,* May 1997.

Allen, William, Jr. "Massasaugas." *Reptile and Amphibian Magazine,* November/December 1990.

Angst, Bim. "The Big Picture: A Profile of Carl Kauffeld." *Reptile and Amphibian Magazine*, September/October 1991.

Attum, Omar. "Snake, Rattle and Roll: Tracing Timber Rattlesnakes." *Kentucky Monthly*, 4, no. 9, 2001.

Balliett, Whitney. "The Talk of the Town: Rattlers." *New Yorker*, April 3, 1954.

Brandt, Buzz. "How the Rattlesnake Became a Symbol of American Unity and Freedom." *Elk Magazine*, June 2012.

Brown, Phil. "Split Rock's Reclusive Rattlers." *Adirondack Explorer*, July/August 2005.

Brown, William S. "The Female Timber Rattlesnake: A Key to Conservation." *Reptile and Amphibian Magazine*, September/October 1995.

Brown, William S. "Hidden Life of the Timber Rattler." *National Geographic*, 172, no. 1, July 1987.

Dickey, James. "Blowjob on a Rattlesnake." *Esquire*, October 1974.

Draud, Matthew Jay. "The Dusky Pigmy Rattlesnake—A Ruby in the Sand." *Reptile and Amphibian Magazine*, May/June 1993.

Grenard, Steve. "Raymond Lee Ditmars." *Reptile and Amphibian Magazine*, May/June 1991.

Grenard, Steve. "Ross Allen (1908–1981)." *Reptile and Amphibian Magazine*, May/June 1990.

Hellman, Geoffrey. "The Talk of the Town: Social Note." *New Yorker*, August 4, 1934.

Holland, Jennifer S. "The Bite That Heals: Scientists Are Unlocking the Medical Potentila of Venom." *National Geographic*, February 2013.

Lang, Susan. "Rattlesnakes Sound Warning on Biodiversity." *Cornell Chronicle*, April 2010.

Levin, Ted. "Ambush." *Audubon*, May 2008.

Levin, Ted. "Castanets on the Ledges." *Vermont Magazine*, March/April 1991.

Levin, Ted. "Rattled in the Blue Hills." *Boston Globe Sunday Magazine*, May 27, 2012.

Levin, Ted. "We Have Rattlesnakes." *OnEarth*, Summer 2008.

Lipske, Michael. "The Private Lives of Pit Vipers." *National Wildlife*, August/September 1995.

Lotz, Darla, and Norman Frank, DVM. "Roger Conant: The Man Who Wrote the Book." *Reptile and Amphibian Magazine*, January/February 1991.

Maloney, Russell. "The Talk of the Town: Hunt." *New Yorker*, November 4, 1939.

Marshall, Jessica. "Rattlesnakes, Avoiding Roads, Become Inbred." *Discovery News*, April 29, 2010.

May, Peter G., Steven T. Heulett, Terence M. Farrell, and Melissa A. Pilgrim. "Live Fast, Love Hard, & Die Young: The Ecology of Pigmy Rattlesnakes." *Reptile and Amphibian Magazine*, January/February 1997.

Meyer, Karl. "In Encounters Where Snake Identity Comes into Question, the Snakes Always Lose." *Sanctuary: Journal of the Massachusetts Audubon Society*, Spring issue, 2010.

Milius, Susan. "Harassing: When a Chipmunk Teases a Rattlesnake." *Science News* 168, August 27, 2005.

Milius, Susan. "Hey, Snake—Rattle This! Some Furry Little Creatures Are Born to Taunt Rattlesnakes." *Science News* 156, October 9, 1999.

Milius, Susan. "The Social Lives of Snakes: From Loner to Attentive Parent." *Science News* 165, March 27, 2004.

Miller, Julie Ann. "Strike!" *Science News* 120, August 29, 1981.

Owen, James. "Rattlesnakes Show Strong Family Bonds, Study Says." *National Geographic News*, February 23, 2004.

Palmer, Thomas. "The Magician and the Rattlesnake." *Discovery*, 14, no. 5, 1993.

Rice, Arthur F. "Vermont Rattlesnakes." *Forest And Stream*, 44, no. 20, 1895.

Roach, John. "Fear of Snakes, Spiders Rooted in Evolution, Study Finds." *National Geographic News*, October 4, 2001.

Rogers, Cameron. "Profiles: Specialist in Snakes." *New Yorker*, July 14, 1928.

Smith, Chris. "Influences: Roger Angell." *New York Magazine*, May 21, 2006.

Specter, Michael. "The Annals of Medicine: The Lyme Wars." *New Yorker*, July 1, 2013.

Weidensaul, Scott. "The Belled Viper." *Smithsonian*, 28, no. 9, 1995.

Welch, Richard F. "The End of the Game: The Timber Rattlesnake on Long Island." *Long Island Forum*, Spring 1995.

White, E. B. "The Talk of the Town: Zoo." *New Yorker*, April 10, 1926.

White, E. B. "The Talk of the Town: Zoo Things." *New Yorker*, May 26, 1928.

Wilkinson, Francis. "Discover Interview: Newt Gingrich." *Discover*, October 2006.

Williams, Ted. "Driving Out the Dread Serpent." *Audubon*, 1988.

Winslow, Joyce. "Who Should Live on Kongscut Mountain?" *Yankee*, August 1987.

Zeidler, Erik. "Snakebite." *Reader's Digest*, March 2012.

Newspapers

Abramson, Rudy. "Hard-Bitten Sackers Get Their Thrills with Rattlers." *San Jose Mercury*, San Jose, CA, July 20, 1981.

Allen, Monica. "Rattlers." *Burlington Free Press*, July 31, 1988.

"Amagansett Plans Museum That Will Memorialize Long Island Off-Shore Whaling." *Suffolk County News*, Sayville, NY, November 13, 1936.

"Ancient History: Forty Years Ago." *Suffolk County News*, Sayville, NY, July 31, 1942.

Anderson, Dave. "Rattlesnakes? Here? Unlikely." *Concord Monitor*, Concord, NH, June 19, 2005.

Anderson, Frank. "Let's Stuff Legislator in a Sack." *Valley News Dispatch*, Tarentum, PA, April 11, 1985.

Angier, Natalie. "The Pit Viper's Complex Life: Bizarre, Gallant, and Complex." *New York Times*, October 20, 1991.

"Are Rattlesnakes Entering Suburbia." *Science Daily*, June 23, 2007.

Bailey, Paul. "Paul Bailey's Historic Long Island Selected and Supplemented by Carl A. Starace." *Islip Bulletin*, July 23, 1970.

Blackwood, Harris. "I Have Enough Faith to Keep Clear of Snakes." *Gainesville Times*, February 11, 2007.

Boucher, Dave. "Timber Rattler Bites, Kills Snake-Handling Preacher." *Charleston Daily Mail*, May 29, 2012.

Braeske, Arnold. "Rattlers Are Labeled 'Rare.'" *New York Times*, May 6, 1973.

Brooke, James. "Fight in a Hartford Suburb: Residents vs. Rattlesnakes." *New York Times*, July 17, 1985.

"C. E. Snyder Eulogized: Tribute Paid Rattlesnake's Victim at Funeral Services." *New York Times*, May 16, 1929.

"Care Needed to Guard against Brush Fire on Long Island." *Suffolk County News*, November 6, 1936.

"Carl F. Kauffeld, Expert On Snakes: Ex-Director of the Staten Island Zoo Dies at 63." *New York Times*, July 11, 1974.

Carroll, Sean B. "As Genes Learn Tricks, Animal Lifestyles Evolve." *New York Times*, August 28, 2012.

Cheuvront, Elmer. "Tioga County Rattlesnake-Roundup." *News*, Lycoming County, PA, June 4, 1966.

Cheuvront, Elmer. "Notes from Nature: At DEATH'S Door! What It Feels Like to be Bitten by a Rattler!" *Aliquippa News*, Aliquippa, PA, August 15, 1963.

"Club Plans Contests." *Johnstown Tribune-Democrat*, Johnstown, PA, August 22, 1981.

Conway, William. "Leave Rattlesnake Alone, They Eat Ticks." *Hartford Courant*, Hartford, CT, December 3, 2013.

Dagle, Shawn R. "Hearing Set for Firing Range." *Glastonbury Citizen*, Glastonbury, CT, May 2, 2013.

Dillon, Karen. "Scaling a Tire Graveyard for the Sake of Snakes: Rod Wittenberg Works to Shed Misunderstandings about Missouri's Largest Venomous Creature." *Kansas City Star*, July 3, 2006.

Ditmars, Raymond L. "Raymond L. Ditmars Urges Suppression of Immoral 'Ragtime.'" *New York Times*, October 13, 1912.

Drakula, Dave. "Rattlesnake Hunting May Be Educational—and Harmful." *Pittsburgh Post-Gazette*, Pittsburgh, PA, July 24, 1981.

Dritschilo, Gordon. "Volunteers Take Humane Rattler Removal Seriously." *Rutland Herald*, Rutland, VT, July 17, 2006.

Dvorchak, Bob. "Snake Sackers Slither around State Ban." *Cortland Standard*, Cortland, PA, September 12, 1985.

"East Moriches Snake Found by Brooklyn Principal." *Brooklyn Daily Eagle*, September 4, 1921.

"Editorial: Don't Tread on Our Rattlers." *Glastonbury Citizen*, Glastonbury, CT, January 23, 2014.

Ecker, Don. "Law Enforcement Agents Put Bite on Snake Poacher." *Record*, Bergen County, NJ, June 2, 1993.

Esch, Mary. "Snakes, Too, Have Their Defenders." *Valley News*, September 4, 1995.

Ek, Derek. "Study to Help Threatened Species." *Courier*, September 9, 2007.

"Evolution of Aversion: Why Even Children Are Fearful of Snakes." *Science Daily*, February 28, 2008.

"Experiment with Snakes." *Corrector*, Sag Harbor, NY, August 20, 1856.

"Fate of Timber Rattlesnake a Cautionary Tale." *Evening Guide*, Port Hope, Ontario, May 19, 2001.

Fernandez, Marry. "Rattlesnake Wranglers, Armed with Gasoline." *New York Times*, March 30, 2014.

"Fossil Found of Earliest Known Snakes: Discovery Suggests Snakes Evolved on Land, Not Sea." *Associated Press*, April 19, 2006.

Frye, Bob. "Running after Rattlers." *Pittsburg Tribune*, December 3, 2006.

Groshong, Kimm. "Oldest Snake Fossil Shows a Bit of Leg." *New Scientist*, April 19, 2006.

Haggerty, Jen. "An Exhibit with 'Bite.'" *New Britain Herald*, January 15, 2006.

Harwig, Stephen H. "Our Readers' Opinions: Snakes Have Defender." *Valley Daily News*, Pittsburgh, PA, October 6, 1969.

"Head of Medusa." *Corrector*, Sag Harbor, NY, August 20, 1856.

"Hidden Treasure in Museum Revealed by Eagle is Extinct Rattlesnake Formerly Plentiful over Long Island." *Brooklyn Eagle*, December 18, 1927.

Holland, Carolyn. "Snake Expert Credits Mountain Upbringing." *Tribune-Review*, August 13, 2006.

"How Snakes Survive Starvation." *Science Daily*, August 27, 2007.

"How to Cure a Rattlesnake Bite." *Corrector*, Sag Harbor, NY, August 20, 1856.

"Hunter Tells How Rattlesnake Was Shot New South Side Club in 1880." *Suffolk County News*, Sayville, NY, April 26, 1935.

"I-86 Rest Area Closed Due to Rattlesnakes." *Associated Press*, July 21, 2006.

Isbell, Lynn A. "Snakes on the Brain." *New York Times*, September 3, 2006.

"It's Snakes 4, Sackers 0 in Bolivar Bagging Event!" *Ligonier Echo*, Ligonier, PA, August 8, 1982.

Jarvis, Peter. "Letters: Just Say No to Rattlesnakes." *Hartford Courant*, Hartford, CT, January 11, 2014.

"Jersey Man Charged in Rattlesnake Attack." *New York Times*, August 2, 1987.

Keller, Thomas W. "Readers' Forum: Supports Rattlesnake Hunts." *Johnstown Tribune Democrat*, Johnstown, PA, July 9, 1991.

"Largest Recorded Rattler Killer at Ludlow." *Forest Press*, Tionesta, PA, August 15, 1968.

Lalo, Julie. "Door to Compromise on Rattlesnake Hunts Closed." *Harrisburg Sunday Patriot*, November 10, 1991.

Levy, Michael. "Survey Says Americans Love Wildlife 108 million Spent $59 Billion Outside during 1991." *Buffalo News*, Buffalo, NY, June 1, 1993.

Loeffler, William. "'Snakes' Alive." *Pittsburgh Tribune-Review*, August 17, 2006.

Loller, Travis. "Death Doesn't Shake Snake Belief." *Valley News* (Associated Press), Lebanon, NH, February 27, 2014.

"Louisiana Drivers Approach Snakes with Tread." *Associated Press*, February 6, 1987.

Marteka, Peter. "Group Sues to Block Project in Snake Habitat." *Hartford Courant*, June 27, 2003.

Marteka, Peter. "Protecting Creatures of Habitat." *Hartford Courant*, May 8. 2000.

Marteka, Peter. "Saving the Timber Rattlesnake: Glastonbury Buying 55-Acre Parcel Home to Dwindling Species." *Hartford Currant*, October 19, 2010.

McFarland, Kent. "When Snakes Hit the Road." *Valley News*, Lebanon, NH, September 5, 2011.

McNall, Carole. "Pennsylvanians Set Snake Hunts." *Olean Times Spring Outdoor Edition*, Olean, NY, May 24, 1985.

McNall, Carole. "Snake Sacking Contest Legality: Slippery Issue." *Olean Times Spring Outdoor Edition*, Olean, NY, May 24, 1985.

McNeil, Donald G., Jr. "Snakes Serve as a Safe Winter Home for Deadly Mosquito-Borne Virus." *New York Times*, October 23, 2012.

Milner, Nathan. "Look for Fewer Rattlers." *Wyoming County Press Examiner*, June 13, 2007.

"Minister Dies after Snake Bite." *Valley News Dispatch*, Tarentum, PA, August 25, 1982.

"Nest of Rattlers near Whitehall Unearthed." *Rutland Daily Herald*, June 24, 1930.

Osborne, Edwin A. "Birds and Animals of Long Island." *Brooklyn Daily Eagle*, April 22, 1928.

Osborne, Edwin A. "Rattlesnake Has Brains or, At Least, Appears to Have, Long Island Expert Believes." *Brooklyn Daily Eagle*, December 16, 1923.

Osgood, George. "Handlers Can't Be Rattled: Longtime Pit Boss to Return after Recovering from Snakebite." *Star Gazette*, June 7, 2007.

"Our Reader's Opinions: Kill Snakes, not Birds." *Valley Daily News*, Pittsburgh, PA, October 6, 1969.

Page, Candace. "For Rattlesnakes in Vermont, We're the Danger." *Burlington Free Press*, June 19, 2011.

"Pennsylvania Moves to Restrict Snake Hunts." *Associated Press*, July 31, 2006.

Pfeiffer, Bryan. "Once Hunted for Bounty, Vt. Rattlers Now Protected." *Times Argus*, May 15, 1990.

Polk, Nancy. "Snakes, A Tough Summer." *Hartford Courant*, August 22, 1999.

Price, Megand. "A Vermont Viper: When It Rattles—Run!" *Rutland Daily Herald*, August 9, 1979.

"Race for Snake Serum Saves S.I. Zoo Worker." *New York Times*, September 7, 1961.

"A Rattle and a Strike, but No Venom." *Maine Times*, August 6, 1998.

"Rattler's Bite Kills Friend of Snakes: C. E. Snyder, Ex-Keeper in the Bronx Zoo, Dies 24 Hours after Accident in Ramapo Hunt." *New York Times*. May 14, 1929.

"Rattlers Still Found Near New York City." *New York Times*, June 2, 1929.

"Rattlesnake at Hooksett." *Telegraph*, July 7, 1933.

"Rattlesnake Ben: And Quaint Stories of Rattlesnake Mountain in Raymond, ME." *Weekly Journal*. Lewiston, ME, May 21, 1891.

"The Rattlesnake Club Makes a Good Haul." *New York Times*, May 21, 1906.

"Rattlesnake Killed in Bayport Yard." *Suffolk County News*, Sayville, NY, May 18, 1934.

"Rattlesnake Oil." *New York Times*, March 22, 1885.

"Rattlesnake Sightings 'Rare, but Reliable.'" *Chronicle*, August 3, 1983.

"A Rattlesnake's Bite: Death from the Poison at the End of Three Months." *New York Times*, February 10, 1887.

Reitz, Stephanie. "Suburbs Invade Rattler Territory." *Hartford Courant*, May 11, 1998.

"Remarkable Discovery." *Poultney Journal*, April 28, 1899.

Revkin, Andrew C. "In a House of Reptiles, Sit Cautiously." *New York Times*, August 24, 1995.

"Robert William's Badly Hurt by Car of Howard Gillette, Bayport." *Suffolk County News*, Sayville, NY, September 26, 1944.

Roddy, Dennis B. "Bolivar Hosts Snake-Sacking." *Greensburg Tribune-Review*, Greensburg, PA, August 22, 1981.

Roop, Lee. "Hunt Paved Way for Future Dreamers." *Huntsville Times*, April 24, 2005.

Rothman, A. "David Shepherd Sure Knows How to Take the Fun Out of Driving." *Wall Street Journal*, February 17, 1987.

"Saw Big Rattlesnake." *Brooklyn Daily Eagle*, May 24, 1910.

Schneck, Marcus. "Commission Gives More Help to Herps." *Patriot-News,*
August 6, 2006.

Schwab, Rick. "Generally Speaking." *Ligonier Echo,* August 3, 1983.

Schwab, Rick. "Generally Speaking." *Ligonier Echo,* Ligonier, PA, November 26,
1983.

"Scientists Are Aiding Timber Rattlesnakes." *Science Daily,* July 26, 2007.

Sellers, Laurin. "Rattlesnake Victim Speaks Out to Warn Wal-Mart Shoppers."
Volusia County News, February 22, 2007.

Shabecoff, Philip. "Snake Steak and Gossip Lead to Aide's Dismissal." *New York
Times,* October 13, 1979.

"Snake Death Ruled Suicide." *Bakersfield Californian,* Bakersfield, July 26, 1955.

"Snake Hunt Unsuccessful." *Valley New Dispatch,* Johnstown, PA, June 15, 1982.

Stacey, Donald. "Letter to the Editor." *Long Islander,* Huntington, NY, November 23, 1972.

Stairs, Kenneth J. "A Differing View on Rattlesnake Hunts." *Johnstown Tribune
Democrat,* Johnstown, PA, July 9, 1991.

"Story of Rattlesnake on Long Island." *Long Islander,* Huntington, NY, March 22,
1935.

Taylor, Dave. "ISU Research Featured on National Geographic TV." *ISU News,*
February 28, 2007.

Thompson, Tim. "Sulphur Powder Helps Keep Rattlesnakes Away." *Gainesville
Times,* August 20, 2006.

True, Morgan. "Lyme Disease Proposal Makes Progress in VT." *Valley News,*
Lebanon, NH, February 27, 2014.

Tucker, Abigail. "Siding with the Snakes." *Post-Star,* Glens Falls, NY, September 14, 2003.

"2 Accused of Using Snakes in Death Plot." *New York Times,* February 28,
1987.

"2 Jersey Men Ordered to Jail in Snake Case." *New York Times,* May 29, 1988.

Untitled. *Brooklyn Daily Eagle,* August 8, 1902.

Untitled. *Corrector,* Sag Harbor, NY, January 21, 1859.

Untitled. *Islip Bulletin,* Islip, NY, August 2, 1962.

Untitled. *Long Island Traveler,* Southold, NY, August 2, 1895.

Untitled. *Port Jefferson Echo,* Port Jefferson, September 2, 1892.

Untitled. *Sag Harbor Express,* Sag Harbor, NY, May 30, 1889.

Untitled. *South Side Signal,* Babylon, NY, July 31, 1901.

Untitled. *South Side Signal,* Babylon, NY, September 9, 1882.

Untitled. *Southside Signal,* Babylon, NY, July 22, 1882.

Untitled. *Suffolk County News,* Sayville, NY, September 2, 1898.

Verhovek, Sam Howe. "State, Den Mother to 5, Takes the Snakes Home." *New
York Times,* July 30.

Walz, Trudy. "'Cobra King' Preaches Heresy?" *Record*, Bergen County, NJ, November 15, 1994.

Watson, Wyndle. "Snake Hunters May Miss at Rules." *Pittsburgh Press*, July 28, 1991.

Watson, Wyndle. "State Tables Proposal on Rattlesnake Hunts." *Pittsburgh Press*, August 1, 1991.

Watson, Wyndle. "Trailblazer: Fish Commission's First Woman Wins over Opponents." *Pittsburgh Press*, February 10, 1985.

"Where to Go for Rattlesnakes: There Are Lots of Them in Western Massachusetts." *New York Times*, August 17, 1895.

Wilson, Larry. "Rattlesnakes Close Rest: Female Timber Rattlers Seek Sunny Spot Off I-86 in Painted Post." *Star-Gazette*, July 20, 2006.

"Yes, Virginia, Rattlers Exist in Vermont." *Fair Haven Gazette*, July 26, 1972.

Zauser, Robert. "Man Find Happiness Hunting for Snakes." *Latrobe Bulletin*, Latrobe, PA, August 14, 1982.

Zimmer, Carl. "Afraid of Snakes? Your Pulvinar May Be to Blame." *New York Times*, October 31, 2013.

Journal Articles, Book Chapters, Papers, and Reports

Agugliaro, Joseph, and Howard K. Reinert. "Comparative Skin Permeability of Neonatal and Adult Timber Rattlesnakes (*Crotalus horridus*)." *Comparative Biochemistry and Physiology* 141, no. 1 (2005).

Ahrenfeldt, R. H. "Two British Anatomical Studies on American Reptiles (1650–1750) II.—Edward Tyson: Comparative Anatomy of the Timber Rattlesnake." *Herpetologica* 11, no. 1 (1955): 49–69.

Aldridge, Robert D., and William S. Brown. "Male Reproductive Cycle, Age at Maturity, and Cost of Reproduction in the Tiber Rattlesnake (*Crotalus horridus*)." *Journal of Herpetology* 29, no. 3 (1995).

Aldridge, Robert D., and David Duvall. "Evolution of the Mating Season in the Pitvipers of North America." *Herpetological Monographs* 16 (2002): 1–25.

Allsteadrt, John, Alan H. Savitzky, Christopher E. Petersen, and Dayanand N. Naik. "Geographic Variation in the Morphology of *Crotalus horridus* (Serpentes: Viperidae)." *Herpetological Monographs* 20, no. 1 (2006): 1–63.

Amato, Christopher A., and Robert Rosenthal. "Endangered Species Protection in New York after State v. Sour Mountain Realty, Inc." *New York University Environmental Law Journal* 10 (2002): 117–145.

Anderson, Corey Devin. "Effects of Movement and Mating Patterns on Gene Flow among Overwintering Hibernacula of the Timber Rattlesnake (*Crotalus horridus*)." *Copeia*, no. 1 (2010): 54–61.

Andrews, Kimberley, and J. Whitfield Gibbons. "How Do Highways Influence Snake Movement? Behavioral Responses to Roads and Vehicles." *Copeia*, no. 4 (2005): 772–782.

Andrews, Kimberly M., J. Whitfield Gibbons, and Denim M. Jochimsen. "Ecological Effects of Roads on Amphibians and Reptiles: A Literature Review." In *Urban Herpetology*, edited by J. C. Mitchell, R. E. Jung Brown, and B. Bartholomew. *Herpetological Conservation* 3 (2008): 121–143.

Anonymous. *Natural Heritage & Endangered Species Program*. Commonwealth of Massachusetts, Division of Fisheries and Wildlife (2007).

Anonymous. "Timber Rattlesnake." *NYNHP Conservation Guide-Timber Rattlesnake (Crotalus horridus)*. New York State Division of Environmental Conservation (2007).

Ashton, Kyle G., and Chris R. Feldman. "Bergmann's Rule in Nonavian Reptiles: Turtles Follow It, Lizards and Snakes Reverse It." *Evolution* 57, no. 5 (2003).

Audubon, John James. "Notes on the Rattlesnake (*Crotalus horridus*)." *Edinburgh New Philosophical Society Journal* 3 (1827): 21–30.

Baarslag, Anton. "The Pilot Blacksnake and the Timber Rattlesnake in Vermont." *Copeia*, no. 4 (1950).

Babcock, Harold L. "Early and Recent Massachusetts Records of *Crotalus horridus* Linne." *Bulletin: Boston Society of Natural History* 68 (1933): 3–4.

Babcock, Harold L. "Food Habits of the Timber Rattlesnake." *Bulletin: Boston Society of Natural History* (1929).

Babcock, Harold L. "Notes on the Colonial Treatment of Rattlesnake Bites during Colonial Times in Massachusetts." *Bulletin of the Antivenin Institute of America* 2, no. 3 (1928): 77–78.

Babcock, Harold L. "Rattlesnakes in Massachusetts." *Bulletin of Boston Society of Natural History* 35 (1925): 5–10.

Bates, Mark C., Gregory P. Wedin, Stafford G. Warren, and David E. Seidler. "Case Report, Treatment Protocol: Approach to the Patient with Severe Rattlesnake Envenomation." *West Virginia Medical Journal* 84 (1988): 398–403.

Bauder, Javan M., Doug Blodgett, Kiley V. Briggs, and Christopher L. Jenkins. "The Ecology of Timber Rattlesnakes (*Crotalus horridus*) in Vermont: A First Year Progress Report Submitted to the Vermont Department of Fish and Wildlife." The Orianne Society, South Clayton, GA (2011).

Berish, Joan E. Diemer. "Characterization of Rattlesnake Harvest in Florida." *Journal of Herpetology* 32, no. 4 (1998).

Billings, Marland P. "Ordovician Cauldron Subsidence of the Blue Hills Complex, Eastern Massachusetts." *Geological Society of America Bulletin* 93 (September 1982).

Bonnet, Xavier, Guy Naulieau, and Richard Shine. "The Dangers of Leaving Home: Dispersal and Mortality in Snakes." *Biological Conservation* 89, no. 1 (1999).

Breen, John F. "Rattlesnakes in Tiverton." *Narragansett Naturalist* 6 (1963): 59–60.

Breen, John F. "Rhode Island's Declining Rattlers." *Narragansett Naturalist* 13, no. 3 (1970): 36–38.

Brick, J. F., L. Gutmann, J. Brick, K. N. Apelgren, and J. E. Riggs. "Timber Rattlesnake Venom-Induced Myokymia: Evidence for Peripheral Nerve Origin." *Neurology* 37 (1987): 1545–1546.

Brown, William S. "Biology and Conservation of the Timber Rattlesnake." In *Conservation of the Timber Rattlesnake in the Northeast*, edited by T. Tyning. Lincoln, MA: Massachusetts Audubon Society, 1992.

Brown, William S. "Emergence, Ingress, and Seasonal Captures at Dens of Northern Timber Rattlesnakes, *Crotalus horridus*." In *Biology of Pitvipers*, edited by J. A. Campbell and E. D. Brodie, Jr. Tyler, TX: Selva Press, 1992.

Brown, William S. "Female Reproductive Ecology in a Northern Population of the Timber Rattlesnake, *Crotalus horridus*." *Herpetologica* 47, no. 1 (1991).

Brown, William S. "Overwintering Body Temperatures of Timber Rattlesnakes (*Crotalus horridus*) in Northeastern New York." *Journal of Herpetology* 16, no. 2 (1982).

Brown, William S., and David B. Greenberg. "Vertical-Tree Ambush Posture in *Crotalus horridus*." *Herpetological Review* 23 (1992).

Brown, William S., Marc Kery, and James E. Hines. "Survival of Timber Rattlesnakes (*Crotalus horridus*) Estimated by Capture-Recapture Models in Relation to Age, Sex, Color Morph, Time, and Birthplace." *Copeia*, no. 3 (2007): 656–671.

Brown, William S., and Frances M. MacLean. "Conspecific Scent-Trailing by Newborn Timber Rattlesnakes, *Crotalus horridus*." *Herpetologica* 39, no. 4 (1983).

Brown, William S., and W. S. Parker. "A Vertical Scale Clipping System for Permanently Marking Snakes (Reptilia, Serpentes)." *Journal of Herpetology* 10, no. 3 (1976).

Brown, William S., Donald W. Pyle, Kimberley R. Green, and Jeffrey B. Friedlaender. "Movements and Temperature Relationships of Timber Rattlesnakes (*Crotalus horridus*)." *Journal of Herpetology* 16, no. 2 (1982).

Bushar, Lauretta M., Howard K. Reinert, and Larry Gelbert. "Genetic Variation and Gene Flow within and between Populations of the Timber Rattlesnake, *Crotalus horridus*." *Copeia*, no. 2 (1998): 411–422.

Capel, Rodrigo O., Fabiola Z. Monica, Marcovan Porto, Samuel Barillas, Marcelo N. Muscara, Simone Teixeira, Andre Arruda, Lorenzo Pissinatti,

Alcides Pissinatti, Andre A. Schenka, Edson Antunes, Cesar Nahoum, Jose C. Cogo, Marco A. de Oliverira, and Gilberto De Nucci. "Role of a Novel Tetrodotoxin-Resistant Sodium Channel in the Nitrergic Relaxation of Corpus Cavernosum from the South American Rattlesnake Crotalus Durissus Terrificus." *International Society for Sexual Medicine* (2011).

Carle, D. H. "Geographic Distribution of Snakes in New Hampshire." *Proceedings of the New Hampshire Academy of Science* 2, no. 1 (1951).

Carle, D. H. "Rattle Snakes on Wantastiquet Mountain." *Proceedings of the New Hampshire Academy of Science* 1, no. 8 (1948).

Chantell, Charles J. "*Crotalus horridus* Remains from Two Caves in Miami Co., Ohio." *Ohio Journal of Science* 70, no. 2 (1970).

Chiszar, David, Robert K. K. Lee, Hobart M. Smith, and Charles W. Radcliffe. "Searching Behaviors of Rattlesnakes following Predatory Strikes." In *Biology of Pitvipers*, edited by J. A. Campbell and E. D. Brodie, Jr. Tyler, TX: Selva Press, 1992.

Clark, A. M., P. E. Moler, E. E. Possardt, A. H. Savitzky, W. S. Brown, and B. W. Bowen. "Phylogeography of the Timber Rattlesnake (*Crotalus horridus*) Based on mtDNA Sequences." *Journal of Herpetology* 37, no. 1 (2003).

Clark, Rulon W. "Diet of the Timber Rattlesnake, *Crotalus horridus*." *Journal of Herpetology* 36, no. 3 (2002).

Clark, Rulon W. "Feeding Experience Modifies the Assessment of Ambush Sites by the Timber Rattlesnake, a Sit-and-Wait Predator." *Ethology* 110 (2004): 471–483.

Clark, Rulon W. "Timber Rattlesnakes (*Crotalus horridus*) Use Chemical Cues to Select Ambush Sites." *Journal of Chemical Ecology* 30, no. 3 (2004).

Clark, Rulon W. "Fixed Videography to Study Predation Behavior of an Ambush Foraging Snake, *Crotalus horridus*." *Copeia*, no. 2 (2006).

Clark, Rulon W. "Kin Recognition in Rattlesnakes." *Royal Society: Biology Letters* (Suppl.) 271, published online February 13, 2004.

Clark, Rulon W. "Post-Strike Behavior of Timber Rattlesnakes (*Crotalus horridus*) during Natural Predation Events." *Ethology* 112 (2006).

Clark, Rulon W. "Public Information for Solitary Foragers: Timber Rattlesnakes Use Conspecific Chemical Clues to Select Ambush Sites." *Behavioral Ecology* 18, no. 2 (2007).

Clark, Rulon W. "Pursuit-Deterrent Communication between Prey Animals and Timber Rattlesnakes (*Crotalus horridus*): The Response of Snakes to Harassment Displays." *Behavioral Ecological Sociobiology* 59 (2005).

Clark, Rulon, William S. Brown, Randy Stechert, and Harry W. Green. "Cryptic Sociability in Rattlesnakes (*Crotalus horridus*) Detected by Kinship Analysis." *Biology Letters* 8, no. 4 (2012): 523–525.

Clark, Rulon W., W. S. Brown, R. Stechert, and K. R. Zamudio. "Integrating

Individual Behavior and Landscape Genetics: The Population Structure of Timber Rattlesnake Hibernacula." *Molecular Ecology* 17, no. 3 (2008).

Clark, Rulon W., William S. Brown, Randy Stechert, and Kelly R. Zamudio. "Roads, Interrupted Dispersal, and Genetic Diversity in Timber Rattlesnakes." *Conservation Biology* 24, no. 4 (2010).

Clark, R. W., Michael N. Marchand, Brendan J. Clifford, Randy Stechert, and Sierra Stephens. "Decline of an Isolated Timber Rattlesnake (*Crotalus horridus*) Population: Interactions between Climate Change, Disease, and Loss of Genetic Diversity. *Biol. Conservation* (2011).

Clark, Rulon W., Sean Tangco, and Matthew A. Barbour. "Field Video Recordings Reveal Factors Influencing Predatory Strike Success of Free-Ranging Rattlesnakes (*Crotalus* spp.)." *Animal Behavior* 84 (2012): 183–190.

Cobb, Vincent A., Jeffery Green, Timothy Worrall, Jake Pruett, and Brad Glorioso. "Initial Den Location Behavior in a Litter of Neonate *Crotalus horridus* (Timber Rattlesnakes)." *Southeastern Naturalist* 4, no. 4 (2005).

Craig, Rudolph D., R. R. Schaefer, D. Saenz, and R. N. Conner. "Arboreal Behavior in the Timber Rattlesnake, *Crotalus horridus*, in Eastern Texas." *Texas Journal of Science*, November 1 (2004).

Dale, Ronald S., and Joseph R. Wielgus. "New Foodplant and Distribution Records for *Megathymus ursus* (Megathymidae)." *Bulletin of the Allyn Museum of Entomology*, Sarasota, FL, 12 (1973).

Duvall, David, Steven J. Arnold, and Gordon W. Schuett. "Pitviper Mating Systems: Ecological Potential, Sexual Selection and Microevolution, in *Biology of Pitvipers*," edited by J. A. Campbell and E. D. Brodie, Jr. Tyler, TX: Selva Press, 1992.

Ebert, J., and G. Westoff. "Behavioural Examination of the Infrared Sensitivity of Rattlesnakes (*Crotalus atrox*)." *Journal of Comparative Physiology* (2006).

Engelhardt, G. P., J. T. Nichols, Roy Latham, and R. C. Murphy. "Long Island Snakes." *Copeia*, no. 17 (1915): 1–5.

Fenton, M. Brock, and Lawrence E. Licht. "Why Rattle Snake?" *Journal of Herpetology* 24, no. 3 (1990).

Fitch, Henry F., and George R. Pisani. "Disappearance of Radio-Monitored Timber Rattlesnakes." *Journal of Kansas Herpetology* 14 (June 2005).

Fitch, Henry F., and George R. Pisani. "Longtime Recapture of a Timber Rattlesnake (*Crotalus horridus*) in Kansas." *Journal of Kansas Herpetology* 3 (September 2002).

Fitch, Henry F., and George R. Pisani. "Rapid Early Growth in Northeastern Kansas Timber Rattlesnake." *Journal of Kansas Herpetology* 20 (December 2006).

Fitch, Henry F., and George R. Pisani. "The Timber Rattlesnake in Northeastern Kansas." *Journal of Kansas Herpetology* 19 (September 2006).

Fitch, Henry F., George R. Pisani, Harry W. Greene, Alice F. Echelle, and Michael Zerwekh. "A Field Study of the Timber Rattlesnake in Leavenworth County, Kansas." *Journal of Kansas Herpetology* 11 (September 2004).

Fry, B. G., and W. Wüster. "Assembling an Arsenal: Origin and Evolution of the Snake Venom Proteome Inferred from Phylogenetic Analysis of Toxin Sequences." *Molecular Biology and Evolution* 21, no. 5 (2004).

Fry, Bryan G., et al. 2006. "Early Evolution of the Venom System in Lizards and Snakes." *Nature: Letters* 439 (February 2, 2004).

Galligan, John H., and William A. Dunson. "Biology and Status of Timber Rattlesnake (*Crotalus horridus*) Populations in Pennsylvania." *Biological Conservation* 15 (1979).

Gardner-Santana, Lynne C., and Steven J. Beaupre. "Timber Rattlesnakes (*Crotalus horridus*) Exhibit Elevated and Less Variable Body Temperatures during Pregnancy." *Copeia*, no. 2 (2009): 363–368.

George, Steven G. "Timber Rattlesnake (*Crotalus horridus*) Swims the Mississippi River." *IRCF Reptiles & Amphibians* 17, no. 1 (2010).

Gibbons, J. Whitefield. "Reproduction, Growth, and Sexual Dimorphism in the Canebrake Rattlesnake (*Crotalus horridus atricaudatus*)." *Copeia*, no. 2 (1972): 222–226.

Glenn, J. L., R. C. Straight, and T. B. Wolt. "Regional Variation in the Presence of Canebrake Toxin in *C. horridus* Venom." *Journal of Comparative Biochemistry and Physiology* 107C (1994): 337–346.

Gold, Barry S., MD, Richard C. Dart, MD, PhD, and Robert A. Barish, MD. "Bites of Venomous Snakes." *New England Journal of Medicine* 347 (2003): 5.

Greene, W. Harry. "The Ecological and Behavioral Context for Pitviper Evolution." In *Biology of Pitvipers*, edited by J. A. Campbell and E. D. Brodie, Jr. Tyler, TX: Selva Press, 1992.

Goris, Richard C. "Infrared Organs of Snakes: An Integral Part of Vision." *Journal of Herpetology* 45 (2011): 1, 2–14.

Guthrie, J. E. "Snakes versus Birds; Birds versus Snakes." *Wilson Bulletin*, June (1932).

Hanlin, Hugh G., Joseph J. Beatty, and Sue W. Hanlin. "An Overlooked Description of a United States Snake from the Eighteen Century Writing of Luigi Castigioni." *Journal of Herpetology* 13, no. 2 (1979).

Hinze, John David, James A. Baker, T. Russell Jones, and Richard E. Winn. "Life-Threatening Upper Airway Edema Caused by a Distal Rattlesnake Bite." *Annals of Emergency Medicine* 38, no. 1 (2001).

Holman, J. Alan. "Paleoclimatic Implications of 'Ecologically Incompatible' Herpetological Species (Late Pleistocene: Southeastern United States)." *Herpetologica* 32, no. 3 (1976).

Hutchinson, Alan, and Malcolm L. Hunter. "The Status of the Timber Rattlesnake in Maine." In *Conservation of the Timber Rattlesnake in the Northeast*, edited by T. Tyning. Lincoln, MA: Massachusetts Audubon Society, 1992.

Johnson, Ralph Gordon. "The Origin and Evolution of the Venomous Snakes." *Evolution* 10 (1955): 56–65.

Kardong, Kenneth V. "The Predatory Strike of the Rattlesnake: When Things Go Amiss." *Copeia*, no. 3 (1986): 816–882.

Kawata, Ken. "Carl and His Rattlesnakes: Herpetology at the Staten Island Zoo." *Herpetological Review* 35, no. 4 (2004).

Keegan, Hugh L., and Ted F. Andrews. "Effects of Venom on North American Snakes." *Copeia*, no. 4 (December 1942).

Keenlyne, K. D. "Sexual Differences in Feeding Habits of *Crotalus horridus horridus*." *Journal of Herpetology* 6, no. 3–4 (1972): 234–237.

Kerns, William II, and Christian Tomaszewski. "Airway Obstruction following Canebrake Rattlesnake Envenomation." *Journal of Emergency Medicine* 20, no. 4, (2001).

Keyler, Daniel E. "Famous Envenomations: Unusual Circumstances and Untimely Deaths of Four illustrious American Herpetologists." *Toxicological History Society Newsletter: Mithridata* 18, no. 1 (2008).

Keyler, Daniel E. "Timber Rattlesnake (*Crotalus horridus*) Envenomations in the Upper Mississippi River Valley." In *Biology of Rattlesnakes*, edited by W. K. Hayes, K. R. Beaman, M. D. Cardwell, and S. P. Bush. Loma Linda, CA: Loma Linda University Press, 2008.

Keyler, Daniel E. "Venomous Snake Bites and Remedies over the Millennia." *Toxicological Society Newsletter: Mithridata* 11, no. 2 (2001).

Keyler, Daniel E., and Kimberly Fuller. "Survey of Timber Rattlesnake (*Crotalus horridus*) Peripheral Range on Southern Minnesota State." Minnesota Department of Natural Resources Nongame Wildlife Program, Minneapolis (1999).

Kitchens, Craig S., Stephen Hunter, and Lodewyk H. S. Van Mierop. "Severe Myonecrosis in a Fatal Case of Envenomation by the Canebrake Rattlesnake (*Crotalus horridus atricaudatus*)." *Toxicon* 25, no. 4 (1987).

Klauber, L. M. "Review of 'The Rattlesnakes, Genera *Sistrurus* and *Crotalus*: A Study in Zoogeography and Evolution' by Howard K. Gloyd." *Copeia*, no. 3 (1940): 206–207.

Krochmal, Aaron R., and George S. Bakken. "Thermoregulation Is the Pits: Use of Thermal Radiation for Retreat Site Selection by Rattlesnakes." *Journal of Experimental Biology* 206 (2003): 2539–2545.

Krochmal, Aaron R., George S. Bakken, and Travis J. LaDuc. "Heat in Evolution's Kitchen: Evolutionary Perspectives on the Functions and Origin of the Facial Pit of Pit Vipers (Viperidae: Crotalinae)." *Journal of Experimental Biology* 207 (2004): 4231–4238.

Langley, William M., Hank W. Lipps, and John F. Theis. "Responses of Kansas Motorists to Snake Models on a Rural Highway." *Transactions of the Kansas Academy of Science* 92, no. 1–2 (1989): 43–48.

Lavonas, Eric, Anne-Michele Ruha, Willaim Banner, Vikhyat Bebarta, Jeffery N. Bernstein, Sean P. Bush, William P. Kerns II, William H. Richardson, Steven A. Seifert, David A. Tanen, and Steve C. Curry. "Unified Treatment Algorithm for the Management of Crotaline Snakebite in the United States: Results of an Evidence-Informed Consensus Workshop." *BMC Emergency Medicine* 11 (2011): 2.

Lewis, James V., and Charles A. Portera, Jr. "Rattlesnake Bite of the Face: Case Report and Review of the Literature." *American Surgeon* 60 (1994): 681–682.

Lillywhite, Harvey B., and Alan Smits. "The Cardiovascular Adaptations of Viperids Snakes." In *Biology of Pitvipers*, edited by J. A. Campbell and E. D. Brodie, Jr. Tyler, TX: Selva Press, 1992.

Madsen, Thomas, Beata Ujvari, and Mats Olsson. "Novel Genes Continue to Enhance Population Growth in Adders (*Vipera berus*)." *Biological Conservation* 120 (2004): 145–147.

Martin, William Henry. "Distribution and Habitat Relationship of the Eastern Diamondback Rattlesnake (*Crotalus adamanteus*)." *Herpetological Natural History* 7, no. 1 (2000).

Martin, William Henry. "Life History Constraints on the Timber Rattlesnake (*Crotalus horridus*) at Its Climatic Limits." In *Biology of Vipers*, edited by G. W. Schuetts, M. Hoggren, M. E. Douglas, and W. H. Greene. Eagle Mountain, UT: Eagle Mountain Publishing, 2002.

Martin, William Henry. "Reproduction of the Timber Rattlesnake (*Crotalus horridus*) in the Appalachian Mountains." *Journal of Herpetology* 27, no. 2 (1993).

Martin, William Henry. "Phenology of the Timber Rattlesnake (*Crotalus horridus*) in an Unglaciated Section of the Appalachian Mountains." In *Biology of Pitvipers*, edited by J. A. Campbell and E. D. Brodie, Jr. Tyler, TX: Selva Press, 1992.

Martin, William Henry. "The Timber Rattlesnake in the Northeast: Its Range, Past and Present." *HERP: Bulletin of the New York Herpetological Society* 17, no. 2 (1982).

Martin, William Henry. "The Timber Rattlesnake: Its Distribution and Natural History." In *The Symposium on the Conservation of the Timber Rattlesnake*, edited by Tom Tyning. December 7, 1991.

Martin, William Henry, William S. Brown, Earl Possardt, and John B. Sealy. "Biological Variation, Management Units, and a Conservation Action Plan for the Timber Rattlesnake (*Crotalus horridus*)." In *The Biology of Rattlesnakes*, edited by W. K. Hayes, K. R Beaman, M. D. Cardwell, and S. P. Bush. Loma Linda, CA: Loma Linda University Press, 2008.

Medden, Rheua Vaughn. "Tales of the Rattlesnakes: From the Works of Early Travelers in America." *Bulletin of the Antivenin Institute of America* 3 (1930): 3.

Medden, Rheua Vaughn. "Tales of the Rattlesnakes: From the Works of Early Travelers in America." *Bulletin of the Antivenin Institute of America* 3 (1930): 4.

Medden, Rheua Vaughn. "Tales of the Rattlesnakes: From the Works of Early Travelers in America." *Bulletin of the Antivenin Institute of America* 4 (1931): 1.

Medden, Rheua Vaughn. "Tales of the Rattlesnakes: From the Works of Early Travelers in America." *Bulletin of the Antivenin Institute of America* 4 (1931): 2.

Medden, Rheua Vaughn. "Tales of the Rattlesnakes: From the Works of Early Travelers in America." *Bulletin of the Antivenin Institute of America* 4 (1931): 3.

Medden, Rheua Vaughn. "Tales of the Rattlesnakes: From the Works of Early Travelers in America." *Bulletin of the Antivenin Institute of America* 4 (1931): 4.

Medden, Rheua Vaughn. "Tales of the Rattlesnakes: From the Works of Early Travelers in America." *Bulletin of the Antivenin Institute of America* 5 (1932): 1.

Merrow, Jed S., and Todd Aubertin. "*Crotalus horridus* (Timber Rattlesnake) Reproduction." *Herpetological Review* 36, no. 2 (2005).

Minton, Sherman A., Jr. "The Feeding Strike of the Timber Rattlesnake." *Journal of Herpetology* 3, no. 3–4 (1969): 121–124.

Mohr, Jeffery R. "Radiotelemetry Study of an Oklahoma Population of Timber Rattlesnakes." Master's thesis, Furman University, 1999.

Moon, Brad R. "Muscle Physiology and the Evolution of the Rattling System in Rattlesnakes." *Journal of Herpetology* 35, no. 3 (2001).

Norton, Arthur H. "The Rattlesnake in Maine." *Maine Naturalist* 9, no. 1 (1929): 25–28.

Oliver, James A. "Amphibians and Reptiles of New Hampshire." In *Biological Survey of the Connecticut River Watershed*. New Hampshire Fish and Game Department, No. 4 (1938).

Otten, Edward J., and Douglas McKimm. "Venomous Snakebite in a Patient Allergic to Horse Serum." *Annals of Emergency Medicine* 12, no. 10 (1983).

Pisani, George R., Joseph P. Collins, and Stephan E. Edwards. "A Re-evaluation of the Subspecies of *Crotalus horridus*." *Transactions of the Kansas Academy of Science* 75, no. 3 (1973).

Place, Aaron J., and Charles I. Abramson. "A Quantitative Analysis of the Ancestral Area of Rattlesnakes." *Journal of Herpetology* 38, no. 1 (2004).

Preisser, Earl L., Daniel I. Bolnick, and Michael F. Bernard. "Scared to Death? The Effects of Intimidation and Consumption in Predator-Prey Interactions." *Ecology* 86 (2005): 501–509.

Reinert, Howard K., David Cundall, and Lauretta M. Bushar. "Foraging Behavior of the Timber Rattlesnake, *Crotalus horridus*." *Copeia*, no. 4 (1984): 976–981.

Reinert, Howard K., Gylia A. MacGregor, Esch Mackenzie, Lauretta M. Bushar, and Robert T. Zappalorti. "Foraging Ecology of Timber Rattlesnakes, *Crotalus horridus*." *Copeia*, no. 3 (2011): 430–442.

Reinert, Howard K., and Robert T. Zappalorti. "Timber Rattlesnakes (*Crotalus horridus*) of the Pine Barrens: Their Movement Patterns and Habitat Preference." *Copeia*, no. 4 (1988): 964–978.

Reinert, Howard K., and Robert T. Zappalorti. "Field Observation of the Association of Adult and Neonatal Timber Rattlesnakes, *Crotalus horridus*, with Possible Evidence for Conspecific Trailing." *Copeia*, no. 4 (1988): 1057–1059.

Rook, John T. "Habitat Protection/Preservation for Timber Rattlesnake Population of Central Connecticut." Master's project, Antioch/New England Graduate School, Keene, NH, October 1989.

Roth, Vincent D. "Ditmars Horned Lizard." *Sonoran Herpetologist*, Tucson Herpetological Society, January (1997).

Rowell, Jeffery. "Timber Rattlesnake (*Crotalus horridus*)." In *Snakes of Ontario: Natural History, Distribution, and Status*. Winnipeg, Manitoba: Art Bookbindry, 2013.

Rudolph, D. Craig, R. R. Schaefer, D. Saenz, and R. N. Conner. "Arboreal Behavior in the Timber Rattlesnake, *Crotalus horridus*, in Eastern Texas." *Texas Journal of Science* (2004).

Savitzky, Alan H. "Embryonic Development of the Maxillary and Prefrontal Bones of Crotaline Snakes." In *Biology of Pitvipers*, edited by J. A. Campbell and E. D. Brodie, Jr. Tyler, TX: Selva Press, 1992.

Schuett, Gordon W. "Is Long-Term Sperm Storage an Important Component of the Reproductive Biology of Temperate Pitvipers?" In *Biology of Pitvipers*, edited by J. A. Campbell and E. D. Brodie, Jr. Tyler, TX: Selva Press, 1992.

Sexton, Owen J., Peter Jacobson, and Judy E. Bramble. "Geographic Variation in Some Activities Associated with Hibernation in Neartic Pitvipers." In *Biology of Pitvipers*, edited by J. A. Campbell and E. D. Brodie, Jr. Tyler, TX: Selva Press, 1992.

Sewell, Henry. "Experiments on the Preventive Inoculations of Rattlesnake Venom." *Journal of Physiology* 8 (1887): 203–210.

Shine, Richard. "Perspective: Tracking Elusive Timber Rattlers with Molecular Genetics." *Molecular Ecology* 17 (2008): 715–718.

Smith, Ronald M. "Spatial Ecology of the Timber Rattlesnake (*Crotalus horridus*) and Northern Pine Snake (*Pituophis melanoleucus*) in the Pine Barrens of New Jersey." A thesis submitted to the faculty of Drexel University, 2013.

Stechert, Randy. "Historical Distribution of Timber Rattlesnake Colonies in New York State." *HERP: Bulletin of the New York Herpetological Society* 17, no. 2 (1982).

Steen, David A., Lora L. Smith, L. Mike Conner, Jean C. Brock, and Shannon K. Hoss. "Habitat Use of Sympatric Rattlesnake Species within the Gulf Coastal Plain." *Journal of Wildlife Management* 71, no. 3 (2005).

Straight, R. C., J. L. Glen, and M. C. Wolfe. "Regional Differences in Content of Small Basic Peptide Toxins in the Venoms of *C. adamanteus* and *C. horridus.*" *Journal of Comparative Biochemistry and Physiology* 100B (1991): 51–58.

Sutherland, Ian D. W. "The 'Combat Dance' of the Timber Rattlesnake." *Herpetologica* 14 (1958): 23–24.

Taylor, James, and Michael Marchand. "Species Profile: Timber Rattlesnake *Crotalus horridus.*" New Hampshire Wildlife Action Plan, Concord, NH, A-226-232 (2004).

Trapido, Harold. "Parturition in the Timber Rattlesnake, *Crotalus horridus horridus* Linne." *Copeia*, no. 4 (1939): 230.

Tyson, Edward. "*Vipera Caudi-Sona Americana*, or The Anatomy of a Rattle-Snake, Dissected at the Repository of the Royal Society in January 1682/3." *Philosophical Transactions* (1683–1775). 1753-01-01. 13 (1683): 25–46.

Van Frank, Richard. "Fossil Rattlesnakes of the Genus *Crotalus* from Northern Massachusetts." *Copeia*, no. 2 (1954): 158–159.

Vonk, Freek J., et al. "Evolutionary Origin and Development of Snake Fangs." *Nature: Letters* 20 (2008).

Waeckerle, Joseph F. "Kiss and Yell,' a Rattlesnake Bit to the Tongue." *Annals of Emergency Medicine* 17, no. 5 (1988).

Walker, Mindy L., Jennifer A. Dorr, Rebecca J. Benjamin, and George R. Pisani. "Successful Relocation of a Threatened Suburban Population of Timber Rattlesnakes (*Crotalus horridus*): Combining Snake Ecology, Politics, and Education." *Reptiles & Amphibians, Conservation and Natural History* 16, no. 4 (2009): 2–13.

Walker, Mindy L., Jennifer A. Dorr, and George R. Pisani. "Observation of Aberrant Growth in a Timber Rattlesnake (*Crotalus horridus*)." *Transactions of the Kansas Academy of Science* 111, no. 1/2 (2008).

Walker, Mindy L., Eric D. Kadlec, Ryan D. Miloshewski, and George Pisani. "Associative Behavior and Affinity for Anthropogenic Habitats in Two Relocated Timber Rattlesnakes." *IRCF Reptiles & Amphibians* 18, no. 4 (2011).

Weaver, Deborah A., Dennis R. Stroup, Beth Ann Slafka, Jane Kuzniewski, and Cathleen Hughes. "Case Review: Timber Rattlesnake Bite to the Hand with Secondary Coagulopathy and Serum Sickness." *Journal of Emergency Nursing* 17, no. 4 (1991).

Weinstein, Scott A. "Snake Venoms: A Brief Treatise on Etymology, Origins of Terminology, and Definitions." *Toxicon* 103 (2015): 188–95.

Weinstein, Scott A., Daniel E. Keyler, and Julian White. "Letter to the Editor: Replies to Fry et al. (*Toxicon* 60, no.4 [2012]: 434–48). Part A. Analysis of Squamate Reptile Oral Glands and Their Products: A Call for Caution in Formal Assignment of Terminology Designating Biological Function." *Toxicon* 60 (2012): 954–63.

Weldon, Paul J., Rudy Ortiz, and Thomas R. Sharp. "The Chemical Ecology of Crotaline Snakes." In *Biology of Pitvipers*, edited by J. A. Campbell and E. D. Brodie, Jr. Tyler, TX: Selva Press, 1992.

Williams, Stephan W., MD. "Rattlesnakes—*Crotalus horridus*." *Boston Medical and Surgical Journal* 37, no. 23 (1848).

Wills, Christina A., and Steven J. Beaupre. "An Application of Randomization for Detecting Evidence of Thermoregulation in Timber Rattlesnake (*Crotalus horridus*) from Northwest Arkansas." *Physiological and Biochemical Zoology* 73, no. 3 (2000).

Wilson, Jeffery A., Dhananjay M. Mohabey, Shanan E. Peters, and Jason J. Head. "Predation upon Hatchling Dinosaurs by a New Snake from the Late Cretaceous of India." *Library of Science Biology* 8, no. 3 (2010).

Yong, Ed. "Scared to Death: How Intimidation Changes Ecosystems." *New Scientist* 2919, June 04, 2013.

Internet

"Are Rattlesnakes Entering Suburbia?" *ScienceDaily*. Retrieved March 11, 2010, from http://www.sciencedaily.com/releases/2007/06/070619125639.htm.

Berg, Christian. "State OKs Rattlesnake Retention Plan." *The Morning Call Online*. Retrieved August 8, 2006, from http://mcall.com/sports/outdoors/all -rattlers0808aug08.0.2342317.story?page=2&coll=all-sportsoutdoors-hed.

Black, Richard. "Snakes in Mysterious Global Decline." *BBC News*. http://news .bbc.uk/gopr/fr/-/2/hi/science/nature/8727863.stm.

Clark, Rulon W., William S. Brown, Randy Stechert, and Harry W. Greene. "Cryptic Sociality in Rattlesnakes (*Crotalus horridus*) Detected by Kinship Analysis." *Biology Letters*, published online February 22, 2012, Doi: 10.1098/ rsbl.2011.1217.

"Crotalus horridus—Homeopathic Remedies." http://abchomeopathy.com/r .php/Crot-h.

"Crotalus horridus (Timber Rattlesnake)." http://iucnredlist.org/redlist/details /64318/0.

Curtis, Brent. "Man Survives Rattlesnake Bite in Fair Haven." *Rutland Herald Online*. 2010. http://rutlandherald.com/article/20100717/NEWSOI/717.

Ecological Misconceptions: Eastern deciduous forest (USA). Focal species: oaks et al. http://ecomiscomceptions.binghamton.edu/posoutlines.htm.

Elena, Ulev. 2008. *Crotalus horridus*. In "Fire Effects Information System" (Online). U.S. Department of Agriculture, Forest Service, Rocky Mountain Research Station, Fire Sciences Laboratory (Producer). October 21, 2009. http://www.fs.fed.us/database/feis.

Fitch, Asa. 2004. "'Cobra King' Is a Problem for Connecticut." http://www

.zwire.com/site/news.cfm?newsid=10742889&BRD=2303&PAG-461&_id
-478976&rfi=6 [January 2].

"Gadsden Flag," *Wikipedia*, the free encyclopedia. http://en.wikipedia.org/wiki
/Gadsden_flag.

Ghose, Tia. "67 Million-Year-Old Snake Fossil Found Eating Baby Dinosaurs." 2010. http://www.wired.com/wiredscience/2010/03/snake-eats
-babydinosaur.

Lauderdale, David. "Quintessentially Lowcountry: Area Was Once Nirvana
for Snake Collectors." 2010. http://www.islandpacket.com/2010/03/26
/1186375/quintessentially.

Miller, Joshua Rhett. "Snake Handling Puts Religious Freedom, Public Safety
in Spotlight after Kentucky Death." February 26, 2014. http://www.foxnews
.com/us/2014/02/26/snake-handling-puts-religious-freedom-public-safety
-in-spotlight-after-kentucky/.

Moyer, Ben. "Outdoors: New Hunting Regulations Test Balance of Conservation and Community." June 17, 2007. http://www.post-gazette.com/pg
/07168/794792–358.stm.

Nunn, Scott. "Rattlesnake Mailed to Ex-Boss." July 22, 1982. Satarnewsonline.com.

"Rattlesnake/Pit Viper: Crotalus horridus." http://herbs2000.com
/homeopathy/crotalus_hor.htm.

"Researchers Track Declining Timber Rattlesnakes." *Newswise Science News*,
August 17, 2007. http://newswise.com/articles/view/532288.

Ripa, Dean. "Degenerated Science Part 4: The Limits of Technology; Snake
Collecting." http:/www.lachesismuta.com/Part%204.htm.

"Study: Rattlesnakes Make Poor Pets." *The Watley Review Science and Medicine*,
February 24, 2004. http://www.watleyreview.com/2004/022404-3.html.

Sweeney, Rory. "Getting Rattled at Noxen Rattlesnake Roundup." June 17, 2007.
http://www.timesleader.com/news/20070617_17snakes_ART.html.

"Timber Rattlesnake (*Crotalus horridus*)." Natural Heritage and Endangered
Species Program. Commonwealth of Massachusetts Division and Fisheries
& Wildlife. www.nhesp.org.

"Timber Rattlesnake *Crotalus horridus*: Identification, Status, Ecology, and
Conservation in the Midwest." Center for Reptile and Amphibian Conservation and Management. http://www.herpcenter.ipfw.edu.

Venesky, Tom. "Bearing Young an Exhausting, Trying Time for Timber Rattlesnakes. *The Times Leader*, July 1, 2007. http://www.timesleader.com/sports
/20070701_01outdoors_ART.html.

Yoshida, Kate Shaw. "Snakes on the Brain." October 31, 2013. http://arttechnica
.com/science/2013/10/snake-on-the-brain/.

Index